NEW-OPATHIES

An Emerging Molecular Reclassification of Human Disease

NEW-OPATHIES

An Emerging Molecular Reclassification of Human Disease

Errol C. Friedberg • Diego H. Castrillon
Rene L. Galindo • Keith A. Wharton, Jr.

University of Texas Southwestern Medical School at Dallas, USA

World Scientific

NEW JERSEY • LONDON • SINGAPORE • BEIJING • SHANGHAI • HONG KONG • TAIPEI • CHENNAI

Published by

World Scientific Publishing Co. Pte. Ltd.
5 Toh Tuck Link, Singapore 596224
USA office: 27 Warren Street, Suite 401-402, Hackensack, NJ 07601
UK office: 57 Shelton Street, Covent Garden, London WC2H 9HE

Library of Congress Cataloging-in-Publication Data
New-opathies : an emerging molecular reclassification of human disease / [edited by] Errol C. Friedberg ... [et al.].
p. ; cm.
Includes bibliographical references and index.
ISBN-13: 978-9814355681 (hardcover : alk. paper)
ISBN-10: 9814355682 (hardcover : alk. paper)
I. Friedberg, Errol C.
[DNLM: 1. Disease--genetics. 2. Disease--classification. 3. Genetic Heterogeneity. QZ 40]

616.001'2--dc23

2012008086

British Library Cataloguing-in-Publication Data
A catalogue record for this book is available from the British Library.

Typeset by Stallion Press
Email: enquiries@stallionpress.com

Printed by FuIsland Offset Printing (S) Pte Ltd Singapore

Contents

Preface

The quest to understand the wealth of diseases that afflict humans has been advanced by scientific and technical revolutions. Early humans attributed alterations in their bodies to external forces — gods and demons of the earth and sky. The first revolution in our understanding of human disease originated in antiquity and applied scientific observation inward — to the examination of diseased tissues, organs, and cadavers. A second revolution, primarily championed by the famous German pathologist Rudolph Virchow and enabled by development of the compound microscope, revealed that diseases ultimately manifest as alterations in the fundamental unit of life — the cell. The current revolution, fueled by genomic, proteomic, and other *-omic* data sets whose complexity approaches that of life itself, is revealing new insights, upsetting long-held dogma and suggesting promising new therapeutic approaches based on an understanding of the complex systems that life, and particularly diseased life, represents.

As a field of study and medical specialty, the discipline of pathology concerns the objective diagnosis of disease. In some cases, diseases thought to be unrelated have been shown to have a common underlying etiology; in other situations, what was thought to be a single disease can now be further subdivided into distinct categories, each with distinct treatment options and prognoses. Cancer is an extreme example, in which the genomic complexity and heterogeneity of each patient's tumor argues that each may represent a distinct disease entity. We refer to "lumpers" and "splitters" to categorize those who aim to minimize or to expand, respectively, the categories of each disease. Each is useful, as prominent examples of lumping and splitting are fresh in our memory.

Emerging cDNA microarray classification of various tumors (particularly diffuse large B cell lymphoma and breast cancer) clearly demonstrate that a population of histologically homogeneous tumors can exhibit molecular heterogeneity with profound significance for prognosis and treatment selection — an example of splitting. Conversely, the analogous signals generated by pathogenic activated Abl and PDGFR kinases, which promote chronic myeloid leukemia and a subset of gastrointestinal stromal tumors, respectively, unite their downstream kinase-driven pathogenesis, and generate uniquely susceptiblity to similar kinase inhibition — an example of lumping.

The language used to describe every disorder, which in turn is driven by newly emerging knowledge, limits our understanding of human disease. Disease classification allows us to keep track of trends, enables communication (or lack thereof) among the biomedical community and the lay public, and shapes our understanding of disease processes in general. A quick perusal of the older medical literature reveals terms such as "carcinoma simplex," "catarrhal jaundice," and "glandular fever," which of course are no longer in use. Despite the existence of disease classification systems such as ICD-10 that designate no less than 14,400 distinct (and until the next revision to ICD-11 in 2015, immutable) disease entities, we must humbly concede that the root cause of many diseases remains controversial, and the relationships between a host of idiopathic diseases remain elusive. The *-omics* revolution is beginning to change all that, with a new round of rational lumping and splitting.

The suffix *-opathies* has been historically used to designate a broad category of diseases whose common feature is an adverse impact on a given organ or tissue. For example, "cardiomyopathies" ultimately lead to heart failure, "nephropathies" to renal failure, "retinopathies" to blindness, and "coagulopathies" to abnormal bleeding. For such *-opathies*, the etiologies are diverse, encompassing genetic, infectious, metabolic, and other causes.

A growing cadre of new *-opathies* is emerging from the modern *-omics* revolution. Each new *-opathy* is characterized by specific defects in a specific organelle, sub-cellular structure, protein complex, or molecular process that can be pathogenic in common ways, but which manifest uniquely in different tissues. The prototype of this class of molecular *-opthies* are the hemoglobinopathies. These represent a compilation of genetic defects in hemoglobin that manifest as altered red blood cell morphology and function, with concomitant alterations in the homeostasis of oxygen transport. The new *-opathies*, examples of which are provided in this text, cross the boundaries of traditional organ system-based disease classifications and herald a new understanding of many diseases previously thought to be unrelated. In this context we hope that this contribution to the literature will enlighten students of pathology at all levels about new ways of thinking about and classifying human diseases.

Errol C. Friedberg, MD, FRCPath
Diego H. Castrillon, MD, Ph.D
Rene L. Galindo, MD, Ph.D
Keith A. Wharton Jr. MD, Ph.D
July, 2011

Acknowledgments

We thank the authors for their willingness to contribute to *New-Opathies*, Mark Smith at the University of Texas Southwestern Medical Center at Dallas for the cover design, and Director of Publishing at World Scientific, Ms. Sook Cheng Lim, for her unflagging assistance and attention.

List of Contributors

Harold Bays, MD, FACP, FACE, FNLA
L-MARC Research Center
3288 Illinois Avenue
Louisville KY 40213
www.lmarc.com
Tel: 502.515.5672
Fax: 502.214.3999
email: hbaysmd@aol.com

Rodrigo T Calado, MD, PhD
Hematology Branch
National Heart, Lung, and Blood Institute
National Institutes of Health
10 Center Drive
Bldg. 10/CRC, Rm. 3E-5140
Bethesda, MD 20892-1202, USA
E-mail: calador@nhlbi.nih.gov

Maria Castella
Divisions of Human Biology and Public
Health Sciences, Fred Hutchinson Cancer
Research Center, Howard Hughes
Medical Institute, 1100 Fairview Avenue
N., Cl-015, Seattle, WA 98109-1024, USA

Ronald S Cheung
Division of Pediatric Hematology/Oncology
Department of Pediatrics, University of
Washington School of Medicine
Seattle, WA 98195-0001, USA

William T Dauer
Department of Neurology and Department of
Cell and Development Biology
University of Michigan
Ann Arbor, MI 48109-2200, USA
E-mail: dauer@umich.edu

Kiranjit K. Dhillon
Divisions of Human Biology and Public
Health Sciences, Fred Hutchinson Cancer
Research Center, Howard Hughes
Medical Institute, 1100 Fairview Avenue
N., Cl-015, Seattle, WA 98109-1024, USA

Morie A. Gertz, MD
Division of Hematology and Internal
Medicine, Mayo Clinic, Professor of Medicine
Mayo Medical School, Rochester, Minnesota, USA
E-mail: gertz.morie@mayo.edu

J. Michael Gonzalez-Campoy
Minnesota Center for Obesity, Metabolism
And Endocrinology, PA, 1185 Town Center Drive
Suite 220, Eagan, MN 55123, USA
E-mail: drmike@mncome.com

Céline Jacquemont
Divisions of Human Biology and Public
Health Sciences, Fred Hutchinson Cancer
Research Center, Howard Hughes
Medical Institute, 1100 Fairview Avenue

Makoto Kuro-o, MD, PhD
Department of Pathology
University of Texas Southwestern Medical Center
at Dallas, 5323 Harry Hines Blvd.
Dallas, TX 75390-9072; USA
makoto.kuro-o@utsouthwestern.edu

Larisa Labzin
Institute of Innate Immunity
Biomedical Center, 1G008
University Hospitals
University of Bonn
Sigmund-Freud-Str. 25
53127 Bonn, Germany

Eicke Latz, MD PhD
Institute of Innate Immunity
Biomedical Center, 1G008
University Hospitals
University of Bonn
Sigmund-Freud-Str. 25
53127 Bonn, Germany
E-mail: eicke.latz@uni-bonn.de

and

University of Massachusetts Medical School
Division of Infectious Diseases & Immunology
364 Plantation St., LRB 360A
Worcester, MA 01605, USA
E-mail: eicke.latz@umassmed.edu

Pia Langhoff
Institute of Innate Immunity
Biomedical Center, 1G008
University Hospitals
University of Bonn
Sigmund-Freud-Str. 25
53127 Bonn, Germany

Dominic De Nardo
Institute of Innate Immunity
Biomedical Center, 1G008
University Hospitals
University of Bonn
Sigmund-Freud-Str. 25
53127 Bonn, Germany

Toshiyasu Taniguchi, MD, PhD
Divisions of Human Biology and Public
Health Sciences, Fred Hutchinson Cancer
Research Center, Howard Hughes
Medical Institute, 1100 Fairview Avenue
N., Cl-015, Seattle, WA 98109-1024, USA
E-mail: ttanigu-tky@umin.ac.jp
ttaniguc@fhcrc.org

Beth L. Thurberg, M.D., Ph.D.
Department of Pathology
Genzyme Corporation
One Mountain Road
Framingham, MA 01701-9322
Tel: 508-271-2739
Fax: 508-820-7664
Beth.Thurberg@genzyme.com

Johanna Vogelhuber
Institute of Innate Immunity
Biomedical Center, 1G008
University Hospitals
University of Bonn
Sigmund-Freud-Str. 25
53127 Bonn, Germany

Howard J. Worman, M.D.
Department of Medicne
College of Physicians and Surgeons
Columbia University
630 West 168th Street, 10th Floor, Room 508
New York, NY 10032
E-mail: hjw14@columbia.edu

Neal S Young, MD, MACP
Hematology Branch
National Heart, Lung, and Blood Institute
National Institutes of Health
10 Center Drive
Bldg. 10/CRC, Rm. 3E-5140
Bethesda, MD 20892-1202, USA
E-mail: youngns@nhlbi.nih.gov

CHAPTER 1

Laminopathies

William T. Dauer and Howard J. Worman†*

NEW "OPATHIES" FOR A NEW MILLENNIUM

The dawn of the 21st century witnessed the emergence of a new "opathy": "laminopathy" or "nuclear envelopathy." The story begins in 1994, when positional cloning of the gene for X-linked Emery–Dreifuss muscular dystrophy identified a novel protein termed "emerin,"[1] which was subsequently found to be an integral membrane protein localized to the nuclear envelope.[2,3] The next significant finding, which launched the field, was the 1999 discovery by Bonne and colleagues[4] that autosomal dominant Emery–Dreifuss muscular dystrophy is caused by a mutation in the *LMNA* gene, which encodes A-type nuclear lamins — proteins that regulate the shape of the nuclear membrane. Mutations in *LMNA* have subsequently been found to cause roughly a dozen distinct, previously described diseases. The roles of the nuclear envelope in maintaining health have expanded, as evidenced by a variety of previously unrelated diseases being linked to mutations in genes encoding other nuclear envelope proteins.

Beyond being fascinating diseases from a clinical standpoint, the so-called "laminopathies" are important because they shed new light on the

*Department of Neurology and Department of Cell and Developmental Biology, University of Michigan, USA.
†Department of Medicine and Department of Pathology and Cell Biology, Columbia University, USA. E-mail: hjw14@columbia.edu.

previously unrecognized complexity and heterogeneity of nuclear envelope function. Prior to the discovery of the laminopathies, it was widely assumed that nuclear envelope proteins were essential, the loss of which would be incompatible with organismal survival. However, most of the laminopathies manifest as tissue-specific diseases (Table 1), selectively affecting striated muscle, adipose tissue, or neural tissue. Some even affect aging, revealing new and crucial functions of the nuclear envelope.

THE NUCLEAR ENVELOPE: NORMAL STRUCTURE AND COMPOSITION

The nuclear membranes, pore complexes, and lamina constitute the nuclear envelope (Fig. 1). Cell biologists divide the nuclear membranes into three interconnected, morphologically distinct domains. The outer nuclear membrane is directly continuous with the rough endoplasmic reticulum and similarly contains ribosomes on its cytoplasmic face. The inner nuclear membrane is separated from the outer nuclear membrane by the perinuclear space, which is continuous with the lumen of the endoplasmic reticulum. Nuclear pore complexes are anchored to pore membranes, the small transcisternal lipid bilayers that connect the inner and outer nuclear membranes. While integral membrane proteins synthesized on endoplasmic-reticulum-bound ribosomes could theoretically diffuse randomly in all three nuclear envelope membrane domains as well as the contiguous endoplasmic reticulum, size limitations on diffusion at the nuclear pore complexes and targeting signals contained within certain integral membrane proteins promote membrane-domain-specific enrichment. For example, emerin and lamin B receptor (LBR) are concentrated in the inner nuclear membrane through binding to lamins, whereas gp210 and POM121 localize to pore membranes by virtue of being components of the nuclear pore complex. A subset of nesprin/syne proteins concentrate in the outer nuclear membrane by binding to the luminal domains of SUN proteins, which themselves are integral to the inner nuclear membrane. These proteins are discussed in this chapter.

Vertebrate nuclear pore complexes, which span the inner and outer nuclear membranes, are macromolecular protein complexes with an aggregate molecular mass of ~125 MDa and are composed of varying copies of approximately 30 different proteins, many of which are termed

Table 1 Laminopathies Resulting from Mutations in Genes Encoding Proteins of Different Components of the Nuclear Envelope

Nuclear Envelope Component	Gene	Protein(s)	Major Protein Function	Disease
Lamina	*LMNA*	Lamin A, lamin C	Mechanical support of nucleus; role in DNA synthesis and transcription	(1) Cardiomyopathy (2) Partial lipodystrophy (3) Peripheral neuropathy (4) Progeria (early aging) affecting skin, bone, arteries (5) Mandibuloacryl dysplasia (bone, fat, some progeroid features)
	LMNB1	Lamin B1	As for *LMNA*	Leukodystophy (white matter degeneration)
	LMNB2	Lamin B2	As for *LMNA*	Possible partial lipodystrophy (adipose)
	ZMPSTE24	ZMPSTE24	Processing of prelamin A	Progeria
Nuclear membranes/ perinuclear space	*EDM*	Emerin	Unknown; binds lamins and other nuclear membrane proteins	Cardlomyopathy with muscular dystrophy
	LBR	Lamin B receptor	Sterol reductase; binds B-type lamins, DNA, and HP1	(1) Pelger–Hüet anomaly (heterozygous loss-of-function) (2) Greenberg skeletal dysplasia (homozygous loss of function)

(*Continued*)

Table 1 (*Continued*)

Nuclear Envelope Component	Gene	Protein(s)	Major Protein Function	Disease
	LEMD3	Emerin	Binds rSmads, antagonizing TGF-beta signaling	Sclerosing bone dysplasia with or without restrictive dermopathy
	SYNE1	Nesprin-1	Element of multiprotein complex ("LINC" complex) that connects nucleus to cytoplasmic cytoskeleton	Cerebellar ataxia
	TOR1A	TorsinA	AAA+ ATPase of ER and perinuclear space; binds lamina-associate polypeptide 1	DYT1 dystonia
	AAAS	Aladin	Nuclear-pore-localized	Triple-A syndrome (achalasia–Addisonian–alacrima syndrome)
Nuclear pore complex	*NUP155*	Nup155	Nucleocytoplasmic transport	Atrial fibrillation
	NUP62	Nup62	Nucleocytoplasmic transport	Infantile striatonigral degeneration
	RANBP2	RanBP2	Nucleocytoplasmic transport	Acute necrotizing encephalopathy

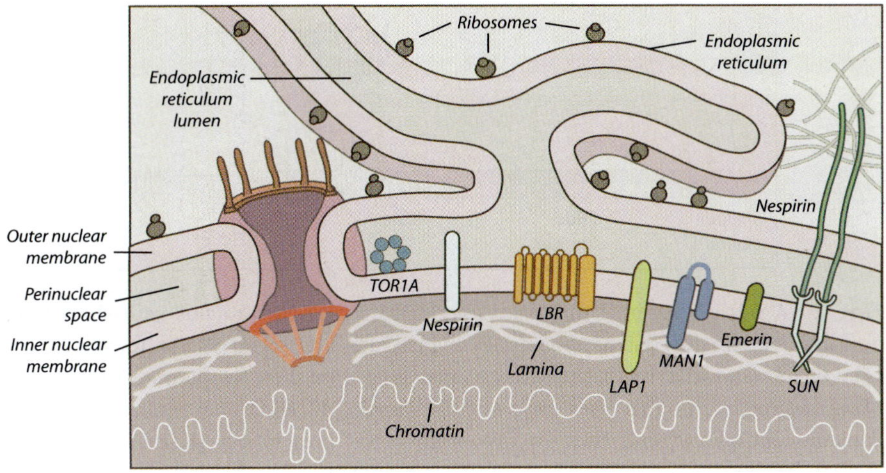

Fig. 1 The nuclear envelope. This schematic diagram shows the major components of the nuclear envelope including locations and topologies of proteins discussed in this chapter.

"nucleoporins." About half of the nucleoporins contain repeats of the diamino acid phenylalanine–glycine ("FG repeats"), which mediate the transport of proteins and ribonucleoprotein complexes through nuclear pore complexes. Nuclear pore complexes regulate both passive and active transport of small and large molecules across the nuclear envelope. Some molecules with molecular masses of less than ~60 kDa can traverse nuclear pore complexes by diffusion, but may use an active transport process to increase the transit rate. Larger proteins, RNAs, and protein–RNA complexes must utilize active transport mechanisms to cross the nuclear pore complexes. Molecules that are actively transported across nuclear pore complexes often contain nuclear import and/or export signals, which are recognized by a family of specialized carrier proteins called "karyopherins" or "importins/exportins." A gradient of GTP-bound Ran (a small GTPase) across the nuclear envelope provides the energy for active transport between the cytoplasm and the nucleus.

The nuclear lamina is a meshwork of 10-nm-diameter intermediate filaments localized primarily on the inner aspect of the inner nuclear membrane. These filaments are composed of proteins called "lamins,"

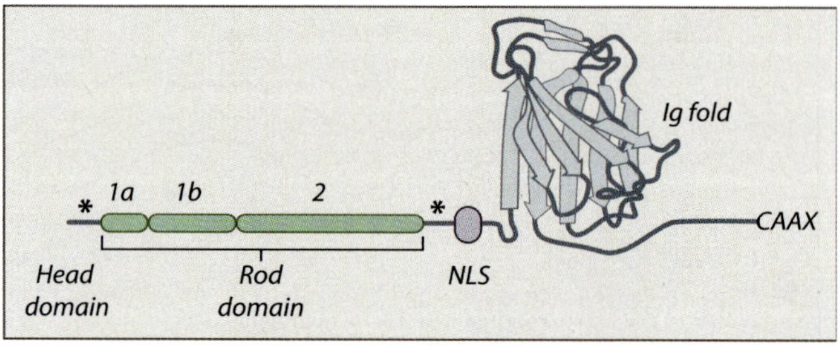

Fig. 2 A nuclear lamin. Lamins are intermediate filament proteins with short head domains, rod domains that are sometimes divided into three segments (1a, 1b, and 2), and tail domains (shown in blue). Vertebrate lamins differ from their cytoplasmic intermediate filament proteins, in that they have slightly longer (42 additional amino acids) rod domains, phosphorylation sites for mitotically active protein kinase (asterisks), and a nuclear localization signal (NLS) in their tail domains. The lamins also contain an immunoglobulin-like (Ig) fold in their tail domains and, except for lamins C and C2, CAAX motifs at their carboxyl termini.

members of the intermediate filament protein family. All intermediate filament proteins contain variable N-terminal head and C-terminal tail domains with a conserved central alpha-helical rod domain; however, the lamins differ from cytoplasmic intermediate filament proteins, such as keratins and neurofilaments, in several ways (Fig. 2). The rod domain of vertebrate lamins is slightly longer than cytoplasmic intermediate filaments, and the tail domain contains the nuclear localization signal. Lamins are phosphorylated by mitotically active protein kinases that regulate polymerization and depolymerization during mitosis. The tail domain contains an immunoglobulin-like fold motif of unclear function, but the importance of which is underscored by the different disease-causing mutations that are located in this region of the molecule.

Mammals have three genes that encode nuclear lamins. *LMNA* encodes the A-type nuclear lamins, the major isoforms being lamin A and lamin C; these proteins are expressed in most differentiated somatic cells. Lamin A is synthesized as a precursor, prelamin A. *LMNB1* and *LMNB2* encode the B-type nuclear lamins, which are expressed in all or most somatic cells. Lamins C2 and B3 are germ-cell-specific isoforms encoded by *LMNA* and *LMNB2*, respectively. Lamins C and C2 differ in their head

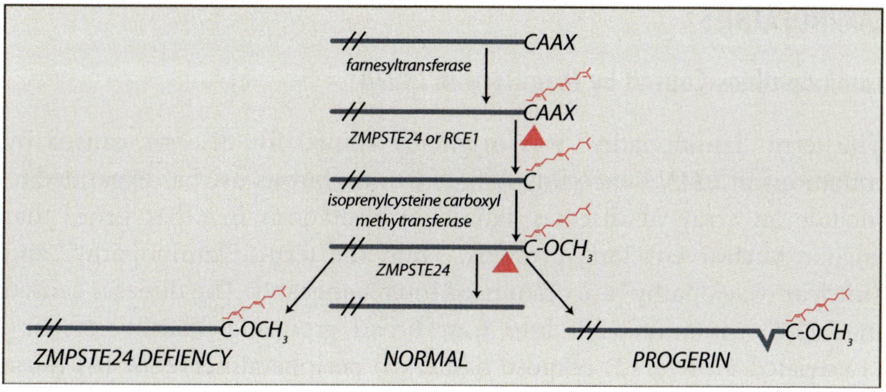

Fig. 3 Modifications at the CAAX motif and subsequent processing of prelamin A. The CAAX motif is the site of three chemical reactions: cysteine farnesylation catalyzed by farnesyltransferase, endoproteolytic cleavage of –AAX catalyzed by RCE1 for B-type lamins and RCE1 and/or ZMPSTE24 for prelamin A and carboxymethylation of the remaining cysteine catalyzed by isoprenylcysteine carboxyl methytransferase. The farnesylated prelamin A is then recognized again by ZMPSTE24, which catalyzes a second endoproteolytic cleavage removing the terminal 15 amino acids to generate normal lamin A, which does not contain a farnesylated cysteine. In cases of ZMPSTE24 deficiency, such as the human disease restrictive dermopathy, the farnesylated prelamin A is not further processed. In Hutchinson–Gilford progeria syndrome, a truncation causes loss of the second ZMPSTE24 endoprotease site, leading to expression of progerin, which contains the farnesylated cysteine. The farnesyl moiety and endoprotease sites are indicated in red.

domain, whereas lamins B2 and B3 differ in their tail domain. All of the lamins except lamins C and C2 contain "CAAX" (cysteine, alanine, alanine, any amino acid) motifs at their carboxyl termini. The CAAX motif is a signal for triggering three sequential chemical reactions (Fig. 3): (1) cysteine farnesylation catalyzed by farnesyltransferase; (2) endopreotolytic cleavage of –AAX catalyzed by RCE1 for B-type lamins and RCE1 and/or ZMPSTE24 for prelamin A; (3) carboxymethylation of the remaining cysteine (following cleavage of –AAX). The terminal 15 amino acids are subsequently cleaved from farnesylated, carboxymethylated prelamin A by the endoprotease ZMPSTE24, generating "mature" lamin A, which polymerizes into the lamina.

LAMINOPATHIES

Laminopathies Caused by Mutations in *LMNA*

The term "laminopathy" was originally coined for diseases caused by mutations in *LMNA* encoding A-type lamins, but its use has expanded to include an array of diseases caused by mutations in other genes that encode nuclear envelope proteins. Thus, the terms "laminopathy" and "nuclear envelopathy" are often used interchangeably. The diseases caused by *LMNA* mutations fall into four broad groups, primarily affecting (1) striated muscle, (2) adipose tissue, (3) peripheral nerve, or (4) those with multisystem manifestations and features of premature aging.[5]

Striated muscle diseases

Autosomal dominant Emery–Dreifuss muscular dystrophy was the first disease to be associated with *LMNA* mutations.[4] The patient typically develops early joint contractures and skeletal muscle wasting in a humeroperoneal distribution in the childhood-to-teenage years and dilated cardiomyopathy with conduction defects during adulthood. The same mutations in *LMNA* can cause dilated cardiomyopathy with an electrical conduction system defect, with minimal-to-no skeletal muscle involvement,[6] or a limb-girdle muscular dystrophy (involving different muscle than Emery–Dreifuss muscular dystrophy) with a predominant dilated cardiomyopathy.[7] Since these early discoveries, *LMNA* mutations have been linked to several other conditions with common cardiac and variable skeletal muscle pathology. The dilated cardiomyopathy in these diseases likely results from a similar cellular abnormality, with modifier genes and other unknown factors dictating the variable skeletal muscle involvement. Myopathy-causing *LMNA* mutations are predominantly single amino substitutions, small deletions, and various splicing abnormalities. Early truncating and promoter mutations that appear to cause haploinsufficiency of A-type lamins have also been described.

The mechanisms whereby mutations in *LMNA* cause tissue-specific abnormalities of striated muscle are poorly understood. Knockout mice deficient in A-type lamins recapitulate most aspects of the Emery–Dreifuss muscular dystrophy phenotype, suggesting that loss of A-type lamin functions is responsible for striated muscle disease.[8] Loss of A-type lamins is associated with a range of abnormalities evident in tissues and in cells cultured *ex vivo*, including increased nuclear deformation, defective mechanotransduction, impaired viability under mechanical strain, and diminished compensatory transcriptional responses to stress and cytokines.[9] These data support a "mechanical stress" hypothesis, which suggests that the selective vulnerability of striated muscles to *LMNA* deficiency derives from the normal role of the nuclear lamina in long-term maintenance of cellular integrity in the face of the constant and repetitive mechanical stress they endure. Cardiomyocytes from knockin mice with an Emery–Dreifuss-causing mutation leading to a single amino acid substitution (H222P) exhibit abnormally enhanced signaling in the JNK and ERK MAP kinase signaling pathways, which are normally activated by stress and extracellular growth factors. These signaling pathways become abnormal prior to the onset of clinically detectable disease[10] and drugs that inhibit JNK and ERK signaling can improve cardiac contractility as well as slow the rate of left ventricular dilatation and cardiac fibrosis in knockin mice with cardiomyopathy.[11] Thus, increased stress-activated cell signaling may contribute to the pathogenesis of *LMNA*-related myopathies. Loss of A-type lamins or the expression of lamin A variants that cause striated muscle disease also leads to abnormal nuclear positioning in migrating cells, suggesting that abnormal nucleo-cytoskeletal dynamics are another consequence of *LMNA* mutations.[12] Such abnormalities may be mechanistically related to the alterations in stress-activated signaling pathways or might constitute an independent mechanism of disease pathogenesis.

Lipodystrophy syndromes

Remarkably, *LMNA* mutations also cause an illness that selectively affects adipose tissue, known as autosomal dominant Dunnigan-type familial partial lipodystrophy.[13–15] This disease, which begins around puberty, is characterized by the simultaneous loss of peripheral subcutaneous adipose tissue and excess adiposity in the face and neck. The loss of peripheral fat leads to insulin resistance, glucose intolerance, and diabetes mellitus, with affected individuals often developing atherosclerosis and hepatic steatosis. Most (missense) partial-lipodystrophy-causing mutations localize to the immunoglobulin-like fold in the tail domain of lamins A and C, where they are believed to act by changing the surface charge,[16,17] but how this defect causes the cellular phenotype is not understood. In contrast, similarly localized amino acid substitutions that disrupt hydrophobic interactions or hydrogen bonds critical for the overall structural integrity of the immunoglobulin-like fold cause striated muscle disease. These observations suggest that a gain of function linked to abnormalities in the immunoglobulin-like fold leads to abnormal differentiation or maintenance of peripheral adipose tissue. In support of this notion, *Lmna*-null mice do not develop lipodystrophy, whereas transgenic mice that overexpress lamin A with a Dunnigan-type partial-lipodystrophy-causing missense mutation do.[18] Overexpression of lamin A and Dunnigan-type partial-lipodystrophy-causing mutants in preadipocytes impedes their differentiation into adipocytes,[19] suggesting that the lipodystrophy is caused by a defect of differentiation. A homozygous missense mutation (R527H) in the immunoglobulin-like fold causes mandibuloacral dysplasia, which, in addition to partial lipodystrophy, features bone, skin, and developmental abnormalities,[20] further linking this domain of A-type lamins to adipocyte biology. A small number of *LMNA* mutations outside of the immunoglobulin-like fold are also reported to cause Dunnigan-type and related forms of partial lipodystrophy, but whether such mutations indirectly affect the structural conformation of the immunoglobulin-like region has not been explored.

Peripheral neuropathy

Charcot–Marie–Tooth (CMT) diseases constitute a group of inherited disorders that predominantly affect peripheral nerves. A recessively inherited form of CMT type 2 is caused by the R298C missense mutation in *LMNA*.[21] As with other type 2 CMTs, this neuropathy is characterized by axonal degeneration, and *LMNA*-null mice (which exhibit a predominant myopathy) also have abnormalities of peripheral nerves, including a reduction in the axon density, an increase in the axon diameter, and the presence of nonmyelinated axons. It is unknown how this specific *LMNA*

R298C mutation disrupts A-type lamin function, or why the phenotypic abnormality is restricted to peripheral nerves.

Progeroid syndromes

Progeroid syndromes are a group of diseases characterized by features of premature aging. Hutchinson–Gilford progeria syndrome is a rare, autosomal dominant disease that arises through sporadic mutations in *LMNA*.[22,23] Children with this disease appear normal at birth but by approximately one year of age develop signs of progeria, including scleroderma, joint contractures, prominent eyes, micrognathia, decreased subcutaneous fat, alopecia, skin dimpling and mottling, prominent vasculature in the skin, fingertip tufting, and growth impairment. Most patients develop arterial abnormalities that mimic atherosclerosis and consequently die before age 20 from myocardial infarctions or strokes.

LMNA mutations that cause Hutchinson–Gilford progeria syndrome create a unique RNA splicing abnormality that results in a 50-amino-acid deletion near the C-terminus of prelamin A that removes a cleavage site for the protease ZMPSTE24. Consequently, this truncated prelamin A (termed "progerin") cannot undergo the final ZMPSTE24 endoprotease cleavage and remains abnormally farnesylated (Fig. 3). The critical role of ZMPSTE24 is highlighted by the disease restrictive dermopathy, caused by recessive loss-of-function mutations in the gene encoding ZMPSTE24.[24] Restrictive demopathy is a neonatal lethal condition characterized by tight skin with prominent superficial vessels, bone mineralization defects, dysplastic clavicles, and arthrogryposis. Mutations in *ZMPSTE24* that retain some enzyme activity cause atypical progeroid syndromes, with some subjects having been given the diagnosis of mandibuloacral dysplasia, which is characterized by partial lipodystrophy with some bone, skin, and developmental abnormalities similar to those in Hutchinson–Gilford progeria syndrome. Interestingly, mandibuloacral dysplasia can also be caused by the recessive *LMNA* R527H mutation. As with the mutations causing Hutchinson–Gilford progeria syndrome, loss of ZMPSTE24 activity leads to abnormal accumulation of stably farnesylated prelamin A, in this case wild-type prelamin A which is not processed (Fig. 3).

Considerable research supports the hypothesis that the farnesylated molecules that accumulate in progeroid syndromes (prelamin A in restrictive dermopathy; progerin in Hutchinson–Gilford progeria) are central to disease pathogenesis. The first supporting experimental evidence came from intercrossing *Lmna*-null and *Zmpste24*-null mice. Strikingly, heterozygosity for the A-type lamin gene ameliorated the progeria-like phenotype characteristic of *Zmpste24*-null mice, supporting a toxic role for farnesylated prelamin A.[25] Further support for this notion came from combined chemical-genetic experiments by Loren Fong, Stephen Young, and colleagues. They administered a farnesytransferase inhibitor (thus blocking farnesylation) to either *Zmpste24*-null mice or a line of *Lmna* knockin mice that were engineered to generate progerin.[26,27] In both cases, treatment with the inhibitor significantly ameliorated the progeroid phenotypes. A separate study using different farnesylation inhibitors in *Zmpste24*-null mice yielded similar results.[28] Moreover, the progeroid phenotype is blocked in knockin mice engineered to express a progerin molecule that cannot undergo farnesylation because of an altered CAAX motif.[29] Pharmacological blockade of protein farnesylation also reversed the microscopic nuclear shape abnormalities from progeroid mouse or human cells.[30] These findings have spurred clinical trials of pharmacological inhibitors of protein farnesylation in human subjects with Hutchinson–Gilford progeria syndrome.

While the accumulation of farnesylated lamin A appears critical for the pathogenesis of progeria, it is unclear how these abnormally expressed proteins alter cellular processes that generate this phenotype. Overexpression of farnesylated lamin A proteins causes abnormal nuclear morphology, but this effect is not specific to progeroid cells as it is present in virtually all A-type lamin-related laminopathies.[30] Abnormalities believed important in the progeroid illness Werner syndrome, such as defects in DNA repair or abnormalities in cell signaling, have also been observed in cells overexpressing progerin or unprocessed prelamin A.[30] Finally, some rare *LMNA* mutations have been reported to cause progeroid phenotypes but apparently do not affect prelamin A processing.[5,30] Hence, progeroid phenotypes can result from some alterations in A-type lamins that do not lead to accumulation of farnesylated prelamin A variants.

Laminopathies Caused by Mutations Affecting B-Type Lamins

No human disease has been linked to the loss of lamin B1 function or alterations in lamin B1 primary structure. In contrast, duplication of *LMNB1* resulting in lamin B1 overexpression causes adult onset autosomal dominant leukodystrophy.[32] This form of leukodystrophy is slowly progressive and is characterized by symmetrical widespread myelin loss in the central nervous system. Increased lamin B1 expression causes premature arrest of oligodendrocyte differentiation, a potential mechanism of disease.[33] Germline loss of lamin B1 yields mice that die at birth with lung and bone abnormalities,[31] precluding an assessment of the normal role of lamin B1 in myelination, which occurs postnatally, in these mice.

Coding polymorphisms in *LMNB2* have been associated with susceptibility to acquired partial lipodystrophy, known as Barraquer–Simons syndrome.[34] Loss of lamin B2 disrupts neuronal migration, leading lamin-B2-null mice to develop a lissencephaly-like phenotype featuring disrupted gyral development of cerebral cortex and cerebellum.[35] This finding is consistent with the emerging role of nuclear lamins in nuclear migration[12] but *LMNB2* mutations have not yet been identified in human neurodevelopmental disorders.

Mutations in Genes Encoding Nuclear Membrane Proteins

Inner nuclear membrane

A mutation in the *EMD* gene which encodes the inner nuclear membrane protein emerin was the first discovered nuclear envelopathy, X-linked Emery–Dreifuss muscular dystrophy.[1] X-linked Emery–Dreifuss muscular dystrophy is phenotypically identical to the dominantly inherited disease of the same name, which can be caused by mutations in *LMNA*. Emerin and A-type lamins appear to be closely associated in the nuclear envelope; expression of A-type lamins is necessary for emerin retention in the inner nuclear membrane.[8] Emerin-null mice exhibit abnormal activation of the ERK but not JNK branch of the MAP kinase pathway in cardiac muscle, whereas both pathways are disrupted in cardiac muscle in the *Lmna*-based model of autosomal Emery–Dreifuss muscular dystrophy.[36]

LBR is an integral protein of the inner nuclear membrane that appears to have two general functions. It contains a domain with eight putative transmembrane segments that is highly homologous to sterol reductases, and a basic N-terminal domain that faces the nucleoplasm and binds to B-type lamins and chromatin components. Heterozygous mutations in *LBR* cause the Pelger–Huët anomaly,[37] a benign abnormality of blood neutrophils, which exhibit abnormally hypolobulated nuclei and coarse chromatin. In contrast, homozygous mutations in *LBR* cause a severe disorder that presents as a neonatal lethal hydrops-ectopic calcification-moth-eaten skeletal dysplasia, known as Greenberg skeletal dysplasia.[37,38] The truncation mutation that causes Greenberg skeletal dysplasia disrupts its 3-beta-hydroxysterol delta(14)-reductase activity, highlighting the crucial requirement for the conversion of lanosterol to cholesterol at the nuclear membrane.[38] Nevertheless, it is unclear whether the loss of this activity in and of itself is sufficient for the disease, as abnormalities in LBR binding to B-type lamins, DNA, and chromatin may also contribute to disease pathogenesis.

The laminopathy that is arguably best understood at a mechanistic level is that caused by mutation in the *LEMD3* gene which encodes MAN1. MAN1 is an integral protein of the inner nuclear membrane with a nucleoplasmic N-terminal region that binds to A-type and B-type lamins. It has two transmembrane segments, and its nucleoplasmic carboxyl terminal region contains a binding domain for rSmads — downstream effectors of transforming growth factor beta signaling. This domain serves to bind and sequester rSmads, thereby blocking transforming growth factor beta-signaling. *Lemd3* knockout mice have severe developmental abnormalities resulting from enhanced transforming growth factor beta signaling and die during embryogenesis. In humans, heterozygous loss-of-function mutations in *LEMD3* cause the sclerosing bone dysplasias osteopoikilosis, Buschke–Ollendorff syndrome, and nonsporadic melorheostosis.[39] These disorders are all characterized by hyperosteotic cortical bone with concurrent hyperproliferation of skin in Buschke–Ollendorff syndrome, phenotypes that can be caused by increased transforming growth factor-beta activity. The clinical features of these diseases also highlight the tissue-selective functions of various nuclear envelope proteins, with MAN1 apparently having a critical role in regulating a signal transduction pathway in bone and skin while other tissues largely withstand alterations in MAN1 function.

Perinuclear space

The perinuclear space is the domain of the rough endoplasmic reticulum lumen that separates the inner from the outer nuclear membrane. Most luminal proteins can freely diffuse between the bulk endoplasmic reticulum and perinuclear space, evenly distributing between the two compartments. One such protein is torsinA, a member of the AAA+ ATPase protein family, a large group of ATPases that have various cellular functions and some structural similarities. A dominantly inherited mutation that removes a single amino acid in torsinA causes DYT1 dystonia, a central nervous system–specific disease.[40] This mutation impairs torsinA function and alters the distribution of the protein in the perinuclear space and the main endoplasmic reticulum,[41] yet how this translates into a selective disorder of movement remains unknown. Additional evidence for a nuclear-membrane-localized role of torsinA comes from the observations that (1) it interacts with LAP1, an integral membrane protein of the inner nuclear membrane of unknown function,[42] and (2) mice null for torsinA or homozygous for the ΔE knockin mutation show tissue-specific disruption of the neuronal nuclear membrane, despite being expressed in both neuronal and nonneuronal tissues.[43] The neural selective phenotype of the ΔE knockin mice might arise because a paralog, torsinB, protects nonneuronal cells from the consequences of torsinA dysfunction.[44]

Outer nuclear membrane

Some isoforms of the transmembrane family of proteins known as nesprins or synes encoded by the *SYNE1* and *SYNE2* genes preferentially localize to the outer nuclear membrane. This localization occurs because the residues in the perinuclear space contain a KASH domain, which binds to the luminal domains of SUNs — integral proteins of the inner nuclear membrane. SUNs in turn bind to nuclear lamins. This transnuclear envelope association of lamins–SUNs–nesprins has been termed the "LINC" complex (<u>li</u>nker of <u>n</u>ucleoskeleton and <u>c</u>ytoskeleton). Mutations in

SYNE1, which encodes nesprin-1, cause a recessively inherited cerebellar ataxia.[45] However, it is not clear if the disease-causing mutations affect nesprin-1 isoforms localized to the LINC complex or elsewhere, as some smaller nesprin-1 splice variants are localized to the inner nuclear membrane. An autosomal recessive form of arthrogryposis multiplex congenita, characterized bybilateral clubfoot, decreased fetal movements, delay in motor milestones, and progressive motor decline has also been reported to result from mutations in *SYNE1*.[46] LINC complex proteins connect the nucleus to cytoskeleton elements and appear to be involved in nuclear migration,[12] a process that is important for normal brain and muscle development.

Mutations in Genes Encoding Nuclear Pore Complex Proteins

Although nuclear pore complexes play a fundamental role in nucleocytoplasmic transport, it is now clear that mutations in genes encoding specific protein components can cause disease. One of these proteins is aladin, encoded by the *AAAS* gene. Mutations in *AAAS* cause achalasia–Addisonianism–alacrima ("triple-A") syndrome, a recessively inherited disorder characterized by adrenal insufficiency, dysfunction of the lower esophageal sphincter, and dry eyes. Wild-type aladin localizes to the nuclear pore complexes, but protein variants that cause triple-A syndromeare found predominantly in the cytoplasm.[47] Nuclear envelope and pore complex morphology appears to be grossly normal in cells from triple-A syndrome patients when examined by electron microscopy, suggesting that tissue-selective abnormalities in pore complex function rather than structure give rise to the disease. However, nuclear envelope ultrastructure has not been examined in cells of all affected tissues. A clear role for nuclear pore function is not established in this disease, as the abnormal mislocalization of aladin variants to the cytoplasm could also alter functions at that location.

Mutations in genes encoding nucleoporins also cause diseases with tissue-selective phenotypes. A homozygous missense mutation in the gene encoding NUP155 causes a familial atrial fibrillation with early sudden cardiac death, and heterozygous null *Nup155* mice develop atrial arrhythmias.[48] Loss of NUP155 appears to alter nucleocytoplasmic transport.

Mutations in genes encoding two other nucleoporins have been associated with central nervous system pathology. Both disorders feature the acute development of necrotic lesions of deep brain structures. One of them, autosomal recessive infantile striatal necrosis, is caused by mutation in the gene encoding NUP62.[49] Autosomal dominant mutations in the gene encoding RanBP2 lead to susceptibility to infection-triggered acute necrotizing encephalopathy.[50] How alterations in specific nuclear pore complex proteins lead to these disorders is not known; however, the tissue specificities of these diseases suggest that different tissues, including possibly cells of the immune system, might have greater dependence on nucleoporins involved in the transport of specific cargoes. These nucleoporins may also have functions other than those in nucleocytoplasmic transport, as RanBP2 binds the Ras-like GTPase Ran, which is involved in other functions such as mitotic spindle assembly and androgen receptor function in addition to nucleocytoplasmic transport.

CONCLUSION

The last decade has witnessed an explosion in genetic discoveries linking mutations in nuclear-envelope-localized proteins to a remarkable variety of human diseases. One common theme of many of these diseases is their marked tissue specificity, despite being due to alterations in proteins that are expressed in a wide variety of tissues, if not ubiquitously. In contrast, progeria-causing mutations, which affect a broad range of tissues and the aging process, disrupt a fundamental aspect of cellular senescence, emphasizing the diverse yet fundamental signaling events that impinge on the nuclear membrane. Clearly, a major focus of research going forward must be to identify how extracellular signals such as those mediated by cytokines or stress and intracellular pathways that transduce these signals are integrated at the nuclear envelope, and to unravel the mechanisms whereby their abnormalities produce characteristic organismal phenotypes. A key element of this effort must be to develop an understanding of the dynamic developmental and adult distribution patterns of nuclear-envelope-related proteins, as such information will likely be essential for understanding the exquisite tissue specificity that so frequently characterizes these illnesses. Only by vigorously pursuing these and related research goals can we hope to begin to develop rational therapeutic strategies for the myriad debilitating "opathies" that are linked to nuclear envelope dysfunction.

REFERENCES

1. Bione S, Maestrini E, Rivella S, *et al.* (1994) Identification of a novel X-linked gene responsible for Emery–Dreifuss muscular dystrophy. *Nat Genet* **8**: 323–327.
2. Nagano A, Koga R, Ogawa M, *et al.* (1996) Emerin deficiency at the nuclear membrane in patients with Emery–Dreifuss muscular dystrophy. *Nat Genet* **12**: 254–259.
3. Manilal S, Nguyen TM, Sewry CA, Morris GE. (1996) The Emery–Dreifuss muscular dystrophy protein, emerin, is a nuclear membrane protein. *Hum Mol Genet* **5**: 801–808.
4. Bonne G, Di Barletta MR, Varnous S, *et al.* (1999) Mutations in the gene encoding lamin A/C cause autosomal dominant Emery–Dreifuss muscular dystrophy. *Nat Genet* **21**: 285–288.
5. Dauer WT, Worman HJ. (2009) The nuclear envelope as a signaling node in development and disease. *Dev Cell* **17**: 626–638.
6. Fatkin D, MacRae C, Sasaki T, *et al.* (1999) Missense mutations in the rod domain of the lamin A/C gene as causes of dilated cardiomyopathy and conduction-system disease. *N Engl J Med* **341**: 1715–1724.
7. Muchir A, Bonne G, van der Kooi AJ, *et al.* (2000) Identification of mutations in the gene encoding lamins A/C in autosomal dominant limb girdle muscular dystrophy with atrioventricular conduction disturbances (LGMD1B). *Hum Mol Genet* **9**: 1453–1459.
8. Sullivan T, Escalante-Alcalde D, Bhatt H, *et al.* (1999) Loss of A-type lamin expression compromises nuclear envelope integrity leading to muscular dystrophy. *J Cell Biol* **147**: 913–920.
9. Lammerding J, Schulze PC, Takahashi T, *et al.* (2004) Lamin A/C deficiency causes defective nuclear mechanics and mechanotransduction. *J Clin Invest* **113**: 370–378.
10. Muchir A, Pavlidis P, Decostre V, *et al.* (2007) Activation of MAPK pathways links LMNA mutations to cardiomyopathy in Emery–Dreifuss muscular dystrophy. *J Clin Invest* **117**: 1282–1293.
11. Wu W, Muchir A, Shan J, Bonne G, Worman HJ. (2011) Mitogen-activated protein kinase inhibitors improve heart function and prevent fibrosis in

cardiomyopathy caused by mutation in lamin A/C gene. *Circulation* **123**: 53–61.

12. Folker ES, Östlund C, Luxton GW, Worman HJ, Gundersen GG. (2011) Lamin A variants that cause striated muscle disease are defective in anchoring transmembrane actin-associated nuclear lines for nuclear movement. *Proc Natl Acad Sci USA* **108**: 131–136.
13. Cao H, Hegele RA. (2000) Nuclear lamin A/C R482Q mutation in Canadian kindreds with Dunnigan-type familial partial lipodystrophy. *Hum Mol Genet* **9**: 109–112.
14. Shackleton S, Lloyd DJ, Jackson SN, *et al.* (2000) LMNA, encoding lamin A/C, is mutated in partial lipodystrophy. *Nat Genet* **24**: 153–156.
15. Speckman RA, Garg A, Du F, *et al.* (2000) Mutational and haplotype analyses of families with familial partial lipodystrophy (Dunnigan variety) reveal recurrent missense mutations in the globular C-terminal domain of lamin A/C. *Am J Hum Genet* **66**: 1192–1198.
16. Krimm I, Östlund C, Gilquin B, *et al.* (2002) The Ig-like structure of the C-terminal domain of lamin A/C, mutated in muscular dystrophies, cardiomyopathy, and partial lipodystrophy. *Structure* **10**: 811–823.
17. Dhe-Paganon S, Werner ED, Chi YI, Shoelson SE. (2002) Structure of the globular tail of nuclear lamin. *J Biol Chem* **277**: 17381–17384.
18. Wojtanik KM, Edgemon K, Viswanadha S, *et al.* (2009) The role of LMNA in adipose: a novel mouse model of lipodystrophy based on the Dunnigan-type familial partial lipodystrophy mutation. *J Lipid Res* **50**: 1068–1079.
19. Boguslavsky RL, Stewart CL, Worman HJ. (2006) Nuclear lamin A inhibits adipocyte differentiation: implications for Dunnigan-type familial partial lipodystrophy. *Hum Mol Genet* **15**: 653–663.
20. Novelli G, Muchir A, Sangiuolo F, *et al.* (2002) Mandibuloacral dysplasia is caused by a mutation in LMNA-encoding lamin A/C. *Am J Hum Genet* **71**: 426–431.
21. De Sandre-Giovannoli A, Chaouch M, Kozlov S, *et al.* (2002) Homozygous defects in LMNA, encoding lamin A/C nuclear-envelope proteins, cause autosomal recessive axonal neuropathy in human (Charcot–Marie–Tooth disorder type 2) and mouse. *Am J Hum Genet* **70**: 726–736.
22. Eriksson M, Brown WT, Gordon LB, *et al.* (2003) Recurrent de novo point mutations in lamin A cause Hutchinson-Gilford progeria syndrome. *Nature* **423**: 293–298.
23. De Sandre-Giovannoli A, Bernard R, Cau P, *et al.* (2003) Lamin A truncation in Hutchinson–Gilford progeria. *Science* **300**: 2055.

24. Navarro CL, Cadiñanos J, De Sandre-Giovannoli A, *et al.* (2005) Loss of ZMPSTE24 (FACE-1) causes autosomal recessive restrictive dermopathy and accumulation of lamin A precursors. *Hum Mol Genet* **14**: 1503–1513.
25. Fong LG, Ng JK, Meta M, *et al.* (2004) Heterozygosity for Lmna deficiency eliminates the progeria-like phenotypes in *Zmpste24*-deficient mice. *Proc Natl Acad Sci USA* **101**: 18111–18116.
26. Fong LG, Frost D, Meta M, *et al.* (2006) A protein farnesyltransferase inhibitor ameliorates disease in a mouse model of progeria. *Science* **311**: 1621–1623.
27. Yang SH, Meta M, Qiao X, *et al.* (2006) A farnesyltransferase inhibitor improves disease phenotypes in mice with a Hutchinson–Gilford progeria syndrome mutation. *J Clin Invest* **116**: 2115–2121.
28. Varela I, Pereira S, Ugalde AP, *et al.* (2008). Combined treatment with statins and aminobisphosphonates extends longevity in a mouse model of human premature aging. *Nat Med* **14**: 767–772.
29. Yang SH, Chang SY, Ren S, *et al.* (2011) Absence of progeria-like disease phenotypes in knock-in mice expressing a non-farnesylated version of progerin. *Hum Mol Genet* **20**: 436–444.
30. Worman HJ, Fong LG, Muchir A, Young SG. (2009) Laminopathies and the long strange trip from basic cell biology to therapy. *J Clin Invest* **119**: 1825–1836.
31. Vergnes L, Peterfy M, Bergo MO, *et al.* (2004) Lamin B1 is required for mouse development and nuclear integrity. *Proc Natl Acad Sci USA* **101**: 10428–10433.
32. Padiath QS, Saigoh K, Schiffmann R, *et al.* (2006) Lamin B1 duplications cause autosomal dominant leukodystrophy. *Nat Genet* **38**: 1114–1123.
33. Lin ST, Fu YH. (2009) miR-23 regulation of lamin B1 is crucial for oligodendrocyte development and myelination. *Dis Model Mech* **2**: 178–188.
34. Hegele RA, Cao H, Liu DM, *et al.* (2006) Sequencing of the reannotated LMNB2 gene reveals novel mutations in patients with acquired partial lipodystrophy. *Am J Hum Genet* **79**: 383–389.
35. Coffinier C, Chang SY, Nobumori C, *et al.* (2010) Abnormal development of the cerebral cortex and cerebellum in the setting of lamin B2 deficiency. *Proc Natl Acad Sci USA* **107**: 5076–5081.
36. Muchir A, Pavlidis P, Bonne G, Hayashi YK, Worman HJ. (2007) Activation of MAPK in hearts of EMD null mice: similarities between mouse models of X-linked and autosomal dominant Emery–Dreifuss muscular dystrophy. *Hum Mol Genet* **16**: 1884–1895.
37. Hoffmann K, Dreger CK, Olins AL, *et al.* (2002) Mutations in the gene encoding the lamin B receptor produce an altered nuclear morphology in granulocytes (Pelger–Huët anomaly). *Nat Genet* **31**: 410–414.

38. Waterham HR, Koster J, Mooyer P, *et al.* (2003) Autosomal recessive HEM/Greenberg skeletal dysplasia is caused by 3 beta-hydroxysterol delta 14-reductase deficiency due to mutations in the lamin B receptor gene. *Am J Hum Genet* **72**: 1013–1017.
39. Hellemans J, Preobrazhenska O, Willaert A, *et al.* (2004) Loss-of-function mutations in LEMD3 result in osteopoikilosis, Buschke–Ollendorff syndrome and melorheostosis. *Nat Genet* **36**: 1213–1218.
40. Ozelius LJ, Hewett JW, Page CE, *et al.* (1997) The early-onset torsion dystonia gene (DYT1) encodes an ATP-binding protein. *Nat Genet* **17**: 40–48.
41. Goodchild RE, Dauer WT. (2004) Mislocalization to the nuclear envelope: an effect of the dystonia-causing torsinA mutation. *Proc Natl Acad Sci USA* **101**: 847–852.
42. Goodchild RE, Dauer WT. (2005) The AAA+ protein torsinA interacts with a conserved domain present in LAP1 and a novel ER protein. *J Cell Biol* **168**: 855–862.
43. Goodchild RE, Kim CE, Dauer WT. (2005) Loss of the dystonia-associated protein torsinA selectively disrupts the neuronal nuclear envelope. *Neuron* **48**: 923–932.
44. Kim CE, Perez A, Perkins G, Ellisman MH, Dauer WT. (2010) A molecular mechanism underlying the neural-specific defect in torsinA mutant mice. *Proc Natl Acad Sci USA* **107**: 9861–9866.
45. Gros-Louis F, Dupré N, Dion P, *et al.* (2007) Mutations in SYNE1 lead to a newly discovered form of autosomal recessive cerebellar ataxia. *Nat Genet* **39**: 80–85.
46. Attali R, Warwar N, Israel A, *et al.* (2009) Mutation of SYNE-1, encoding an essential component of the nuclear lamina, is responsible for autosomal recessive arthrogryposis. *Hum Mol Genet* **18**: 3462–3469.
47. Cronshaw JM, Matunis MJ. (2003) The nuclear pore complex protein ALADIN is mislocalized in triple A syndrome. *Proc Natl Acad Sci USA* **100**: 5823–5827.
48. Zhang X, Chen S, Yoo S, *et al.* (2008) Mutation in nuclear pore component NUP155 leads to atrial fibrillation and early sudden cardiac death. *Cell* **135**: 1017–1027.
49. Basel-Vanagaite L, Muncher L, Straussberg R, *et al.* (2006) Mutated nup62 causes autosomal recessive infantile bilateral striatal necrosis. *Ann Neurol* **60**: 214–222.
50. Neilson DE, Adams MD, Orr CM, *et al.* (2009) Infection-triggered familial or recurrent cases of acute necrotizing encephalopathy caused by mutations in a component of the nuclear pore, RANBP2. *Am J Hum Genet* **84**: 44–51.

CHAPTER 2

Inflammasomopathies: Diseases Linked to the NLRP3 Inflammasome

Dominic De Nardo, Johanna Vogelhuber,* Larisa Labzin,* Pia Langhoff* and Eicke Latz*,†,‡*

INTRODUCTION AND OVERVIEW

The innate immune system evolved via diversification of families of germline-encoded pattern recognition receptors (PRRs) that mediate inflammatory responses to microbes by recognizing evolutionarily conserved pathogen-associated molecular patterns (PAMPs). In addition, some of these receptors can sense exogenous and endogenous structures such as nucleic acids associated with infection or inflammation which are called "danger-associated" molecular patterns (DAMPs). PRRs are classified into four major families based on genetic and functional properties: C-type lectins (CLRs), Toll-like receptors (TLRs), rentinoic-acid-inducible gene (RIG)-I-like receptors (RLRs), and Nod-like receptors (NLRs).

*Institute of Innate Immunity, Biomedical Center 1OG008, University Hospitals, Sigmund-Freud-str. 25, University of Bonn, 53173 Bonn, Germany.
†Division of Infectious Diseases and Immunology, University of Massachusetts Medical School, Worcester, MA 01605, USA.
‡E-mail: eicke.latz@uni-bonn.de.

Members of the NLR family have been shown to form large multimeric complexes upon activation, termed "inflammasomes," which mediate the maturation of IL-1β family cytokines, which in turn initiate and amplify inflammatory responses. The most-studied and perhaps the most important of these, the NLRP3 inflammasome, assembles in response to a plethora of PAMPs and DAMPs and is implicated in a variety of disease states, ranging from genetic, infection, environmental, to metabolically driven conditions. This chapter will summarize the current understanding of NLRP3 and its function, examining the role of NLRP3 in diverse diseases which we term "inflammasomopathies," and discuss the current therapeutic approaches targeting these diseases.

THE NLRP3 INFLAMMASOME: A PLATFORM FOR MATURATION OF IL-1β FAMILY CYTOKINES

NOD-Like Receptors (NLRs)

NLRs constitute a large family of intracellular PRRs, with 23 human and 34 mouse members currently identified. They have a tripartite structure, including a C-terminal leucine-rich repeat (LRR) domain, a centralized nucleotide binding and oligomerization domain (NOD or NACHT), and a variable N-terminal pyrine domain (PYD).[1] LRR domains are a common component of many proteins and display a range of variable biological functions, including ligand binding.[2] Indeed, within the TLR family of innate immune receptors the LRR region has been shown to mediate pathogen recognition.[3] As in TLRs, the NLR LRR region is thought to act primarily by sensing and autoregulating in response to stimuli, and possibly by mediating protein–protein interactions. The only domain common to all the NLRs, the NBD (nucleotide-binding domain) or NACHT (N̲AIP, C̲IITA, H̲ET-E, and T̲P1) domain, is thought to enable activation by ATP-dependent self-oligomerization.[4] The unique N-terminal domain of each of the NLRs, which initiates downstream signaling via protein–protein interactions, further classifies the NLRs into subgroups.[5] In humans the largest of these consists of 14 proteins (including NLRP1 and NLRP3) containing an N-terminal P̲YD domain and are hence called NLRPs. NOD1 (also known as NLRC1), NOD2 (also known as NLRC2), and

NLRC4 (also known as IPAF) contain a caspase recruitment domain (CARD) at the N-terminus and are therefore classified as NLRCs, whilst other NLR family members, containing a baculoviral inhibitory repeat-like (BIR) domain, form NLRBs. Finally, NLRs having an N-terminal acidic transactivation domain are classified as NLRAs.

Inflammasome-Forming Proteins

To date, three members of the NLR family, NLRP1, NLRP3, and NLRC4, have been shown to assemble into large multimeric complexes upon activation, termed "inflammasomes."[5–7] Inflammasome formation modulates the activation of inflammatory caspases and ultimately the production of two members of the IL-1 family of cytokines, IL-1β and IL-18. The bacterial peptidylglycan derivative muramyl dipeptide (MDP) and anthrax lethal toxin have been shown to induce the formation of NLRP1 inflammasome,[8,9] whilst assembly of the NLRC4 inflammasome can be triggered by bacterial and intracellular flagellin.[10–12] The NLRP3 inflammasome appears to form in response to a variety of diverse exogenous and endogenous stimuli, including crystalline material (e.g. silica), pore-forming toxins, bacteria, and viruses (see Table 1 for details). Interestingly, to date no direct binding between NLRP3 and any of its ligands has been demonstrated, suggesting that a common upstream receptor probably activates the NLRP3 inflammasome. Another inflammasome-inducing receptor is the PYHIN family protein absence in melanoma 2 (AIM2), following its recognition of cytoplasmic double-stranded DNA.[13–16]

Inflammasome formation results in the recruitment of inactive procaspase-1 via a homotypic CARD–CARD domain interaction. During the formation of the NLRP3, NLRC4, and AIM2 inflammasomes, procaspase-1 is recruited indirectly via the adaptor molecule, apoptosis-associated speck-like protein containing a CARD (ASC), through a homotypic PYD–PYD interaction. Interaction of ASC molecules results in formation of large, ordered ASC speck-like structures (up to 2 μm in diameter) and recruitment of the "pro-"catalytically inactive zymogen form of caspase-1.[17] Accumulation of procaspase-1 on the inflammasome then induces its cleavage, producing the p10 and p20 caspase-1 subunits, which subsequently reassemble to form active caspase-1 heterodimers.[18] Active caspase-1 then

Table 1 Activators of the NLRP3 Inflammasome

PAMPs	References
Adenovirus	25
Bacteria: *Listeria monocytogenes*, *Neisseria gonorrhoeae*, *Staphylococcus aureus*	26–28
Bacterial pore-forming toxins: hemolysin, listerolysin O, maitotoxin, nigericin, saponin, streptolysin	26, 28–30
Encephalomyocarditis virus	31
Malarial hemozoin	32
Influenza A virus: M2 channel protein	33–35
Sendai virus	33
Yeast (fungi): *Candida albicans*, *Saccharomyces cerevisiae*	36, 37
DAMPs	**References**
Alum (aluminum salts)	38–42
Amyloid-β aggregates	43
Asbestos fibers	44, 45
Calcium pyrophosphate dehydrate (CPPD) crystals	46
Ceramide	47
Cholesterol crystals	48, 49
Extracellular ATP, P2X7	26
Hyaluronan	50
Islet amyloid polypeptide (IAPP)	51
Microparticles: TiO_2, SiO_2, ZnO	
Monosodium urate (uric acid) crystals	46
Palmitate	52
Silica	39, 44, 45
Skin irritants: trinitrophenylchloride, trinitrochlorobenzene, dinitrofluorobenzene	53

processes pro-IL-1β and pro-IL-18 into their bioactive "mature" forms by cleavage of an N-terminal leader sequence.[19] The mature form of IL-1β is a potent endogenous pyrogen well known for its ability to induce fever as well as provoke a variety of proinflammatory responses, including activation of lymphocytes and endothelial cells. In addition, IL-1β increases the expression of adhesion molecules and chemokines at sites of inflammation, which together promote the extravasation of immune cells from the circulation into the tissues to destroy pathogens.[20]

By virtue of the potentially damaging proinflammatory effects of uncontrolled IL-1β release, its production is tightly regulated at multiple levels. Firstly, unlike pro-IL-18, which is constitutively expressed in resting cells,[21] the expression of the IL-1β precursor is induced following activation of NF-κB pathways, such as via TLRs. Secondly, IL-1β maturation requires processing by active caspase-1, whose production is also under the tight control of the inflammasome, as described above. Finally, as IL-1β lacks a conventional secretory signal its release from cells has been demonstrated to occur via a nonconventional route involving a highly inflammatory form of caspase-1-induced programmed cell death termed "pyroptosis." Pyroptosis involves rapid caspase-1-dependent DNA fragmentation and formation of 1–2 nm pores in the plasma membrane leading to cytoplasmic swelling and osmotic lysis via plasma membrane rupture.[22,23] The formation of these large pores and cellular lysis also facilitates the release of cytosolic IL-1β and IL-18 into the extracellular space from immune cells such as macrophages.[24]

Focus on the NLRP3 Inflammasome

The most extensively characterized inflammasome to date is that assembled around NLRP3 (also known as cryopyrin or NALP3). NLRP3 is expressed in a variety of myeloid cells, including macrophages, neutrophils, and dendritic cells, and is induced upon activation by various inflammatory stimuli.[54–57] Assembly of the NLRP3 inflammasome involves the formation of a large signaling complex containing NLRP3, ASC, and caspase-1, ultimately leading to release of IL-1-family cytokines, as described above. In order to prevent accidental NLRP3 activation which could potentially cause inflammatory damage in the surrounding tissues, a two-step process has evolved to mediate formation of the NLRP3 inflammasome. Firstly, activation of NF-κB via PRR or cytokine receptors (a priming signal) is critical for concomitantly increasing expression of pro-IL-1β as well as NLRP3 to a functional level. Indeed, in immune cells endogenous NLRP3 is insufficient to permit inflammasome activation.[55] Following increased expression of both NLRP3 and pro-IL-1β protein in the cytosol, a second signal (activation signal) is required to initiate assembly of the inflammasome.

Presently, the precise process leading to NLRP3 activation in not clearly understood; however, several potential mechanisms for explaining the initiation of NLRP3 inflammasome assembly have been proposed. By virtue of its C-terminal LRR domains, NLRP3 has been suggested to act as a cytosolic receptor that can directly engage its stimuli, allowing subsequent formation of the inflammasome. However, given the large number of PAMPs and DAMPs that can activate NLRP3 and their varied physical and chemical natures, it is highly unlikely that this model is correct. Furthermore, to date no concrete evidence has been obtained to suggest that NLRP3 can directly bind to any of the ligands that can activate it. Other, more favored models suggest that NLRP3 recognizes a common upstream signal following disruption of cellular homeostasis to initiate activation indirectly. One hypothesis posits that intracellular reactive oxygen species (ROS), which are commonly produced in response to uptake of a number of NLRP3 activators, provide this signal. This notion is supported by observations that ROS inhibition or scavenging can suppress NLRP3 inflammasome activation.[58] A number of sources of ROS are implicated in this process, including ROS derived from NADPH oxidases during phagocytosis[44] and those produced from the mitochondria.[59,60] However, ROS do not appear essential for NLRP3 activation, since mice deficient in NADPH oxidase subunits and humans with defects in NADPH oxidase (chronic granulomatous) display normal NLRP3 inflammasome function.[61] Another hypothesis proposes that perturbation of lysosomal membranes by large crystals or protein aggregates is sensed by NLRP3 as a danger signal.[39,43] The uptake of such molecules results in lysosomal damage and successive rupture, leading to spillage of the lysosomal proteases such as cathepsins B and L into the cytosol. In this model it is hypothesized that in the steady state NLRP3 exists in a closed/inactive conformation in which the NACHT domain is guarded by the LRR.[62] The release of such proteases could potentially mediate NLRP3 inflammasome activation by cleavage of the receptor or an intermediate protein, resulting in conformational changes that facilitate binding of ASC and formation of a mature inflammasome. Finally, pore formation in the plasma membrane via either P2X7 receptor or by pore-forming toxins results in cellular potassium (K^+) efflux, which has been found

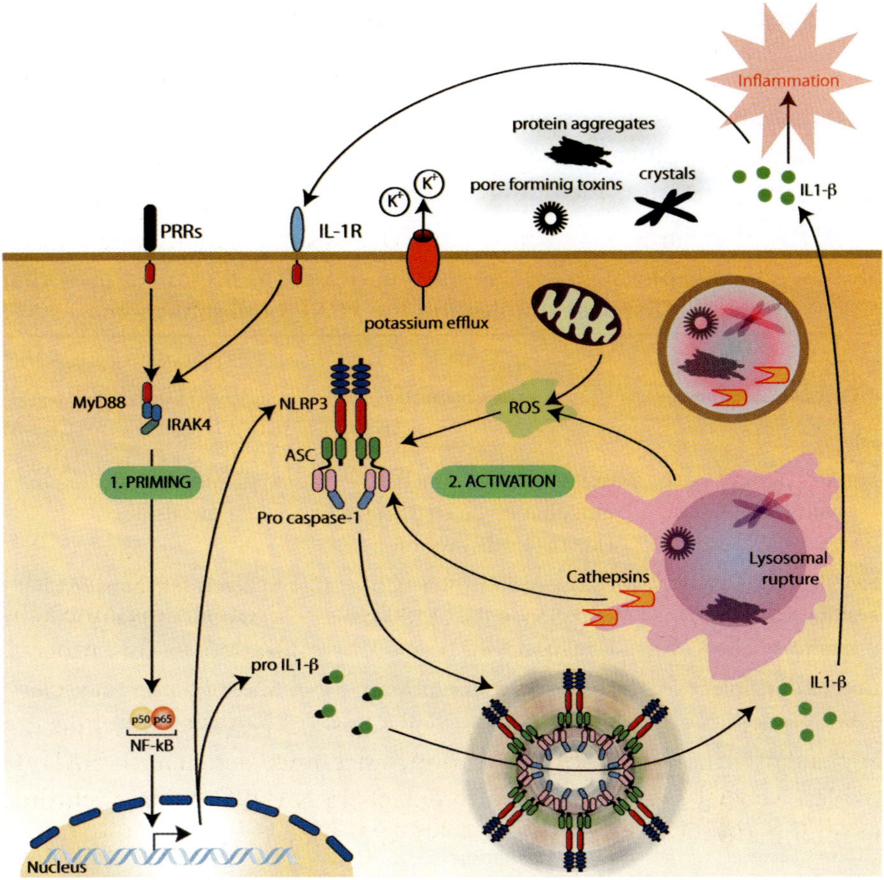

Fig. 1 Models of NLRP3 inflammasome activation. Activation of NLRP3 inflammasome requires an NF-κB priming step (signal 1) to upregulate NLRP3 expression and pro-IL-1β, followed by a subsequent activation step (signal 2) resulting in maturation of IL-1β. A number of mechanisms are implicated in NLRP3 activation, including production of ROS during phagocytosis and from mitochondria, cathepsin release into the cytoplasm following lysosomal damage, and intracellular potassium efflux.

to be required for inflammasome activation in response to an array of NLRP3 stimuli.[17,29,44,63,64] It is highly likely that elements of all three models or their interaction are required for full NLRP3 activation. A representation of the current models of NLRP3 inflammasome activation is provided in Fig. 1.

INFLAMMASOMOPATHIES: THE NLRP3 INFLAMMASOME AND DISEASE

The NLRP3 inflammasome assembles in response to a wide array of stimuli, ranging from pathogens to environmental and metabolic danger signals; therefore it is not surprising that NLRP3 is implicated in a variety of disease states, including those arsing from genetic variants of NLRP3 (Table 2).

Table 2 Diseases Implicating the NLRP3 Inflammasome

Disease Category	Causes/Stimuli	Primary Disease Location
Genetic:		
Muckle–Wells syndrome	Missense (gain-of-function) mutation in the NLRP3 gene, Chr 1q44, NBD region	Systemic inflammation in various tissues
Familial cold autoinflammatory syndrome	Missense (gain-of-function) mutation in the NLRP3 gene, Chr 1q44, NBD region	Systemic inflammation in various tissues
Chronic infantile neurological cutaneous and articular syndrome (CINCA, NOMID)	Missense (gain-of-function) mutation in the NLRP3 gene, Chr 1q44, NBD region	Systemic inflammation in various tissues
Crohn's disease	Missense mutation in the NLRP3 gene, Chr 1q44, 3′UTR (decreased NLRP3 expression)	Gastrointestinal tract
Environmental:		
Asbestosis	Asbestos fibers	Lung, bronchoalveolar space
Silicosis	Silica crystals	Lung, bronchoalveolar space
Nanoparticles	TiO_2, SiO_2	Skin, epithelial tissue
Infectious:		
Virus infection	Influenza A virus	Nose, throat, lungs
	Adenovirus	Upper respiratory tract
	Encephalomyocarditis virus (EMCV)	Heart, liver, kidney
Malaria/cerebral malaria	Malarial hemozoin/uric acid	Bloodstream/brain

(Continued)

Table 2 *(Continued)*

Disease Category	Causes/Stimuli	Primary Disease Location
Metabolic:		
Early type 2 diabetes and obesity	Palmitate	Adipose tissue, liver
Type 2 diabetes progression	Ceramide	Adipose tissue
	Islet amyloid polypeptide (IAPP)	Pancreas
	Thioredoxin-interacting protein (TXNIP)	Pancreas
Atherosclerosis	Cholesterol crystals	Large arteries
Gout	Monosodium urate (MSU) crystals	Articulation, joints (Primarily big toe and ankle)
Pseudogout	Calcium pyrophosphate dehydrate (CPPD) crystals	Articulation, joints
Alzheimer's disease	Amyloid beta	Brain
Other sterile inflammation:		
Contact hypersensitivity	Sodium dodecylsulfate Trinitro-chlorobenzene	Skin
Injury	Hyaluronan, ATP, uric acid, mitochondria	Various tissues, liver

Genetic NLRP3 Inflammasome Disorders

Cryopyrin-associated periodic syndromes (CAPS)

"Cryopyrinopathies," otherwise known as cryopyrin-associated periodic syndromes (CAPS), are rare autoinflammatory diseases characterized by recurrent, systemic inflammation in the absence of infection. They manifest in a disease continuum of increasing severity, with three identifiable phenotypes. The mildest phenotype is familial cold autoinflammatory syndrome (FCAS). First reported in 1940 by Kile and Rusk,[65] FCAS is characterized by inflammation (including rash, fever, and joint pain) in response to cold exposure. Muckle–Wells syndrome (MWS) was first described in 1962 and represents the intermediate phenotype with similar systemic inflammation to FCAS, but has no association with cold exposure.[66] The most severe clinical manifestation of this recurrent inflammation phenotype is observed in neonatal onset multisystem inflammatory disease (NOMID) also known as chronic infantile neurological cutaneous articular (CINCA) syndrome.[67] This disorder includes symptoms involving the central nervous system, bones, and joints. There is extensive overlap of clinical symptoms between the cryopyrinopathies, often making diagnosis within the CAPS spectrum difficult.[68] Similar to other autoinflammatory disorders such as familial Mediterranean fever (FMF) and tumor necrosis factor receptor-associated periodic syndrome (TRAPS), CAPS are monogenic disorders with Mendelian inheritance. The gene identified as dysregulated in CAPS, *NLRP3* (*CIAS1*), was initially termed "cryopyrin," literally translated to "ice–fire," because of the recurrent cold-induced fever associated with these disorders.[69] The integral role of the proinflammatory cytokine IL-1β in CAPS has been confirmed by the efficacy of anti-IL-1β therapy in the treatment of CAPS.[70,71]

Clinical manifestation of CAPS diseases

CAPS patients generally present with symptoms of systemic inflammation including fever, a hives-like skin rash similar to but distinct from urticaria, and joint pain without apparent trigger or infection. A hallmark of the inflammation associated with CAPS is a massive neutrophil influx into various sites. For instance, influx into the joints leads to joint inflammation (arthralgia), whilst influx of neutrophils into the perivascular dermis leads to the urticaria-like rash. This polymorphonuclear cell infiltration is unlike the typical lymphocytic and eosinophilic infiltration associated with classical urticaria.[68] In addition, a strong acute phase inflammatory response is initiated in CAPS, with release of typical proinflammatory mediators such as C-reactive protein and serum amyloid A from the liver into the bloodstream.[68] Cuisset *et al.* reported that the most common symptoms of CAPS are elevated levels of C-reactive protein and the urticaria-like rash with an age of onset less than 20 years.[72] As a prelude to genetic screening for *CIAS1* mutations for definitive diagnosis of CAPS, they also recommend testing after at least three recurrent bouts to exclude inflammation due to infection.[72]

Familial cold autoinflammatory syndrome (FCAS)

The mildest phenotype along the CAPS spectrum is FCAS. FCAS usually presents as short bouts of recurrent inflammation upon exposure to cold.[68] The most common symptoms include the urticaria-like rash, fever, joint pain, conjunctivitis with profuse sweating, drowsiness, headache, extreme thirst, and often nausea.[73] The severity of the symptoms generally correlates with the intensity of the cold trigger.[68] The symptoms usually begin 2 h after exposure to cold and last, on average, up to 12 h.[73] Hoffman *et al.* demonstrated that by artificially inducing FCAS episodes, the urticaria-like rash can actually develop within the first hour after cold challenge, whilst joint pain and fever develop in the first 2–3 h.[70] The symptoms correlated with increased serum IL-6 and blood neutrophilia, with resolution within 12–18 h.[70] In the majority of cases, FCAS presents within the first six months of life.[74]

Muckle–Wells syndrome (MWS)

The intermediate phenotype, MWS, is also characterized by the urticaria-like rash, recurrent fever, as well as joint and eye inflammation. However, unlike FCAS, episodes of MWS are not preceded by exposure to cold or any other apparent trigger, and indeed in some patients the symptoms do not abate.[68] In approximately 70% of cases, perceptive deafness develops from childhood or early adulthood, but the mechanism by which this develops is not understood.[68] The most severe complication observed in MWS, seen in ~25% of cases, is the deposition of serum amyloid A (amyloidosis) in the kidneys, leading to renal impairment and nephrotic syndrome.[75] However, the severity of other inflammatory symptoms does not predict the likelihood of amyloidosis in MWS patients.[75]

Neonatal onset multisystem inflammatory disease (NOMID)

The most severe phenotype of the CAPS spectrum, NOMID, presents with the typical urticaria-like rash, joint inflammation, fever, and chronic aseptic meningitis.[76] Two-thirds of NOMID patients are born with a cutaneous rash, and in the remaining cases this rash develops within the following 1–2 weeks of life.[77] The second-most-common feature of NOMID is neutrophilic central nervous system (CNS) inflammation, resulting in headaches, seizures, and transient episodes of limb paralysis and leg spasticity.[77] As in the milder phenotypes of CAPS, patients display joint inflammation, but one-third of NOMID sufferers also show more severe arthropathy, with overgrowth of the long bones, patella, and cartilaginous growth plates, resulting in joint deformities, pain, and limited motion.[68,77] Loss of vision and hearing is also reported in older patients,[77] and approximately 20% of patients die before reaching adulthood.[78]

Genetics of CAPS

In 1999 genetic linkage studies of three families with MWS mapped the susceptibility locus for the disease to chromosome 1q44,[79] and in 2000 FCAS was mapped to the same region.[80] The gene was later identified as *CIAS1* (NLRP3).[74] To date, 127 distinct heterozygous coding mutations in the NLRP3 gene have been reported from CAPS patients.[81,82] These mutations are very rare; a recent study reported the prevalence of NLRP3 mutations in France as approximately 1 in 400,000.[72] In FCAS and MWS, mutations in NLRP3 are primarily inherited in an autosomal dominant fashion, while in NOMID most of the mutations are spontaneous.[68,83] While nearly all the mutations are located in exon 3 (which encodes the NACHT domain), there is no clear correlation between mutational position and disease severity: L353P and L305P are associated with FCAS and milder phenotypes, F309S and Y570C are associated with NOMID, while other mutations such as R260W show a variable phenotypic severity.[84,85] Furthermore, many CAPS patients, including up to 40%–60% of NOMID patients, are negative for *CIAS1* mutations,[86] suggesting additional genetic mechanisms. Saito *et al.* showed that three NOMID patients, originally identified as *CIAS1*-mutation-negative, actually had a subset of monocytes with disease-associated mutations in *CIAS1*.[87] While somatic mosaicism may explain why some CAPS patients show no mutations in *CIAS1*, there are significant obstacles to directly associating mutational position with phenotypic severity, including the rarity of the disease as well as imprecise diagnostic criteria.[84]

Mechanism of inflammasome activation in CAPS

Most of the mutations in *CIAS1* are located in exon 3, which encodes the NACHT domain of NLRP3. Modeling of the NACHT domain suggests that these mutations occur on a single solvent-exposed surface of the protein.[84] It is hypothesized that the mutations in *CIAS1* interfere with inhibitory protein–protein interactions, leading to a "gain of function" NLRP3 inflammasome which is constitutively active or hyperactive.[69] A model proposed by Aksentijevich *et al.* suggests that in the steady state NLRP3 assumes an inactive conformation whereby the bound LRR and NACHT domains are mutually inhibitory. Mutations within these domains disrupt these interactions, which leads to a more "open" active conformation, requiring a lower threshold for inflammasome activation.[84] Peripheral blood mononuclear cells isolated from FCAS patients,[88] and monocytes isolated from a NOMID patient[83] showed constitutive IL-1β release and increased IL-1β production in response to the proinflammatory stimulus LPS. Moreover, knockin mouse models of FCAS and MWS showed corresponding phenotypes of underlying inflammation, skin abscesses, neutrophilia, and a Th17-cytokine dominant immune response.[89,90] (Th17-driven immune responses are often associated with chronic inflammation and tissue injury.)[91] Crossing mice harboring mutations that cause FCAS or MWS with ASC-deficient mice completely suppressed the disease phenotype, suggesting that these mutations do indeed act through the NLRP3 inflammasome.[89] Similarly, macrophages from FCAS knockin mice produce active IL-1β in response to TLR stimuli alone, suggesting that only the priming step (signal 1) but not the second signal is required for production of pro-IL-1β.[90] While these experimental data support the hypothesis that CAPS mutations lead to hyperactive inflammasome activation resulting in excessive IL-1β production, the exact mechanism remains to be determined.

Another unanswered question is how cold exposure can activate the NLRP3 inflammasome in FCAS but not in other NLRP3 genetically driven diseases, such as MWS. A role for cryoglobulins (which also display cold-sensitive behaviors) in cold-dependent inflammasome activation has not yet been investigated. Monocytes isolated from FCAS patients and then incubated at 32°C for 4 h or 16 h produced IL-1β in response to cold exposure alone, while monocytes from healthy donors did not.[92] Similarly, cells from mice expressing an FCAS-associated mutation released IL-1β when exposed to 32°C, while cells from mice with the MWS mutation did not.[89] Metabolic labeling assays demonstrated that NLRP3 with an FCAS-associated mutation (L353P) has an increased half-life at 32°C, but it is not clear whether the cold-associated phenotype is due to stabilized NLRP3 or altered conformation or binding partners associated with the mutant NLRP3.[89]

The NLRP3 Inflammasome and Environmental Diseases

Silicosis and asbestosis

For many years asbestos was a common material utilized for industrial and building purposes; however, until recently very little was known about its potentially hazardous side effects. At the beginning of the 20th century, studies by Cooke and colleagues first revealed that extended exposure to asbestos fibers could lead to changes in the lung and alveolar space.[93] Such changes were termed "asbestosis."[94] Today, asbestosis is defined as bronchoalveolar inflammation due to the inhalation of small crystalline asbestos particles that culminate in an irreversible lung fibrosis known as pneumoconiosis. Pneumoconiosis is a restrictive pulmonary disease, often leading to mesothelioma and cancer.

Silicosis is probably the most common lung disease worldwide, especially in developing countries, and shows a significant association with industrial development. It is typically diagnosed in patients after prolonged contact with crystalline silicon-dioxide-containing dust, as is commonly the case in miners, as well as individuals working in cement factories, on farms, or in manufacturing. As with asbestosis, silicosis is caused by inhalation of small crystals (silica crystals) contained in dust, leading to progressive pulmonary inflammation and fibrosis which ultimately results in pneumoconiosis or lung cancer. In addition, silicosis can lead to chronic bronchitis and recurrent interstitial lung infections, such as mycobacterial and fungal infections.

Until recently the precise pathway by which asbestos and silica crystals trigger inflammation was not fully understood. Although other crystals such as monosodium urate (MSU) were known to activate the NLRP3 inflammasome,[46] the involvement of NLRP3 in the pathogenesis of asbestosis and silicosis was recently demonstrated by several groups.[39,44,45] In response to either silica or asbestos crystals, LPS-primed human peripheral blood mononuclear cells (PBMCs) secreted IL-1β and cleaved caspase-1. Consequently, macrophages from mice deficient in NLRP3 or ASC showed no detectable caspase-1 cleavage or IL-1β maturation following silica or asbestos crystal stimulation.[39,44] Furthermore, utilizing an *in vivo* model of lung inflammation following intranasal administration of silica, NLRP3- or ASC-deficient mice had reduced granuloma formation, collagen deposition, and fibrosis in comparison with wild-type mice.[45]

Phagocytosis was found to be critical for the activation of NLRP3 in response to silica or asbestos crystals. PBMCs pretreated with the phagocytosis inhibitor cytochalasin D showed severely reduced IL-1β secretion in response to crystals.[39] Using confocal reflection microscopy, it was revealed that large amounts of phagocytosed silica crystals destabilized lysosomes and spilled phagolysosomal contents into the cytoplasm.[39] In particular, release of the lysosomal protease, cathepsin B, is important in NLRP3 inflammasome activation in response to silica, as NLRP3 activation was markedly reduced upon inhibition of cathepsin B or in cathepsin-B-deficient mice.[39,43] In addition, ROS production induced by phagocytosis of crystals was required for NLRP3 activation.[44]

Nanoparticles

Nanoparticles such as inorganic metal oxides, including titanium dioxide (TiO_2) and silica dioxide (SiO_2), are increasingly used in diverse products and manufacturing processes, including cosmetics, biomedicine, and electronics.[95] For example, by virtue of its opacifying effects, TiO_2 became a common additive in cosmetic products. In addition, nanoparticles are commonly used in paints, plastics, and in the production of various industrial items. Initially, nanoparticles such as TiO_2 were purported to be innocuous and biologically inert; however, within the last 30 years it has become increasingly clear that extended exposure to and subsequent inhalation of nanoparticles can initiate and exacerbate pulmonary diseases.[96–98] Nevertheless, the mechanisms by which nanoparticles cause disease were not well characterized until recently. In a study by Yazdi *et al.*, macrophages and primary human keratinocytes incubated with TiO_2 or SiO_2 displayed caspase-1 cleavage and IL-1β production, implicating activation of the NLRP3 inflammasome.[99] The necessity for NLRP3 was confirmed by shRNA knockdown of NLRP3 in a human macrophage cell line and also in macrophages from NLRP3- or ASC-deficient mice, each of which is unresponsive to nanoparticles. Furthermore, mice deficient in inflammasome components (NLRP3, ASC, or IL-1R) had reduced peritonitis following intraperitoneal injection of TiO_2 as compared to wild-type mice. In a similar *in vivo* model, the effects of TiO_2 on bronchial tissue showed results comparable to those of the peritonitis model.[99]

In order to investigate the molecular mechanism of NLRP3 inflammasome activation by nanoparticles, experiments utilizing the ROS inhibitor, (2R,4R)-4-aminopyrrolidine-2,4-dicarboxylate (APDC), and the pharmacological inhibitor, glibenclamide, which blocks ATP-dependent potassium efflux, were recently reported.[99] In both cases, nanoparticle-induced NLRP3 activation was decreased, suggesting the involvement of ROS production and potassium efflux in this process.

In contrast to crystals, NLRP3 activation by nanoparticles does not appear to involve lysosomal disruption. Firstly, NLRP3 inflammasome activity is detectable in primary human keratinocytes, which are non-phagocytic. Secondly, inhibition of actin-mediated phagocytosis by cytochalasin D did not affect TiO_2 and SiO_2-induced caspase-1 cleavage and IL-1β secretion.[99] Electron microscopy has also suggested that nanoparticle uptake is independent of phagocytosis by showing a random localization of nanoparticles in the cell cytoplasm. This data is perhaps not surprising, as nanoparticles are single molecules generally smaller than 100 nm. However, the route by which nanoparticles enter the cell is still unclear, since disruption of lipid rafts by methyl-β-cyclodextrin (MCD) or inhibition of caveolin-dependent or clathrin-dependent uptake did not perturb nanoparticle ingestion.[99]

The NLRP3 Inflammasome in Infectious Diseases

Influenza A virus

The influenza virus is one of the most prevalent human pathogenic viruses, comprising three different strains: types A, B, and C.[100] In annual epidemics, the avian and human influenza A virus infects millions of people worldwide, causing a respiratory infection commonly known as the flu, which is characterized by high fever, cough, head and limb pains, dyspnea, and interstitial or hemorrhagic pneumonia. Influenza A is a negative single-stranded (ss) RNA virus with an RNA genome segmented into eight parts, with each segment containing one or two genes, which differs from common viral genomes, which consist of a single nucleic acid molecule. This "genome segmentation" enables antigenic variation and mutation, explaining the frequent appearance each year of new highly contagious influenza A variants as the evolving virus escapes recognition by the adaptive immune system.

There are three different classes of PRRs involved in the recognition of PAMPs generated during influenza A infection.[101] Once internalized by cells such as plasmacytoid dendritic cells (DCs), endosomal membrane-bound TLR7 or TLR3 recognizes influenza A via its ssRNA and dsRNA respectively, resulting in a strong type I interferon response.[102–104] In addition, the intracellular retinoic-acid-inducible gene-I (RIG-I) senses 5′-triphosphates on virus genomic ssRNA, triggering a type I IFN response in respiratory bronchial epithelial cells.[103,105,106]

Recent studies revealed the NLRP3 inflammasome as an essential mediator in driving proinflammatory and protective response to bacterial infection. This prompted investigation into NLRP3 in viral recognition and a protective inflammatory response to the influenza A virus. *In vivo* studies showed that mice deficient in NLRP3 or caspase-1 infected with a high (sublethal) dose of the influenza A virus had reduced survival (less than 40%) in comparison with wild-type mice, indicating that the NLRP3 inflammasome is protective against influenza A virus–induced mortality.[34,107] Moreover, measurement of cytokines in the bronchoalveolar lavage fluid (BALF) following intranasal challenge with influenza A revealed a significant reduction of IL-1β and IL-18 in either NLRP3- or caspase-1-deficient mice. The levels of other inflammatory cytokines and the keratinocyte chemoattractant KC were also reduced in KO mice, which explains the appearance of fewer immune cells such as neutrophils and DCs in the KO BALF. These findings demonstrate a central role for the NLRP3 inflammasome in recruitment of inflammatory innate immune cells during influenza A infection to drive an adequate immune response.

Although the influenza A virus can be sensed by TLR-7 and RIG-I, the role of the NLRP3 inflammasome in influenza remained unclear until recently. Two studies demonstrated that influenza A can provide both the initial priming signal that upregulates transcription of pro-IL-1β and NLRP3 through TLR-7 and the second signal necessary for NLRP3 inflammasome activation.[35,107] In a study by Thomas *et al.*, NLRP3- or caspase-1-deficient DCs showed a significant decrease in IL-1β secretion in response to influenza A ssRNA, suggesting that viral nucleic acid is sufficient to activate NLRP3.[107] In a separate study by Ichinohe and colleagues this finding was not replicated, and the authors suggested the virally encoded M2 protein to be the genuine NLRP3 activator involved in influenza.[35] M2 is a homotetrameric integral membrane proton ion channel that is essential for influenza infection and replication. It promotes viral entry into cells via endosome neutralization, thus enabling uncoating of the virus within the cytoplasm.[108] It also forms proton channels in the trans-Golgi-network — normally an acidic environment — in order to neutralize the luminal pH, thereby ensuring proper production and assembly of virion particles.[109] In bone-marrow-derived macrophages and DCs transduced with a lentivirus carrying the M2 channel protein, IL-1β and IL-18 secretion were significantly increased in comparison with nonprimed and/or nontransduced cells.[35] The Na^+H^+-antiporter monensin can export protons from the trans-Golgi-network and has channel-mediated effects similar to M2 *in vitro*, as primed immune cells incubated with monensin produced IL-1β and IL-18. Localization studies revealed that the M2 protein channel localized in the trans-Golgi-network is able to activate the NLRP3 inflammasome.[35] This finding suggests that the change in cellular homeostasis, i.e. the pH imbalance mediated by M2, is the key danger signal sensed by the NLRP3 inflammasome during influenza infection.

As with NLRP3 inflammasome activation in response to other stimuli, lysosomal destabilization and burst with subsequent release of proteolytic enzymes into the cytoplasm, production of ROS, and potassium efflux are all thought to be important for fighting influenza A infection.[34] Whether the NLRP3 inflammasome is necessary and sufficient for adequate adaptive immune responses to influenza A infection remains to be clarified, given several discordant results in the literature. As multiple pathways are involved in influenza A recognition and defense, the crosstalk between the different molecular pattern recognition systems, namely RIG-I, TLR, and NLRP3 inflammasomes, must be tightly regulated to avoid overproduction of cytokines and to prevent a possible cytokine storm that can exacerbate influenza A infection.[101]

Malaria

With 300–600 million infections and over 2 million deaths worldwide per year, malaria is a serious problem, particularly in developing nations within Asia, Africa (especially sub-Saharan Africa), and the Americas.[110] Malaria is caused by infection with protozoan parasites of the *Plasmodium* genus resulting in periodic episodes of cyclic fevers and chills.[111] The disease is transmitted via protozoan-infected mosquitoes and is therefore most endemic in the tropics. The life cycle of parasitemia involves various phases, including the liver and blood stages. During the blood stage, parasites released from the liver colonize erythrocytes and replicate therein, causing widespread destruction of red blood cells and associated anemia.[112] Erythrocyte rupture with subsequent release of *Plasmodia* in the merozoite cell stage mediates a strong proinflammatory response known as the "cytokine storm," which causes the periodic fevers associated with malaria.[113,114] This excessive production of proinflammatory cytokines, including increased TNF and IL-1β, results in many of the typical symptoms of malaria, including fever, anorexia, fatigue, arthralgia, and myalgia. In 1%–2% of patients, malaria can progress into a severe and often lethal form, cerebral malaria (CM), leading to impaired consciousness, coma, and persistent neurological defects. Placental malaria can also manifest, leading to low birth weight, premature delivery, or loss of the fetus, as well as a shock-like syndrome resulting in systemic organ failure.[111] It is now widely accepted that the cytokine storm produced by the host immune response is both beneficial and damaging; it is responsible not only for parasite clearance but also for much of the pathology and clinical manifestations of the disease.[111,115,116]

The molecular mechanisms initiating the proinflammatory cytokine release in malaria have been the subject of extensive investigation. PRRs have been implicated in sensing several parasite byproducts during the progression of malaria. Activation of TLR2 has been demonstrated in response to glycosylphosphatidylinositols (GPIs), which are glycolipids from *P. falciparum*.[117,118] TLR9 is thought to induce inflammation by recognizing haemozoin, an insoluble heme crystal formed by the parasite from host blood to protect against the oxidative damage caused by free heme.[119–123] More recently, NLRP3 has been implicated in the sensing of hemozoin. A study by Dostert *et al.* revealed that components of the NLRP3 inflammasome help recruit proinflammatory immune cells to hemozoin and intiate a proinflammatory response.[32] There are, however, conflicting reports concerning the proinflammatory capacities of hemozoin, as it may not directly activate NLRP3; rather, it might stimulate the production of uric acid (released from dead cells), which can form monosodium urate (MSU) crystals and activate the inflammasome.[124] Specifically, high amounts of uric acid are observed in the peritoneal cavity of mice after intraperitoneal injection of hemozoin. Furthermore, pretreatment with allopurinol and uricase to deplete uric acid showed a decreased neutrophil influx into the peritoneal cavity following hemozoin injection.[124]

The molecular mechanism by which hemozoin induces NLRP3 inflammasome activation remains unclear and a matter of debate. Phagocytosis of hemozoin crystals, potassium efflux, and the production of ROS are crucial events in driving proinflammatory NLRP3-inflammasome-dependent responses.[32] However, lysosomal destabilization and leakage did not appear to play a role during hemozoin-mediated activation of the NLRP3 inflammasome.[114] There is evidence that Syk, the widely expressed spleen tyrosine kinase, is necessary for a lysosomal-destabilization-independent release of cathepsin B into the cytoplasm.[125] Indeed, Syk is activated upon hemozoin stimulation (upstream of NLRP3), and inhibiton of Syk in mouse macrophages diminished IL-1β maturation in response to hemozoin.[114] Interestingly, Syk interacts directly with the NLRP3 inflammasome complex by binding to the PYD domain, which also mediates the interaction between NLRP3 and ASC. These results suggest that malarial hemozoin can activate NLRP3 via two separate pathways.

Whilst malarial hemozoin can activate NLRP3 and appears to play an important role in malaria, the role of the NLRP3 inflammasome in the pathogenesis of the more severe CM remains contentious. On the one hand, longer survival and reduced development of CM have been observed in NLRP3-deficient mice,[32] but mice deficient in ASC, caspase-1, or IL-1R showed no differences in CM development in comparison with wild-type mice.[126] However, in the latter study, NLRP3-deficient mice showed a delayed onset of CM, suggesting a role for NLRP3. The role of the NLRP3 inflammasome in the pathogenesis of CM requires further investigation.

The NLRP3 Inflammasome in Metabolic Diseases

Gout and pseudogout

Gout is an ancient disease that has been well documented throughout history — in the Egyptian (27th century BC) and Grecian (5th century BC) civilizations — as a condition distinct from other arthritic diseases.[127] During the past century, the prevalence of gout and pseudogout increased concomitantly with the increasing rates of obesity, diabetes, and metabolic syndrome.[128] By definition, gout is associated with increased levels of circulating uric acid (hyperuricemia) and the deposition of MSU crystals within the joints and periarticular tissues.[129] Like gout, pseudogout is also an arthritic disease that affects the joints. However, in pseudogout the inflammation is coupled with deposition of salt crystals formed from calcium pyrophosphate dihydate (CPPD) rather than from uric acid. These crystals are easily distinguished by the extent of birefringence (double refraction) during microscopic analysis: MSU crystals are strongly birefringent and easily seen under polarized light, whereas CPPD crystals are nonrefringent and better visualized without polarizing filters.[130] In both diseases the joints affected are most commonly the big toe and ankle, in which the crystalline deposits cause severe swelling and immense pain. Gout begins with asymptomatic hyperuricemia, followed by periodic attacks of acute gout, with patients often presenting with fevers, chills, and severe pains leading to joint swelling and skin rashes (erythema). Acute attacks of gouty inflammation over many years lead to chronic disease in which large crystal deposits cause substantial joint damage with chronic inflammation and immobility.[131]

Although it had long been accepted that the causative agents are the crystalline materials specific to each disease, the molecular mechanisms underlying both MSU- and CPPD-induced inflammation have only more recently become clear. A study by Martinon *et al.* demonstrated that both MSU and CPPD crystals elicited proinflammatory effects on LPS-primed macrophages, primarily resulting from caspase-1-dependent release of IL-1β.[46] Utilizing strains of mice deficient in various inflammasome components (NLRP3, ASC, or caspase-1), it was shown that macrophages from such mice were incapable of crystal-induced IL-1β production. Furthermore, the finding that mice deficient in NLRP3, ASC, caspase-1, or the IL-1 receptor have impaired influx of neutrophils into the peritoneum following injection of MSU crystals supports the proinflammatory nature of such crystals *in vivo*.[46,132,133]

Whilst the exact mechanism of MSU- and CPPD-induced assembly of the NLRP3 inflammasome still remains to be elucidated, it is likely to involve phagosomal disruption and/or increased oxidative stress (i.e. ROS), as observed following the uptake of other crystalline materials such as cholesterol crystals and silica.[39,48] Additionally, MSU crystals have been shown to directly engage cholesterol components of cell membranes to elicit Th2 cellular responses via activation of Syk kinase.[134] However, given that these effects were found to be independent of NLRP3, it is unlikely that MSU or CPPD crystals activate the inflammasome via this route.[135]

Alzheimer's disease

Alzheimer's disease is a progressive neurodegenerative disease — most commonly seen in the elderly — that is characterized by synapse dysfunction and neuronal cell death that manifests as progressive dementia. The accumulation of extracellular amyloid-β (Aβ) has long been associated with the disease. Indeed, prominent extracellular deposits of insoluble Aβ fibrils are clearly visible upon microscopic examination of the affected brain.[136,137] This association, coupled with the identification of several disease-causing mutations in the amyloid precursor protein (APP) which is processed to form Aβ, initiated the "amyloid hypothesis" of Alzheimer's disease, which suggests that Aβ accumulation in senile plaques is the primary event in the pathogenesis of the disease.[138,139] Inflammatory processes involving microglia and invading mononuclear phagocytes are also central to the pathogenesis of Alzheimer's disease. Specifically, microglia are recruited to senile plaques, where they phagocytose Aβ, resulting in activation and secretion of inflammatory cytokines, primarily IL-1β.[140,141] As expected, high levels of IL-1β can be found in the cerebrospinal fluid of Alzheimer's patients.[142]

A potential molecular mechanism responsible for the inflammatory response to Aβ was recently shown to be mediated by the NLRP3 inflammasome in a study by Halle *et al.*[43] Fibrillar Aβ was shown to elicit IL-1β production from microglia in a caspase-1-dependent manner. Furthermore, microglia obtained from NLRP3- and ASC-deficient mice were unresponsive to Aβ stimulation, demonstrating that Aβ is sensed by the NLRP3 inflammasome. Phagocytosis of fibrillar Aβ appeared to activate NLRP3 inflammasomes by facilitating lysosomal damage and the release of cathepsin B. Previous studies have also linked cathepsin B to Alzheimer's disease: large amounts of the protease have been reported around senile plaques, and inhibition of cathepsin B is somewhat protective in a mouse model of Alzheimer's disease.[143,144] Finally, the inflammatory response to Aβ was also shown *in vivo* to be mediated by the NLRP3 inflammasome, as mice deficient in NLRP3, ASC, or caspase-1 exhibited a reduced accumulation of microglia adjacent to stereotactically introduced Aβ.[43]

Atherosclerosis

Atherosclerosis is an underlying cause of about 50% of all deaths in the western world. It is a disease predominantly of the large arteries, which results in the formation of atherosclerotic plaques within the arterial walls (atheroma). Rupture of these plaques can lead to a thrombus or blood clot, which often progresses to the clinical complications of myocardial infarction or stroke.[145] Although the disease displays many hallmarks of an inflammatory disease and is characterized by the recruitment of immune cells to artery walls, and whilst infectious agents have been found in human lesions,[146] the appearance of atherosclerosis in germ-free mice suggests that the chronic inflammation associated with atherosclerosis is initiated by endogenous components, rather than by infectious agents.[147] Cholesterol is one such endogenous molecule that has long been associated with atherosclerosis. Large amounts of cholesterol — either in the form of cholesteryl esters within macrophage foam cells or in the intra- and extracellular environment as crystalline cholesterol — can be found within atherosclerotic plaques in close proximity to activated immune cells.[145] Furthermore, elevated cholesterol levels in blood are closely linked to disease development, and dietary administration of increased amounts of cholesterol is sufficient to cause atherosclerosis in a variety of animal models.

Cholesterol has very low solubility in aqueous solutions, and cholesterol crystals can be detected by standard histology as so-called "cholesterol crystal clefts" in advanced atherosclerotic lesions. However, utilizing reflection-microscopic techniques, Duewell *et al.* showed that in addition to the large crystals that leave clefts in tissues, a great number of much smaller cholesterol crystals accumulate in the extracellular space as well as within macrophages in atherosclerotic lesions.[48] Cholesterol crystals can induce secretion of large amounts of IL-1β in an NLRP3-inflammasome-dependent manner in mouse- or human-primed macrophages.[48,49] Moreover, when bone marrow from mice deficient in either NLRP3, ASC, or IL-1α/β was transferred into irradiated atherosclerosis-prone low-density lipoprotein (LDL) receptor–deficient mice and the animals were fed a high-cholesterol "Western" diet, they displayed markedly decreased atheromata as well as reduced levels of circulating IL-18 in comparison with mice transplanted with wild-type bone marrow.[48]

LDL is an important particle in the progression of atherosclerosis that can be found within the innermost layer of the artery wall early in the disease process, where it can become modified and promote recruitment of monocytes as well as the accumulation of cholesterol-engorged macrophage foam cells.[145] The oxidized form of LDL, whose formation is dependent on ROS produced from macrophages and adjacent cells, is tightly linked to atherosclerotic disease progression.[148] Interestingly, oxidized LDL itself also induces the production of ROS[149] and causes lysosomal damage,[150] both of which are implicated as possible mechanisms of NLRP3 inflammasome activation. Moreover, the reduced secretion of IL-1β observed in cathepsin-B- or cathepsin-L-deficient mouse macrophages stimulated with cholesterol crystals suggests that phagosomal leakage is required to activate the NLRP3 inflammasome in atherosclerosis.[48] In an *in vivo* setting the priming step required for NLRP3 inflammasome activation is likely to be mediated by a modified form of LDL. Indeed, several studies have found that modified LDL induces the expression of pro-IL-1β in macrophages.[48,51] This is due to the ability of oxidized LDL to activate NF-κB following recognition by a receptor complex involving TLR4/6,[151] CD14,[152] and various scavenger receptors (e.g. CD36, SR-A).[153,154] Activation of NF-κB by oxidized LDL also requires MyD88-dependent signaling[155,156] and the kinase activity of IRAK-4,[157] both components of canonical intracellular TLR signaling. Perhaps not surprisingly, mice deficient in any of these receptors or signaling components displayed reduced atherosclerotic lesions. Within the atherosclerotic lesion, oxidized LDL also provides cells with the activation step of the NLRP3 inflammasome as it facilitates the formation of cholesterol crystals.[48,158]

Obesity and early type 2 diabetes (T2D)

The prevalence of obesity is constantly rising, currently affecting around 30% of the world's adult population.[159] Obesity has long been associated with an underlying chronic low-level inflammation characterized by abnormal cytokine production (e.g. increased TNF), increased acute phase reactants, and activation of inflammatory signaling pathways.[160] Obese mice display highly enriched gene networks for controlling inflammation, immunity, and macrophage activation.[161] Accompanying the rise in obesity is increased inflammation-driven metabolic diseases, including atherosclerosis and T2D.[160,162] Several recent studies have found that obesity-related risk factors induce the production of IL-1β in a NLRP3-inflammasome-dependent manner, leading to insulin resistance in early T2D.[47,52] In a study by Vandanmagsar *et al.* mice fed on a normal chow diet had increased expression of both NLRP3 and IL-1β in visceral adipose tissue, which correlated directly with body weight and adiposity (fat content) when compared to mice fed on a calorie-restricted diet.[47] In the same study a similar association was found in humans; weight loss in obese T2D individuals correlated with reduced NLRP3 and IL-1β expression in subcutaneous adipose tissue. The direct involvement of NLRP3 in obesity was suggested by studies on gene-deficient mice. Firstly, NLRP3-deficient mice fed on a high-fat diet displayed reduced caspase-1 activation and pro-IL-1β expression in adipose tissue as well as a loss of IL-18 in serum when compared to their wild-type counterparts. In addition, NLRP3- or caspase-1-deficient mice are more protected from high-fat diet-induced insulin resistance than are wild-type mice.[47,163] Decreased insulin sensitivity was found to be a consequence of NLRP3-inflammasome-mediated activation of effector adipose T cells which, through release of interferon-gamma, mediate downstream pathways, resulting in insulin resistance.

In some forms of obesity, elevated circulating levels of free fatty acids (FFAs) result in their accumulation in adipose tissue. FFAs might act as a sterile "danger signal" to activate adipose-tissue-resident immune cells via assembly of the NLRP3 inflammasome.[164,165] Vandanmagsar *et al.* tested whether ceramide, which is composed of sphingoside and fatty acid, might serve as a danger signal.[166] Although ceramide-stimulated macrophages displayed NLRP3-dependent caspase-1 activation and production of IL-1β, the output levels were very low.[47] Another candidate danger signal is palmitate, one the most abundant FFAs in plasma that is further elevated in obesity. Recently Wen *et al.* tested this lipid's ability to activate the NLRP3 inflammasome.[52] When stimulated with palmitate, LPS-primed macrophages showed NLRP3-dependent caspase-1 activation and IL-1β maturation. Interestingly, palmitate seems to mediate NLRP3 activation via a unique mechanism involving reduction of AMP-protein kinase (AMPK) activity which results in defective autophagosomal processes and subsequently the accumulation of mitochondrial ROS. Other studies have suggested that mitochondrial ROS is required for NLRP3 activation.[59,60] Wen *et al.* suggest that subsequent IL-1β production via NLRP3 mediates insulin resistance both directly via inhibition of insulin signaling and indirectly via increased production of TNF, a known inducer of insulin resistance.[167,168]

Whilst FFAs appear to be able to mediate the activation of NLRP3 inflammasome in obesity, these molecules may also be responsible for priming immune cells in adipose tissues via NF-κB activation. Previously, FFAs were shown to induce NF-κB signaling via TLR4,[165] the expression of which is increased in adipose tissue during obesity.[169] Furthermore, TLR4-deficient mice are protected from fat-induced inflammation and insulin resistance.[170–173]

Type 2 diabetes progression

Another disease that is potentially mediated by a sterile endogenous danger signal is type 2 diabetes (T2D), some mechanisms of which have only recently been described. Disease characteristics include obesity-induced insulin resistance and dysfunction of islet beta cells in the pancreas.[174] An additional feature is pancreatic accumulation of islet amyloid polypeptide (IAPP), which is readily taken up by both dendritic cells (DCs) and macrophages.[175] A recent study showed that stimulation of primed bone marrow-derived DCs and macrophages with human IAPP induced cleavage of caspase-1, formation of ASC specks, and production of IL-1β in an NLRP3-dependent manner.[51] Consistent with prior observations, activation of the NLRP3 inflammasome required ROS production and the activity of the phagosomal protease cathepsin B. Mouse IAPP does not form active amyloid aggregates; thus, a transgenic mouse model overexpressing human IAPP was utilized to demonstrate that IAPP induces macrophages to produce IL-1β in pancreatic islets *in vivo*.[51] It is possible that the IL-1-induced inflammation produced in the pancreas results in death of beta cells, T2D disease progression, and development of insulin-dependent diabetes.

Thioredoxin-interacting protein (TXNIP) has previously been implicated in NLRP3 inflammasome activation,[176] and glucose-induced TXNIP has been shown to correlate with activation of caspase-1 and IL-1β production in human and mouse adipose tissue.[177] In contrast, the study by Masters *et al.* suggests that TXNIP is not involved in T2D progression, since normal NLRP3-induced IL-1β production was observed in cells from TXNIP-deficient mice. These discrepancies may be explained by different experimental approaches, which may result in the induction of NLRP3 by TXNIP only under specific conditions or in specific cells.

In the study by Masters *et al.* the authors also investigated the endogenous priming mechanism of the NLRP3 inflammasome in T2D. Population-based studies have found a correlation between obesity, T2D, and increased concentrations of modified forms of LDL, including oxidized LDL.[178] In the setting of atherosclerosis, modified LDLs have the capacity to prime macrophages for NLRP3 activation via TLR2 and TLR6.[48]

Here, modified LDL was also demonstrated to prime cells, this time via TLR4 signaling, for IAPP-induced inflammasome activation in T2D.[51]

CURRENT THERAPEUTIC STRATEGIES TARGETING INFLAMMASOMOPATHIES

No therapeutics are currently available that directly target the NLRP3 inflammasome. Available therapies are instead aimed at the neutralization of the major NLRP3 product, IL-1β. Presently a range of antibodies and inhibitors exist to target IL-1β,[179] some of which have been administered in NLRP3-associated diseases. Anakinra, a recombinant protein almost identical to endogenous interleukin-1 receptor antagonist, resolved the inflammatory symptoms associated with a number of genetic NLRP3 disorders (CAPS), including those associated with FCAS and MWS, as well as improving the hearing problems in some MWS cases and decreasing CNS symptoms in NOMID patients.[70,180,181] Furthermore, more recent studies with new biologic IL-1 inhibitors such as Rilonacept (often referred to as IL-1 trap; IL-1α/β neutralizer)[71] and Canakinumab[182] also show efficacy in the treatment of CAPS.

Anakinra has also produced positive results in a number of metabolic NLRP3-related disease trials, including those for gout, pseudogout, and T2D.[183–186] Furthermore, Rilonacept can suppress chronic gouty arthritis.[187] Inhibition of IL-1β with the high-affinity monoclonal antibody XOMA 052 has recently been demonstrated to protect against atherosclerosis in mice.[188] In phase 2b clinical trials, XOMA 052 showed potential in the treatment of cardiovascular disease; however, the antibody was not effective in reducing hyperglycemia in T2D patients (XOMA Limited press release, March 2011). In addition, the IL-1 receptor agonist, MRC-ILA-HEART, has previously been shown in preclinical studies to improve vascular function in patients with acute coronary syndrome arising from atherosclerosis-driven inflammation.[189]

The observation that NLRP3-driven metabolic diseases are linked to obesity suggests that more upstream, preventive therapeutic approaches are required, in contrast to the current strategies, which are aimed at blocking the outcomes of inflammasome activation. A recent study showed that direct inhibition of caspase-1 was beneficial in reducing obesity and improving insulin sensitivity in mice.[163] In addition, the small molecule caspase-1 inhibitor, pralnacasan (VX-765), blocked excess IL-1β and IL-18 in PBMCs derived from CAPS patients.[88] Unfortunately, clinical trials with VX-765 were abandoned due to safety issues.[190]

CONCLUDING REMARKS

The formation of the NLRP3 inflammasome is a multifaceted process that is regulated on a number of levels, although the precise mode of activation remains unclear. The NLRP3 inflammasome assembles in response to a diverse range of stimuli and has been implicated in numerous diseases. These so-called "inflammasomopathies" span the range of disease classifications to include genetic disorders, infectious and environmentally induced conditions, and diseases driven by alterations in metabolism. To date, the treatment of inflammasomopathies is limited to inhibition of the downstream effector IL-1β. A better understanding of the precise mechanisms by which NLRP3 is activatied should reveal strategies that aid the design of superior, inflammasome-targeted therapeutics. The recent findings presented in this chapter clearly highlight the wide-ranging influence that NLRP3 has on the inflammatory response and in disease.

REFERENCES

1. Bryant C, Fitzgerald KA. (2009) Molecular mechanisms involved in inflammasome activation. *Trends Cell Biol* **19**(9): 455–464.
2. Bella J, Hindle KL, McEwan PA, Lovell SC. (2008) The leucine-rich repeat structure. *Cell Mol Life Sci* **65**(15): 2307–2333.
3. Bell JK, Mullen GE, Leifer CA, *et al.* (2003) Leucine-rich repeats and pathogen recognition in Toll-like receptors. *Trends Immunol* **24**(10): 528–533.
4. Duncan JA, Bergstralh DT, Wang Y, *et al.* (2007) Cryopyrin/NALP3 binds ATP/dATP, is an ATPase, and requires ATP binding to mediate inflammatory signaling. *Proc Natl Acad Sci USA* **104**(19): 8041–8046.
5. Ting JP, Lovering RC, Alnemri ES, *et al.* (2008) The NLR gene family: a standard nomenclature. *Immunity* **28**(3): 285–287.
6. Martinon F, Burns K, Tschopp J. (2002) The inflammasome: a molecular platform triggering activation of inflammatory caspases and processing of proIL-beta. *Mol Cell* **10**(2): 417–426.

7. Stutz A, Golenbock DT, Latz E. (2009) Inflammasomes: too big to miss. *J Clin Invest* **119**(12): 3502–3511.
8. Faustin B, Lartigue L, Bruey JM, *et al.* (2007) Reconstituted NALP1 inflammasome reveals two-step mechanism of caspase-1 activation. *Mol Cell* **25**(5): 713–724.
9. Boyden ED, Dietrich WF. (2006) Nalp1b controls mouse macrophage susceptibility to anthrax lethal toxin. *Nat Genet* **38**(2): 240–244.
10. Sutterwala FS, Flavell RA. (2009) NLRC4/IPAF: a CARD carrying member of the NLR family. *Clin Immunol* **130**(1): 2–6.
11. Franchi L, Amer A, Body-Malapel M, *et al.* (2006) Cytosolic flagellin requires Ipaf for activation of caspase-1 and interleukin-1 beta in salmonella-infected macrophages. *Nat Immunol* **7**(6): 576–582.
12. Lightfield KL, Persson J, Brubaker SW, *et al.* (2008) Critical function for Naip5 in inflammasome activation by a conserved carboxy-terminal domain of flagellin. *Nat Immunol.* **9**(10): 1171–1178.
13. Burckstummer T, Baumann C, Bluml S, *et al.* (2009) An orthogonal proteomic-genomic screen identifies AIM2 as a cytoplasmic DNA sensor for the inflammasome. *Nat Immunol* **10**(3): 266–272.
14. Fernandes-Alnemri T, Yu JW, Datta P, *et al.* (2009) AIM2 activates the inflammasome and cell death in response to cytoplasmic DNA. *Nature* **458**(7237): 509–513.
15. Hornung V, Ablasser A, Charrel-Dennis M, *et al.* (2009) AIM2 recognizes cytosolic dsDNA and forms a caspase-1-activating inflammasome with ASC. *Nature* **458**(7237): 514–518.
16. Roberts TL, Idris A, Dunn JA, *et al.* (2009) HIN-200 proteins regulate caspase activation in response to foreign cytoplasmic DNA. *Science* **323**(5917): 1057–1060.
17. Fernandes-Alnemri T, Wu J, Yu JW, *et al.* (2007) The pyroptosome: a supramolecular assembly of ASC dimers mediating inflammatory cell death via caspase-1 activation. *Cell Death Differ* **14**(9): 1590–1604.
18. Wilson KP, Black JA, Thomson JA, *et al.* (1994) Structure and mechanism of interleukin-1 beta converting enzyme. *Nature* **370**(6487): 270–275.
19. Dinarello CA. (2009) Immunological and inflammatory functions of the interleukin-1 family. *Annu Rev Immunol* **27**: 519–550.
20. Dinarello CA. (2009) Interleukin-1 beta and the autoinflammatory diseases. *N Engl J Med* **360**(23): 2467–2470.
21. Puren AJ, Fantuzzi G, Dinarello CA. (1999) Gene expression, synthesis, and secretion of interleukin 18 and interleukin-1 beta are differentially regulated

in human blood mononuclear cells and mouse spleen cells. *Proc Natl Acad Sci USA* **96**(5): 2256–2261.
22. Bergsbaken T, Fink SL, Cookson BT. (2009) Pyroptosis: host cell death and inflammation. *Nat Rev Microbiol* **7**(2): 99–109.
23. Kepp O, Galluzzi L, Zitvogel L, Kroemer G. (2010) Pyroptosis — a cell death modality of its kind? *Eur J Immunol* **40**(3): 627–630.
24. Fink SL, Cookson BT. (2006) Caspase-1-dependent pore formation during pyroptosis leads to osmotic lysis of infected host macrophages. *Cell Microbiol* **8**(11): 1812–1825.
25. Muruve DA, Petrilli V, Zaiss AK, *et al.* (2008) The inflammasome recognizes cytosolic microbial and host DNA and triggers an innate immune response. *Nature* **452**(7183): 103–107.
26. Mariathasan S, Weiss DS, Newton K, *et al.* (2006) Cryopyrin activates the inflammasome in response to toxins and ATP. *Nature* **440**(7081): 228–232.
27. Warren SE, Mao DP, Rodriguez AE, *et al.* (2008) Multiple Nod-like receptors activate caspase 1 during *Listeria monocytogenes* infection. *J Immunol* **180**(11): 7558–7564.
28. Craven RR, Gao X, Allen IC, *et al.* (2009) *Staphylococcus aureus* alpha-hemolysin activates the NLRP3-inflammasome in human and mouse monocytic cells. *PLoS One* **4**(10): e7446.
29. Walev I, Reske K, Palmer M, *et al.* (1995) Potassium-inhibited processing of IL-1 beta in human monocytes. *EMBO J* **14**(8): 1607–1614.
30. Gurcel L, Abrami L, Girardin S, *et al.* (2006) Caspase-1 activation of lipid metabolic pathways in response to bacterial pore-forming toxins promotes cell survival. *Cell* **126**(6): 1135–1145.
31. Poeck H, Bscheider M, Gross O, *et al.* (2010) Recognition of RNA virus by RIG-I results in activation of CARD9 and inflammasome signaling for interleukin-1 beta production. *Nat Immunol* **11**(1): 63–69.
32. Dostert C, Guarda G, Romero JF, *et al.* (2009) Malarial hemozoin is a Nalp3 inflammasome activating danger signal. *PLoS One* **4**(8): e6510.
33. Kanneganti TD, Ozoren N, Body-Malapel M, *et al.* (2006) Bacterial RNA and small antiviral compounds activate caspase-1 through cryopyrin/Nalp3. *Nature* **440**(7081): 233–236.
34. Allen IC, Scull MA, Moore CB, *et al.* (2009) The NLRP3 inflammasome mediates *in vivo* innate immunity to influenza A virus through recognition of viral RNA. *Immunity* **30**(4): 556–565.
35. Ichinohe T, Pang IK, Iwasaki A. (2010) Influenza virus activates inflammasomes via its intracellular M2 ion channel. *Nat Immunol* **11**(5): 404–410.

36. Gross O, Poeck H, Bscheider M, *et al.* (2009) Syk kinase signalling couples to the Nlrp3 inflammasome for anti-fungal host defence. *Nature* **459**(7245): 433–436.
37. Hise AG, Tomalka J, Ganesan S, *et al.* (2009) An essential role for the NLRP3 inflammasome in host defense against the human fungal pathogen *Candida albicans. Cell Host Microbe* **5**(5): 487–497.
38. Franchi L, Nunez G. (2008) The Nlrp3 inflammasome is critical for aluminium hydroxide-mediated IL-1 beta secretion but dispensable for adjuvant activity. *Eur J Immunol* **38**(8): 2085–2089.
39. Hornung V, Bauernfeind F, Halle A, *et al.* (2008) Silica crystals and aluminum salts activate the NALP3 inflammasome through phagosomal destabilization. *Nat Immunol* **9**(8): 847–856.
40. Kool M, Petrilli V, De Smedt T, *et al.* (2008) Cutting edge: alum adjuvant stimulates inflammatory dendritic cells through activation of the NALP3 inflammasome. *J Immunol* **181**(6): 3755–3759.
41. Li H, Willingham SB, Ting JP, Re F. (2008) Cutting edge: inflammasome activation by alum and alum's adjuvant effect are mediated by NLRP3. *J Immunol* **181**(1): 17–21.
42. Eisenbarth SC. (2008) Use and limitations of alum-based models of allergy. *Clin Exp Allergy* **38**(10): 1572–1575.
43. Halle A, Hornung V, Petzold GC, *et al.* (2008) The NALP3 inflammasome is involved in the innate immune response to amyloid-beta. *Nat Immunol* **9**(8): 857–865.
44. Dostert C, Petrilli V, Van Bruggen R, *et al.* (2008) Innate immune activation through Nalp3 inflammasome sensing of asbestos and silica. *Science* **320**(5876): 674–677.
45. Cassel SL, Eisenbarth SC, Iyer SS, *et al.* (2008) The Nalp3 inflammasome is essential for the development of silicosis. *Proc Natl Acad Sci USA* **105**(26): 9035–9040.
46. Martinon F, Petrilli V, Mayor A, *et al.* (2006) Gout-associated uric acid crystals activate the NALP3 inflammasome. *Nature* **440**(7081): 237–241.
47. Vandanmagsar B, Youm YH, Ravussin A, *et al.* (2011) The NLRP3 inflammasome instigates obesity-induced inflammation and insulin resistance. *Nat Med* **17**(2): 179–188.
48. Duewell P, Kono H, Rayner KJ, *et al.* (2010) NLRP3 inflammasomes are required for atherogenesis and activated by cholesterol crystals. *Nature* **464**(7293): 1357–1361.
49. Rajamaki K, Lappalainen J, Oorni K, *et al.* (2010) Cholesterol crystals activate the NLRP3 inflammasome in human macrophages: a novel link between cholesterol metabolism and inflammation. *PLoS One* **5**(7): e11765.

50. Yamasaki K, Muto J, Taylor KR, *et al.* (2009) NLRP3/cryopyrin is necessary for interleukin-1 beta (IL-1 beta) release in response to hyaluronan, an endogenous trigger of inflammation in response to injury. *J Biol Chem* **284**(19): 12762–12771.
51. Masters SL, Dunne A, Subramanian SL, *et al.* (2010) Activation of the NLRP3 inflammasome by islet amyloid polypeptide provides a mechanism for enhanced IL-1 beta in type 2 diabetes. *Nat Immunol* **11**(10): 897–904.
52. Wen H, Gris D, Lei Y, *et al.* (2011) Fatty acid-induced NLRP3-ASC inflammasome activation interferes with insulin signaling. *Nat Immunol* **12**(5): 408–415.
53. Sutterwala FS, Ogura Y, Szczepanik M, *et al.* (2006) Critical role for NALP3/CIAS1/cryopyrin in innate and adaptive immunity through its regulation of caspase-1. *Immunity* **24**(3): 317–327.
54. O'Connor W, Jr., Harton JA, Zhu X, *et al.* (2003) Cutting edge: CIAS1/cryopyrin/PYPAF1/NALP3/CATERPILLER 1.1 is an inducible inflammatory mediator with NF-κB suppressive properties. *J Immunol* **171**(12): 6329–6333.
55. Bauernfeind FG, Horvath G, Stutz A, *et al.* (2009) Cutting edge: NF-κB activating pattern recognition and cytokine receptors license NLRP3 inflammasome activation by regulating NLRP3 expression. *J Immunol* **183**(2): 787–791.
56. Franchi L, Eigenbrod T, Nunez G. (2009) Cutting edge: TNF-alpha mediates sensitization to ATP and silica via the NLRP3 inflammasome in the absence of microbial stimulation. *J Immunol* **183**(2): 792–796.
57. Guarda G, Zenger M, Yazdi AS, *et al.* (2011) Differential expression of NLRP3 among hematopoietic cells. *J Immunol* **186**(4): 2529–2534.
58. Meissner F, Molawi K, Zychlinsky A. (2008) Superoxide dismutase 1 regulates caspase-1 and endotoxic shock. *Nat Immunol* **9**(8): 866–872.
59. Zhou R, Yazdi AS, Menu P, Tschopp J. (2011) A role for mitochondria in NLRP3 inflammasome activation. *Nature* **469**(7329): 221–225.
60. Nakahira K, Haspel JA, Rathinam VA, *et al.* (2011) Autophagy proteins regulate innate immune responses by inhibiting the release of mitochondrial DNA mediated by the NALP3 inflammasome. *Nat Immunol* **12**(3): 222–230.
61. Meissner F, Seger RA, Moshous D, *et al.* (2010) Inflammasome activation in NADPH oxidase defective mononuclear phagocytes from patients with chronic granulomatous disease. *Blood* **116**(9): 1570–1573.
62. Latz E. (2010) The inflammasomes: mechanisms of activation and function. *Curr Opin Immunol* **22**(1): 28–33.

63. Petrilli V, Papin S, Dostert C, *et al.* (2007) Activation of the NALP3 inflammasome is triggered by low intracellular potassium concentration. *Cell Death Differ* **14**(9): 1583–1589.
64. Perregaux D, Gabel CA. (1994) Interleukin-1 beta maturation and release in response to ATP and nigericin. Evidence that potassium depletion mediated by these agents is a necessary and common feature of their activity. *J Biol Chem* **269**(21): 15195–15203.
65. Kile RI, Rusk HA. (1940) A case of cold urticaria with an unusual family history. *J Am Med Assoc* **114**(12): 1067–1068.
66. Muckle TJ, Wells M. (1962) Urticaria, deafness, and amyloidosis: a new heredo-familial syndrome. *Q J Med* **31**(2): 235–248.
67. Prieur AM, Griscelli C. (1981) Arthropathy with rash, chronic meningitis, eye lesions, and mental retardation. *J Pediatr* **99**(1): 79–83.
68. Neven B, Prieur AM, Quartier dit Maire P. (2008) Cryopyrinopathies: update on pathogenesis and treatment. *Nat Clin Pract Rheumatol* **4**(9): 481–489.
69. Hoffman HM, Brydges SD. (2011) Genetic and molecular basis of inflammasome-mediated disease. *J Biol Chem* **286**(13): 10889–10896.
70. Hoffman HM, Rosengren S, Boyle DL, *et al.* (2004) Prevention of cold-associated acute inflammation in familial cold autoinflammatory syndrome by interleukin-1 receptor antagonist. *Lancet* **364**(9447): 1779–1785.
71. Hoffman HM, Throne ML, Amar NJ, *et al.* (2008) Efficacy and safety of rilonacept (interleukin-1 Trap) in patients with cryopyrin-associated periodic syndromes: results from two sequential placebo-controlled studies. *Arthritis Rheum* **58**(8): 2443–2452.
72. Cuisset L, Jeru I, Dumont B, *et al.* (2011) Mutations in the autoinflammatory cryopyrin-associated periodic syndrome gene: epidemiological study and lessons from eight years of genetic analysis in France. *Ann Rheum Dis* **70**(3): 495–499.
73. Hoffman HM, Wanderer AA, Broide DH. (2001) Familial cold autoinflammatory syndrome: phenotype and genotype of an autosomal dominant periodic fever. *J Allergy Clin Immunol* **108**(4): 615–620.
74. Hoffman HM, Mueller JL, Broide DH, *et al.* (2001) Mutation of a new gene encoding a putative pyrin-like protein causes familial cold autoinflammatory syndrome and Muckle–Wells syndrome. *Nat Genet* **29**(3): 301–305.
75. Shinkai K, McCalmont TH, Leslie KS. (2008) Cryopyrin-associated periodic syndromes and autoinflammation. *Clin Exp Dermatol* **33**(1): 1–9.
76. Feldmann J, Prieur AM, Quartier P, *et al.* (2002) Chronic infantile neurological cutaneous and articular syndrome is caused by mutations in CIAS1, a

gene highly expressed in polymorphonuclear cells and chondrocytes. *Am J Hum Genet* **71**(1): 198–203.
77. Prieur AM. (2001) A recently recognised chronic inflammatory disease of early onset characterised by the triad of rash, central nervous system involvement and arthropathy. *Clin Exp Rheumatol* **19**(1): 103–106.
78. Prieur AM, Griscelli C, Lampert F, *et al.* (1987) A chronic, infantile, neurological, cutaneous and articular (CINCA) syndrome. A specific entity analysed in 30 patients. *Scand J Rheumatol Suppl* **66**: 57–68.
79. Cuisset L, Drenth JP, Berthelot JM, *et al.* (1999) Genetic linkage of the Muckle–Wells syndrome to chromosome 1q44. *Am J Hum Genet* **65**(4): 1054–1059.
80. Hoffman HM, Wright FA, Broide DH, *et al.* (2000) Identification of a locus on chromosome 1q44 for familial cold urticaria. *Am J Hum Genet* **66**(5): 1693–1698.
81. Touitou I, Lesage S, McDermott M, *et al.* (2004) Infevers: an evolving mutation database for auto-inflammatory syndromes. *Hum Mutat* **24**(3): 194–198.
82. Infevers: an online database for autoinflammatory mutations [database on the Internet]. (2011) Available from http://fmf.igh.cnrs.fr/ISSAID/infevers/.
83. Aksentijevich I, Nowak M, Mallah M, *et al.* (2002) *De novo* CIAS1 mutations, cytokine activation, and evidence for genetic heterogeneity in patients with neonatal-onset multisystem inflammatory disease (NOMID): a new member of the expanding family of pyrin-associated autoinflammatory diseases. *Arthritis Rheum* **46**(12): 3340–3348.
84. Aksentijevich I, C DP, Remmers EF, *et al.* (2007) The clinical continuum of cryopyrinopathies: novel CIAS1 mutations in North American patients and a new cryopyrin model. *Arthritis Rheum* **56**(4): 1273–1285.
85. Jesus AA, Silva CA, Segundo GR, *et al.* (2008) Phenotype–genotype analysis of cryopyrin-associated periodic syndromes (CAPS): description of a rare non-exon 3 and a novel CIAS1 missense mutation. *J Clin Immunol* **28**(2): 134–138.
86. Henderson C, Goldbach-Mansky R. (2010) Monogenic autoinflammatory diseases: new insights into clinical aspects and pathogenesis. *Curr Opin Rheumatol* **22**(5): 567–578.
87. Saito M, Nishikomori R, Kambe N, *et al.* (2008) Disease-associated CIAS1 mutations induce monocyte death, revealing low-level mosaicism in mutation-negative cryopyrin-associated periodic syndrome patients. *Blood* **111**(4): 2132–2141.
88. Stack JH, Beaumont K, Larsen PD, *et al.* (2005) IL-converting enzyme/caspase-1 inhibitor VX-765 blocks the hypersensitive response to an

inflammatory stimulus in monocytes from familial cold autoinflammatory syndrome patients. *J Immunol* **175**(4): 2630–2634.

89. Brydges SD, Mueller JL, McGeough MD, *et al.* (2009) Inflammasome-mediated disease animal models reveal roles for innate but not adaptive immunity. *Immunity* **30**(6): 875–887.
90. Meng GX, Zhang FP, Fuss I, *et al.* (2009) A mutation in the Nlrp3 gene causing inflammasome hyperactivation potentiates Th17 cell–dominant immune responses. *Immunity* **30**(6): 860–874.
91. Steinman L. (2007) A brief history of T(H)17, the first major revision in the T(H)1/T(H)2 hypothesis of T cell-mediated tissue damage. *Nat Med* **13**(2): 139–145.
92. Rosengren S, Mueller JL, Anderson JP, *et al.* (2007) Monocytes from familial cold autoinflammatory syndrome patients are activated by mild hypothermia. *J Allergy Clin Immunol* **119**(4): 991–996.
93. Cooke WE. (1924) Fibrosis of the lungs due to the inhalation of asbestos dust. *Br Med J* **2**(3317): 142, 147.
94. Cooke WE. (1927) Pulmonary asbestosis. *Br Med J* **2**(3491): 1024–1025.
95. Klein J. (2007) Probing the interactions of proteins and nanoparticles. *Proc Natl Acad Sci USA* **104**(7): 2029–2030.
96. Garabrant DH, Fine LJ, Oliver C, *et al.* (1987) Abnormalities of pulmonary function and pleural disease among titanium metal production workers. *Scand J Work Environ Health* **13**(1): 47–51.
97. Inoue K, Takano H, Ohnuki M, *et al.* (2008) Size effects of nanomaterials on lung inflammation and coagulatory disturbance. *Int J Immunopathol Pharmacol* **21**(1): 197–206.
98. Bermudez E, Mangum JB, Asgharian B, *et al.* (2002) Long-term pulmonary responses of three laboratory rodent species to subchronic inhalation of pigmentary titanium dioxide particles. *Toxicol Sci* **70**(1): 86–97.
99. Yazdi AS, Guarda G, Riteau N, *et al.* (2010) Nanoparticles activate the NLR pyrin domain containing 3 (Nlrp3) inflammasome and cause pulmonary inflammation through release of IL-1alpha and IL-1beta. *Proc Natl Acad Sci USA* **107**(45): 19449–19454.
100. Pang IK, Iwasaki A. (2011) Inflammasomes as mediators of immunity against influenza virus. *Trends Immunol* **32**(1): 34–41.
101. Owen DM, Gale M, Jr. (2009) Fighting the flu with inflammasome signaling. *Immunity* **30**(4): 476–488.
102. Lund JM, Alexopoulou L, Sato A, *et al.* (2004) Recognition of single-stranded RNA viruses by Toll-like receptor 7. *Proc Natl Acad Sci USA* **101**(15): 5598–5603.

103. Guillot L, Le Goffic R, Bloch S, *et al.* (2005) Involvement of Toll-like receptor 3 in the immune response of lung epithelial cells to double-stranded RNA and influenza A virus. *J Biol Chem* **280**(7): 5571–5580.
104. Le Goffic R, Balloy V, Lagranderie M, *et al.* (2006) Detrimental contribution of the Toll-like receptor (TLR)3 to influenza A virus–induced acute pneumonia. *PLoS Pathog* **2**(6): e53.
105. Pichlmair A, Schulz O, Tan CP, *et al.* (2006) RIG-I-mediated antiviral responses to single-stranded RNA bearing 5′-phosphates. *Science* **314**(5801): 997–1001.
106. Hornung V, Ellegast J, Kim S, *et al.* (2006) 5′-triphosphate RNA is the ligand for RIG-I. *Science* **314**(5801): 994–997.
107. Thomas PG, Dash P, Aldridge JR, Jr., *et al.* (2009) The intracellular sensor NLRP3 mediates key innate and healing responses to influenza A virus via the regulation of caspase-1. *Immunity* **30**(4): 566–575.
108. Pinto LH, Lamb RA. (2006) The M2 proton channels of influenza A and B viruses. *J Biol Chem* **281**(14): 8997–9000.
109. Sakaguchi T, Leser GP, Lamb RA. (1996) The ion channel activity of the influenza virus M2 protein affects transport through the Golgi apparatus. *J Cell Biol* **133**(4): 733–747.
110. Greenwood BM, Bojang K, Whitty CJ, Targett GA. (2005) Malaria. *Lancet* **365**(9469): 1487–1498.
111. Schofield L, Grau GE. (2005) Immunological processes in malaria pathogenesis. *Nat Rev Immunol* **5**(9): 722–735.
112. Silvie O, Mota MM, Matuschewski K, Prudencio M. (2008) Interactions of the malaria parasite and its mammalian host. *Curr Opin Microbiol* **11**(4): 352–359.
113. Kwiatkowski D, Nowak M. (1991) Periodic and chaotic host–parasite interactions in human malaria. *Proc Natl Acad Sci USA* **88**(12): 5111–5113.
114. Shio MT, Eisenbarth SC, Savaria M, *et al.* (2009) Malarial hemozoin activates the NLRP3 inflammasome through Lyn and Syk kinases. *PLoS Pathog* **5**(8): e1000559.
115. Hunt NH, Golenser J, Chan-Ling T, *et al.* (2006) Immunopathogenesis of cerebral malaria. *Int J Parasitol* **36**(5): 569–582.
116. Miller LH, Baruch DI, Marsh K, Doumbo OK. (2002) The pathogenic basis of malaria. *Nature* **415**(6872): 673–679.
117. Schofield L, Hackett F. (1993) Signal transduction in host cells by a glycosylphosphatidylinositol toxin of malaria parasites. *J Exp Med* **177**(1): 145–153.
118. Naik RS, Branch OH, Woods AS, *et al.* (2000) Glycosylphosphatidylinositol anchors of *Plasmodium falciparum*: molecular characterization and naturally

elicited antibody response that may provide immunity to malaria pathogenesis. *J Exp Med* **192**(11): 1563–1576.

119. Egan TJ. (2008) Haemozoin formation. *Mol Biochem Parasitol* **157**(2): 127–136.
120. Jaramillo M, Godbout M, Olivier M. (2005) Hemozoin induces macrophage chemokine expression through oxidative stress-dependent and -independent mechanisms. *J Immunol* **174**(1): 475–484.
121. Jaramillo M, Plante I, Ouellet N, *et al.* (2004) Hemozoin-inducible proinflammatory events *in vivo*: potential role in malaria infection. *J Immunol* **172**(5): 3101–3110.
122. Coban C, Ishii KJ, Kawai T, *et al.* (2005) Toll-like receptor 9 mediates innate immune activation by the malaria pigment hemozoin. *J Exp Med* **201**(1): 19–25.
123. Coban C, Ishii KJ, Uematsu S, *et al.* (2007) Pathological role of Toll-like receptor signaling in cerebral malaria. *Int Immunol* **19**(1): 67–79.
124. Griffith JW, Sun T, McIntosh MT, Bucala R. (2009) Pure hemozoin is inflammatory *in vivo* and activates the NALP3 inflammasome via release of uric acid. *J Immunol* **183**(8): 5208–5220.
125. He J, Tohyama Y, Yamamoto K, Kobayashi M, *et al.* (2005) Lysosome is a primary organelle in B cell receptor-mediated apoptosis: an indispensable role of Syk in lysosomal function. *Genes Cells* **10**(1): 23–35.
126. Reimer T, Shaw MH, Franchi L, *et al.* (2010) Experimental cerebral malaria progresses independently of the Nlrp3 inflammasome. *Eur J Immunol* **40**(3): 764–769.
127. Nuki G, Simkin PA. (2006) A concise history of gout and hyperuricemia and their treatment. *Arthritis Res Ther* **8** (Suppl 1): S1.
128. Martinon F. (2010) Mechanisms of uric acid crystal-mediated autoinflammation. *Immunol Rev* **233**(1): 218–232.
129. Dalbeth N, Haskard DO. (2005) Mechanisms of inflammation in gout. *Rheumatology (Oxford)* **44**(9): 1090–1096.
130. Lumbreras B, Pascual E, Frasquet J, *et al.* (2005) Analysis for crystals in synovial fluid: training of the analysts results in high consistency. *Ann Rheum Dis* **64**(4): 612–615.
131. Mandell BF. (2008) Clinical manifestations of hyperuricemia and gout. *Cleve Clin J Med* **75** (Suppl 5): S5–S8.
132. Shi Y, Evans JE, Rock KL. (2003) Molecular identification of a danger signal that alerts the immune system to dying cells. *Nature* **425**(6957): 516–521.
133. Mayor A, Martinon F, De Smedt T, *et al.* (2007) A crucial function of SGT1 and HSP90 in inflammasome activity links mammalian and plant innate immune responses. *Nat Immunol* **8**(5): 497–503.

134. Ng G, Sharma K, Ward SM, *et al.* (2008) Receptor-independent, direct membrane binding leads to cell-surface lipid sorting and Syk kinase activation in dendritic cells. *Immunity* **29**(5): 807–818.
135. Kool M, Willart MA, van Nimwegen M, *et al.* (2011) An unexpected role for uric acid as an inducer of T helper 2 cell immunity to inhaled antigens and inflammatory mediator of allergic asthma. *Immunity* **34**(4): 527–540.
136. Tanzi RE, Bertram L. (2005) Twenty years of the Alzheimer's disease amyloid hypothesis: a genetic perspective. *Cell* **120**(4): 545–555.
137. Weiner HL, Frenkel D. (2006) Immunology and immunotherapy of Alzheimer's disease. *Nat Rev Immunol* **6**(5): 404–416.
138. Glenner GG, Wong CW, Quaranta V, Eanes ED. (1984) The amyloid deposits in Alzheimer's disease: their nature and pathogenesis. *Appl Pathol* **2**(6): 357–369.
139. Meyer-Luehmann M, Spires-Jones TL, Prada C, *et al.* (2008) Rapid appearance and local toxicity of amyloid-beta plaques in a mouse model of Alzheimer's disease. *Nature* **451**(7179): 720–724.
140. Simard AR, Soulet D, Gowing G, *et al.* (2006) Bone marrow-derived microglia play a critical role in restricting senile plaque formation in Alzheimer's disease. *Neuron* **49**(4): 489–502.
141. Akama KT, Van Eldik LJ. (2000) Beta-amyloid stimulation of inducible nitric-oxide synthase in astrocytes is interleukin-1 beta- and tumor necrosis factor-alpha (TNFalpha)-dependent, and involves a TNFalpha receptor-associated factor- and NFκB-inducing kinase-dependent signaling mechanism. *J Biol Chem* **275**(11): 7918–7924.
142. Blum-Degen D, Muller T, Kuhn W, *et al.* (1995) Interleukin-1 beta and interleukin-6 are elevated in the cerebrospinal fluid of Alzheimer's and *de novo* Parkinson's disease patients. *Neurosci Lett* **202**(1–2): 17–20.
143. Mueller-Steiner S, Zhou Y, Arai H, *et al.* (2006) Antiamyloidogenic and neuroprotective functions of cathepsin B: implications for Alzheimer's disease. *Neuron* **51**(6): 703–714.
144. Hook VY, Kindy M, Hook G. (2008) Inhibitors of cathepsin B improve memory and reduce beta-amyloid in transgenic Alzheimer disease mice expressing the wild-type, but not the Swedish mutant, beta-secretase site of the amyloid precursor protein. *J Biol Chem* **283**(12): 7745–7753.
145. Lusis AJ. (2000) Atherosclerosis. *Nature* **407**(6801): 233–241.
146. Kol A, Santini M. (2004) Infectious agents and atherosclerosis: current perspectives and unsolved issues. *Ital Heart J* **5**(5): 350–357.
147. Wright SD, Burton C, Hernandez M, *et al.* (2000) Infectious agents are not necessary for murine atherogenesis. *J Exp Med* **191**(8): 1437–1442.

148. Koller E, Volf I, Gurvitz A, Koller F. (2006) Modified low-density lipoproteins and high-density lipoproteins. From investigation tools to real *in vivo* players. *Pathophysiol Haemost Thromb* **35**(3–4): 322–345.
149. Napoli C, de Nigris F, Palinski W. (2001) Multiple role of reactive oxygen species in the arterial wall. *J Cell Biochem* **82**(4): 674–682.
150. Yuan XM, Li W, Olsson AG, Brunk UT. (1997) The toxicity to macrophages of oxidized low-density lipoprotein is mediated through lysosomal damage. *Atherosclerosis* **133**(2): 153–161.
151. Stewart CR, Stuart LM, Wilkinson K, *et al.* (2010) CD36 ligands promote sterile inflammation through assembly of a Toll-like receptor 4 and 6 heterodimer. *Nat Immunol* **11**(2): 155–161.
152. Miller YI, Viriyakosol S, Binder CJ, *et al.* (2003) Minimally modified LDL binds to CD14, induces macrophage spreading via TLR4/MD-2, and inhibits phagocytosis of apoptotic cells. *J Biol Chem* **278**(3): 1561–1568.
153. Suzuki H, Kurihara Y, Takeya M, *et al.* (1997) A role for macrophage scavenger receptors in atherosclerosis and susceptibility to infection. *Nature* **386**(6622): 292–296.
154. Febbraio M, Podrez EA, Smith JD, *et al.* (2000) Targeted disruption of the class B scavenger receptor CD36 protects against atherosclerotic lesion development in mice. *J Clin Invest* **105**(8): 1049–1056.
155. Bjorkbacka H, Kunjathoor VV, Moore KJ, *et al.* (2004) Reduced atherosclerosis in MyD88-null mice links elevated serum cholesterol levels to activation of innate immunity signaling pathways. *Nat Med* **10**(4): 416–421.
156. Michelsen KS, Wong MH, Shah PK, *et al.* (2004) Lack of Toll-like receptor 4 or myeloid differentiation factor 88 reduces atherosclerosis and alters plaque phenotype in mice deficient in apolipoprotein E. *Proc Natl Acad Sci USA* **101**(29): 10679–10684.
157. Kim TW, Febbraio M, Robinet P, *et al.* (2011) The critical role of IL-1 receptor-associated kinase 4-mediated NF-{kappa}B activation in modified low-density lipoprotein-induced inflammatory gene expression and atherosclerosis. *J Immunol* **186**(5): 2871–2880.
158. Klinkner AM, Waites CR, Kerns WD, Bugelski PJ. (1995) Evidence of foam cell and cholesterol crystal formation in macrophages incubated with oxidized LDL by fluorescence and electron microscopy. *J Histochem Cytochem* **43**(10): 1071–1078.
159. Flegal KM, Carroll MD, Ogden CL, Curtin LR. (2010) Prevalence and trends in obesity among US adults, 1999–2008. *JAMA* **303**(3): 235–241.

160. Hotamisligil GS. (2006) Inflammation and metabolic disorders. *Nature* **444**(7121): 860–867.
161. Chen Y, Zhu J, Lum PY, *et al.* (2008) Variations in DNA elucidate molecular networks that cause disease. *Nature* **452**(7186): 429–435.
162. Shoelson SE, Lee J, Goldfine AB. (2006) Inflammation and insulin resistance. *J Clin Invest* **116**(7): 1793–1801.
163. Stienstra R, Joosten LA, Koenen T, *et al.* (2010) The inflammasome-mediated caspase-1 activation controls adipocyte differentiation and insulin sensitivity. *Cell Metab* **12**(6): 593–605.
164. Shah C, Yang G, Lee I, *et al.* (2008) Protection from high fat diet-induced increase in ceramide in mice lacking plasminogen activator inhibitor 1. *J Biol Chem* **283**(20): 13538–13548.
165. Shi H, Kokoeva MV, Inouye K, *et al.* (2006) TLR4 links innate immunity and fatty acid-induced insulin resistance. *J Clin Invest* **116**(11): 3015–3025.
166. Boden G. (2008) Ceramide: a contributor to insulin resistance or an innocent bystander? *Diabetologia* **51**(7): 1095–1096.
167. Hotamisligil GS, Peraldi P, Budavari A, *et al.* (1996) IRS-1-mediated inhibition of insulin receptor tyrosine kinase activity in TNF-alpha- and obesity-induced insulin resistance. *Science* **271**(5249): 665–668.
168. Uysal KT, Wiesbrock SM, Marino MW, Hotamisligil GS. (1997) Protection from obesity-induced insulin resistance in mice lacking TNF-alpha function. *Nature* **389**(6651): 610–614.
169. Vitseva OI, Tanriverdi K, Tchkonia TT, *et al.* (2008) Inducible Toll-like receptor and NF-kappaB regulatory pathway expression in human adipose tissue. *Obesity (Silver Spring)* **16**(5): 932–937.
170. Davis JE, Gabler NK, Walker-Daniels J, Spurlock ME. (2008) Tlr-4 deficiency selectively protects against obesity induced by diets high in saturated fat. *Obesity (Silver Spring).* **16**(6): 1248–1255.
171. Suganami T, Mieda T, Itoh M, *et al.* (2007) Attenuation of obesity–induced adipose tissue inflammation in C3H/HeJ mice carrying a Toll-like receptor 4 mutation. *Biochem Biophys Res Commun* **354**(1): 45–49.
172. Suganami T, Tanimoto-Koyama K, Nishida J, *et al.* (2007) Role of the Toll-like receptor 4/NF-kappaB pathway in saturated fatty acid-induced inflammatory changes in the interaction between adipocytes and macrophages. *Arterioscler Thromb Vasc Biol* **27**(1): 84–91.
173. Song MJ, Kim KH, Yoon JM, Kim JB. (2006) Activation of Toll-like receptor 4 is associated with insulin resistance in adipocytes. *Biochem Biophys Res Commun* **346**(3): 739–745.

174. Kahn SE, Hull RL, Utzschneider KM. (2006) Mechanisms linking obesity to insulin resistance and type 2 diabetes. *Nature* **444**(7121): 840–846.
175. de Koning EJ, van den Brand JJ, Mott VL, *et al.* (1998) Macrophages and pancreatic islet amyloidosis. *Amyloid* **5**(4): 247–254.
176. Zhou R, Tardivel A, Thorens B, (2010) Thioredoxin-interacting protein links oxidative stress to inflammasome activation. *Nat Immunol* **11**(2): 136–140.
177. Koenen TB, Stienstra R, van Tits LJ, *et al.* (2011) Hyperglycemia activates caspase-1 and TXNIP-mediated IL-1 beta transcription in human adipose tissue. *Diabetes* **60**(2): 517–524.
178. Njajou OT, Kanaya AM, Holvoet P, *et al.* (2009) Association between oxidized LDL, obesity and type 2 diabetes in a population-based cohort — the Health, Aging and Body Composition Study. *Diabetes Metab Res Rev* **25**(8): 733–739.
179. Deal watch: (2011) XOMA and Servier to develop anti-IL-1 beta antibody for inflammatory diseases. *Nat Rev Drug Discov* **10**(3): 166.
180. Leslie KS, Lachmann HJ, Bruning E, *et al.* (2006) Phenotype, genotype, and sustained response to anakinra in 22 patients with autoinflammatory disease associated with CIAS-1/NALP3 mutations. *Arch Dermatol* **142**(12): 1591–1597.
181. Goldbach-Mansky R, Dailey NJ, Canna SW, *et al.* (2006) Neonatal-onset multisystem inflammatory disease responsive to interleukin-1 beta inhibition. *N Engl J Med* **355**(6): 581–592.
182. Lachmann HJ, Kone-Paut I, Kuemmerle-Deschner JB, *et al.* (2009) Use of canakinumab in the cryopyrin-associated periodic syndrome. *N Engl J Med* **360**(23): 2416–2425.
183. McGonagle D, Tan AL, Madden J, *et al.* (2008) Successful treatment of resistant pseudogout with anakinra. *Arthritis Rheum* **58**(2): 631–633.
184. So A, De Smedt T, Revaz S, Tschopp J. (2007) A pilot study of IL-1 inhibition by anakinra in acute gout. *Arthritis Res Ther* **9**(2): R28.
185. Larsen CM, Faulenbach M, Vaag A, *et al.* (2007) Interleukin-1-receptor antagonist in type 2 diabetes mellitus. *N Engl J Med* **356**(15): 1517–1526.
186. Osborn O, Brownell SE, Sanchez-Alavez M, *et al.* (2008) Treatment with an interleukin-1 beta antibody improves glycemic control in diet-induced obesity. *Cytokine* **44**(1): 141–148.
187. Terkeltaub R, Sundy JS, Schumacher HR, *et al.* (2009) The interleukin 1 inhibitor rilonacept in treatment of chronic gouty arthritis: results of a placebo-controlled, monosequence crossover, non-randomised, single-blind pilot study. *Ann Rheum Dis* **68**(10): 1613–1617.

188. Bhaskar V, Yin J, Mirza AM, *et al.* (2011) Monoclonal antibodies targeting IL-1 beta reduce biomarkers of atherosclerosis *in vitro* and inhibit atherosclerotic plaque formation in apolipoprotein E-deficient mice. *Atherosclerosis* Feb. 24.
189. Crossman DC, Morton AC, Gunn JP, *et al.* (2008) Investigation of the effect of interleukin-1 receptor antagonist (IL-1ra) on markers of inflammation in non-ST elevation acute coronary syndromes (the MRC-ILA-HEART Study). *Trials* **9**: 8.
190. Cornelis S, Kersse K, Festjens N, *et al.* (2007) Inflammatory caspases: targets for novel therapies. *Curr Pharm Des* **13**(4): 367–385.

CHAPTER 3

Amyloidosis

*Morie A. Gertz**

INTRODUCTION

Amyloidosis is defined as the presence, in histologic tissue, of amorphous eosinophilic deposits that bind Congo red and demonstrate green birefringence under polarized light. These deposits are found extracellularly and lead to pathology because of interference with the function of the organs in which they are deposited. The disorder is quite heterogeneous and is found in multiple forms. In clinical medicine, the term "amyloid" was first used by Rudolph Virchow, in 1854.[1] Originally, the deposit of amyloid was felt to be starch-like or lardaceous; both of these views were incorrect in recognizing the proteinaceous nature of all amyloid deposits.[1] The first reports of amyloidosis were almost certainly on secondary or AA amyloidosis due to the high prevalence of tuberculosis, osteomyelitis, and suppurative pulmonary infections in the patient population seen in that era. Congo red as the identifying stain for amyloid was introduced in 1927,[2] and amyloid was found to have a fibrillar ultrastructure under the electron microscope in 1959.[3]

*Division of Hematology and Internal Medicine, Mayo Clinic; Mayo Clinic Medical School; Rochester, Minnesota, USA. E-mail: gertz.morie@mayo.edu.

PATHOGENESIS

All amyloid deposits have in common nonbranching fibrils of indeterminate length, with a width of approximately 10 nm. Each fibril is composed of protofilaments 3 nm in diameter which run along the longitudinal axis of the amyloid fibril[4] and twist, creating a helical repeat that forms the β-pleated sheet structure, which runs perpendicular to the long axis of the fibril.[5]

In spite of the ultrastructural similarity at the level of the protofilament, the amyloid proteins have been identified as belonging to at least 25 different precursors (Table 1). Immunoglobulin light chains are the precursors for AL amyloidosis, previously referred to as primary amyloidosis and now appropriately known as immunoglobulin light chain amyloidosis. Most patients have an underlying plasma cell dyscrasia; amyloidosis A, previously known as secondary amyloidosis, is related to the deposition of a fragment of the normal serum acute phase reactant (SAA). Inherited amyloidosis has been described from point mutations of transthyretin, apolipoprotein, lysozyme, and fibrinogen. The most common inherited amyloidosis syndromes are related to mutations of transthyretin, and the largest pedigrees have a substitution of TTR Val30Met. Over 100 point mutations in the transthyretin molecule have been described as being responsible for inherited forms of amyloidosis.

The formation of the amyloid fibril has been thought to be a result of protein misfolding with a dynamic equilibrium between a soluble state and the insoluble fibrillar protein complex.[6,7] Nonprotein components of the amyloid fibril include the glycoprotein amyloid P component found in all forms of amyloidosis, heparan sulfate proteoglycans, and apolipoprotein E.[8] One characteristic of amyloidogenesis is an overproduction of precursor protein permissive for deposition, although exceptions exist, such as in senile systemic amyloidosis when a native form of transthyretin deposits in the myocardium. It is thought that in inherited forms of amyloidosis misfolding is promoted by specific genetic mutations that reduce solubility and make deposition possible. It is likely that there is an element of proteolytic cleavage necessary for amyloid formation. In most forms of light chain amyloidosis, the entire immunoglobulin light chain is not present but a proteolytic fragment.[9]

Table 1 Amyloidosis: Some of the Proteins Known to Cause Clinical Amyloid Disease in Humans

Precursor Protein	Human Disease	Major Causative Association	Major Clinical Manifestation
Immunoglobulin light chain	Primary (AL) amyloidosis	Plasma cell disorders	Renal, cardiac, GI, peripheral nervous system
Immunoglobulin heavy chain	Primary (AH) amyloidosis	Plasma cell disorders	Much less frequent than AL amyloidosis
Serum amyloid A	Secondary (AA) amyloidosis	Inflammation	Renal
Transthyretin	Familial	Mutation of ATTR	Peripheral nervous system, heart
	Senile	Wild-type ATTR, aging	Multiple organs, cardiac most clinically prominent
Apolipoprotein AI	Familial	Mutation of Apo-AI	Very slowly progressive disease
Apolipoprotein AII	Familial	Point mutation of stop codon leading to additional 20 amino acid residues	Renal
Lysozyme	Familial	Mutation of lysozyme	Kidney, liver, lung, spleen
Fibrinogen A α-chain	Familial	Mutation of fibrinogen	Renal
Gelsolin	Familial (Finnish type)	Mutation of gelsolin	Corneal dystrophy, cranial neuropathy, cutis laxa
Cystatin C	Familial (Icelandic type)	Mutation of cystatin	Amyloid angiopathy, cerebral hemorrhage
B-2-microglobulin	Aβ2M	Hemodialysis	Joints
LECT2	ALECT2		Renal

In amyloidosis A, cleavage invariably occurs from the original 76-residue precursor. In Alzheimer's disease, the amyloid protein represents the N-terminal fragment of the amyloid precursor protein, either A β 1:40 or A β 1:42.[10]

AMYLOID ORGAN INVOLVEMENT

Little is understood about why different amyloidogenic proteins have a specific tropism for specific organs such as TTR amyloidosis, which primarily affects the peripheral nervous system and the heart, where fibrinogen forms of amyloidosis have a preferential deposition in the kidneys.[11] Some patients with light chain amyloidosis present with renal involvement that does not progress over the course of the disease. Other patients with AL develop widespread involvement. Table 2 outlines the most common organs involved with amyloid deposits among the various forms of systemic amyloidosis.

It is unclear how amyloid causes organ dysfunction after deposition. Whether the inert amyloid deposits are directly responsible for organ dysfunction or soluble toxic intermediates that are codeposited with the visible amyloid deposits are responsible remains uncertain.[12] There is no correlation between the extent of organ dysfunction and the amount of amyloid visible on biopsy. After successful therapy in light chain amyloidosis, organ improvement is regularly seen. However, anatomic resolution of the congophilic deposit does not occur. In certain situations, the protofibril and oligomeric intermediates are toxic rather than the mature

Table 2 Organ Systems Commonly Involved Clinically by Various Forms of Amyloidosis

Organ System	Primary (AL) Amyloidosis	Secondary (AA) Amyloidosis	Dialysis-associated β_2-Microglobulin (β_2-m) Amyloidosis	Hereditary Amyloidosis*
Heart	X			X
Kidney	X	X		X
Vascular	X			
Peripheral nerves	X			X
Autonomic nerves	X			X
Liver	X			
Gastro-intestinal tract	X	X		
Joints	X		X	

*Organ involvement varies according to the specific amyloid precursor protein.

amyloid fibril itself. Nonfibrillar oligomers and protofibrillar intermediates have been shown to directly induce neurotoxicity.[13] In TTR amyloidosis, administration of TTR to cell lines as immature amyloid which has not aggregated leads to cell death where the mature amyloid fibril is not toxic.[14]

DIAGNOSIS

A challenge in amyloidosis is when to suspect the diagnosis of such a rare disorder.[15] The signs and symptoms of the disease are nonspecific and are shared by other more common disorders. Involvement of the kidney, which is the most commonly involved organ in amyloid, presents with nephrotic range proteinuria, which causes edema, fatigue, and weight loss. Peripheral nerve involvement causes paresthesias and pain. Cardiac involvement results in dyspnea and edema. Intestinal involvement can result in weight loss, diarrhea, or vomiting. In any case, all of these symptoms are quite common in the general medical population, and amyloidosis can be expected to be the etiology in only a very small proportion. Pathognomonic findings exist in amyloidosis such as glossomegaly and periorbital purpura. These are actually uncommon findings in patients with amyloidosis, seen in no more than 15% of patients.[16]

A high index of suspicion is needed by a clinician faced with a patient with unexplained weight loss, edema, dyspnea, fatigue, or paresthesias of uncertain etiology. Amyloidosis needs to be considered when patients are seen with these symptoms that are not readily explainable by diabetes mellitus, or ischemic or valvular heart disease.[16] Ten percent of adult nephrotic syndrome in nondiabetics after renal biopsy is found to be renal amyloidosis.[17] Subtle changes in the electrocardiogram, which would include low voltage or an infarction pattern without a history of ischemic heart disease or echocardiographic evidence of thickening of the cardiac walls without a history of significant hypertension, should be considered potential clues. Unexplained hepatomegaly, enlargement of the tongue, and steatorrhea would all be consistent with amyloidosis and justify screening. Occasionally, patients present with a syndrome that is labeled atypical multiple myeloma with a small clonal plasmacytosis in the bone marrow, and these patients may have amyloid.

SCREENING FOR AMYLOIDOSIS

A tissue biopsy should not be the first method of screening, because the prevalence of amyloidosis is low, and the majority of patients with the symptoms referred to above will not turn out to have amyloidosis, and so noninvasive screening would be desirable. The majority of patients with systemic amyloidosis have immunoglobulin light chain amyloidosis, and there will generally be a characteristic presence of a monoclonal protein in the serum or urine, or an immunoglobulin free light chain abnormality with screening.[18] When patients have a compatible clinical syndrome, immunofixation of the serum and urine and an immunoglobulin free light chain assay will detect virtually all patients with light chain amyloidosis.[19] A simple serum or urine protein electrophoresis is insufficient, since a high proportion of patients have only light chain proteinemia, and as a consequence a demonstrable spike will not be seen on the electrophoretic pattern. Only 10% of patients will have an M spike in the serum >1.3 g/dL, and only 25% will have an M spike >0.5 g/dL, historically the lower limit of ability to visualize a discrete peak.

The single most sensitive test abnormality when amyloidosis is suspected is the immunoglobulin free light chain assay.[20] This commercial test will detect an abnormal κ or λ light chain population in 99% of patients. Even in patients with amyloidosis and a negative serum and urine immunofixation, seen in 14% of patients, the immunoglobulin light chain assay will detect 86% of patients with κ and 30% of patients with λ light chain amyloidosis.

CONFIRMING THE DIAGNOSIS

Amyloidosis can only be confirmed with biopsy demonstrating green birefringence with Congo red staining. There are a number of diagnostic pitfalls in interpretation of the Congo red stain. Overfixation and trapping of the stain can result in a false positive result. Collagen and elastin in skin and subcutaneous fat can often be birefringent.[21] The detection of Congo-red-positive deposits, however, does not validate the type of the amyloid protein. Congo red positivity does not distinguish localized from systemic, nor does it differentiate among the various forms of systemic amyloidosis outlined in Table 1. Efforts to identify the specific protein involved require either immunohistochemical staining, immunogold, or mass-spectroscopic analysis which can be performed on Congo-red-stained sections. Particular caution is required in determining the type of amyloidosis, since 3%–5% of patients over the age of 70 have monoclonal gammopathies, and a monoclonal protein can be detected in patients with localized as well as hereditary forms of amyloidosis. Therefore, specific diagnostic testing that confirms the form of amyloid is imperative.[22–24]

Although organ biopsy will always demonstrate the diagnosis, it is generally not required. In our experience, performing both a bone marrow biopsy and fat aspiration on 378 patients, both tests were positive for amyloid in 53.4%; one was positive in 32% and both were negative or equivocal in only 14.6%, representing only 1 in 7 patients. As a consequence, the biopsy diagnosis of amyloid can generally be done noninvasively, and only for the minority does a biopsy of the kidney, heart, liver, or nerve become necessary. Moreover, Congo red deposits in fat and in bone marrow can be examined by mass-spectrometry-based proteomic analysis.[25] Other groups have reported very high yields using endoscopic biopsies.[26] A diagnostic pathway to confirm AL, the assessment of a patient with AL amyloidosis, is given in Table 3.

Table 3 Diagnostic Pathway for AL Amyloidosis

1. Consider AL amyloidosis in patients with:
   - Nondiabetic nephrotic syndrome
   - Nonischemic cardiomyopathy with an echocardiogram showing "hypertrophy"
   - Hepatomegaly or alkaline phosphatase elevation without imaging abnormality
   - Peripheral neuropathy with MGUS or CIDP with autonomic features
   - Atypical myeloma with monoclonal light chains and modest marrow plasmacytosis
2. Screen suspicious patients with electrophoresis and immunofixation of serum and urine and serum-free light chain assay. If negative, inquire about the possibility of familial or localized amyloidosis.
3. Biopsy patients with a monoclonal protein; fat and bone marrow should be sampled first.
4. Assess prognosis with echocardiography including Doppler and strain echocardiography; evaluate serum troponin and NT-proBNP levels.
5. Refer for anti–plasma cell therapy.

Abbreviations (not used in text): CIDP — chronic inflammatory demyelinating polyradiculoneuropathy; MGUS — monoclonal gammopathy of unknown significance.

IMAGING OF AMYLOIDOSIS

Echocardiography, magnetic resonance imaging (MRI), radionuclide imaging, and serum amyloid P component scanning have all been used to detect, quantify, and diagnose amyloidosis. Echocardiography is routinely used to differentiate amyloid heart disease from other forms of cardiomyopathy. The typical findings are thickening of the left ventricular wall with diastolic dysfunction, which is measured by Doppler as reduced compliance of the chambers with slowing of inflow of blood return.[27] Later phases result in restrictive cardiomyopathy. Systolic dysfunction manifested by a reduction in the ejection fraction is a late finding. Amyloid involvement of the conduction system can lead to significant rhythm disturbances. The use of strain echocardiography has enhanced the sensitivity of echocardiography in its ability to detect amyloidosis.[28] Cardiac MRI has shown characteristic patterns of gadolinium enhancement in subendocardial tissue and has an important role in the diagnosis of the disease. Late gadolinium enhancement is common in cardiac amyloidosis and detects interstitial expansion from amyloid deposition. Global transmural or subendocardial enhancement is most common, but focal patchy enhancement is also observed. These findings occur in patients with a normal left ventricular thickness and are associated with clinical markers of prognosis.[29] The radioiodinated SAP component localizes in amyloid infiltrated tissues in proportion to the amount of disease. Patients with amyloidosis clear the radioiodinated SAP from the plasma at an accelerated rate in proportion to the amyloid body burden. The technique has been useful for prognosis, for its ability to identify potential biopsy sites to confirm the diagnosis, and for serialized monitoring of response to therapy.[30]

CLINICAL CLASSIFICATIONS OF AMYLOIDOSIS

Immunoglobulin Light Chain Amyloidosis (AL)

Patients with AL have, by definition, a monoclonal light chain in the serum or urine or a detectable circulating free light chain in addition to a clonal plasma cell disorder in the bone marrow. The percentage of plasma cells in the bone marrow is, on average, approximately 5%. Only a small fraction of these patients will ultimately develop overt multiple myeloma,[31] reflecting the differing pathogenesis of these two disorders. Approximately 80% of patients with amyloidosis have λ light chain disease. This is in distinct contrast to monoclonal gammopathy of undetermined significance and multiple myeloma, where only one-third of patients have λ light chains. This suggests that λ light chains have a higher intrinsic amyloidogenicity than κ light chains. The λ 6 subgroup of light chains is almost always associated with amyloid.[32]

Two-thirds of patients with AL are male and the median age of patients seen at Mayo Clinic is 67 years. Half of patients have echocardiographic evidence of cardiac involvement, and a third actually have symptomatic cardiac insufficiency. Nephrotic range proteinuria is seen in approximately one-third of patients, and peripheral neuropathy in approximately 15%. The neuropathy, by electromyographic studies, is usually mixed, axonal, and demyelinating. Hepatomegaly is a feature seen in 15% of patients but is actually a dominant symptom in only approximately 5%.[33] Gastrointestinal involvement is rare and can present with intestinal pseudo-obstruction or malabsorption. Enlargement of the tongue is seen in 15%. Periarticular infiltration of the shoulders with amyloid is a rare event but is recognized by its production of the "shoulder pad sign."

Of all the known prognostic factors, cardiac involvement is the most important. The median survival of patients with cardiac failure is measured in months. For patients who have no significant cardiac involvement, the median survival is approximately 26 months. The most sensitive predictors of survival are a combination of serum troponin T and N-terminal probrain natriuretic peptide. Patients can be separated into three stages: whether both are normal (stage 1), both are abnormal (stage 3), and one of the two is abnormal (stage 2). This results in median survival for stages 1, 2, and 3 of 26.4, 10.5, and 3.5 months, respectively.[34] The pretransplant value of the immunoglobulin free light chain is also prognostic for survival.[35]

Secondary Systemic Amyloidosis (AA)

Secondary amyloidosis results from the deposition of fragments of serum amyloid A protein. Precursor protein elevations are seen in any chronic inflammatory state.[36]

The serum amyloid A protein is an apolipoprotein bound to high-density lipoprotein particles during inflammation.[37] The serum amyloid A–rich HDL may be involved in cholesterol metabolism during inflammation. Macrophages appear to have a central role in the pathogenesis of AA amyloidosis, since proteolytic cleavage is required to convert the 104-amino-acid SAA to the 76-amino-acid amyloid A protein, the subunit of the secondary amyloid fibril. There are isoforms of SAA, with SAA1 and SAA2 being involved with amyloidogenesis since they function as acute phase reactants in humans.

AA amyloidosis was the form most frequently seen in the 19th century and most likely encountered in the original pathological descriptions by Virchow and Rokitansky.[38] In that era, tuberculosis, leprosy, osteomyelitis, and suppurative infections including lung abscesses were common causes. Today, difficult-to-treat chronic inflammatory noninfectious conditions appear to dominate and include psoriatic arthritis, anklyosing spondylitis,[39] juvenile rheumatoid arthritis, paraplegia, cutaneous abscesses related to elicit drug injections, and granulomatous colitis. Amyloidosis is a common accompaniment of long-standing cystic fibrosis in those who survive to adulthood.[40] There is an inherited form of secondary amyloidosis. There are a number of inherited periodic fever syndromes, which include familial Mediterranean fever and inherited mutations of the tumor necrosis factor receptor[41] that result in intermittent febrile illnesses often associated with polyarthritis and skin rash which will lead to secondary systemic amyloidosis on an autosomal recessive basis. In the West, inflammatory spondyloarthropathies are the most common cause of AA amyloidosis, and even these have become increasingly rare causes of AA in the past two decades.[42]

Familial Mediterranean fever is the most frequent periodic fever syndrome worldwide that is associated with AA amyloid. Its presentation is virtually always the kidney presenting with nephrotic range proteinuria leading to end-stage renal failure. The target organs are usually the kidney, the GI tract, and the thyroid. Large reported groups of patients with familial Mediterranean fever exist in Turkey and Israel. Cardiac involvement is rare and is generally seen only years after the development of AA amyloid, usually in patients who are sustained with chronic hemodialysis. Liver involvement is seen in 9% of patients, although scanning evidence of amyloid in the liver is seen far more frequently.[43]

The survival for AA amyloidosis is significantly better than that for AL, in large part due to the availability of support for end-stage renal disease and the low frequency of significant cardiac involvement. The development of AA amyloid is directly related to the ability to suppress the underlying inflammatory condition. Cardiac involvement, when it does occur, predicts shorter overall survival: 31% vs. 63%. Glomerular involvement with proteinuria does worse than for patients who have interstitial involvement with azotemia only. The best indicator of a successful outcome is suppression of the production of serum amyloid A protein.[44] When the serum amyloid A protein is normalized, 10-year survivals were typical. When the serum amyloid A protein was reduced by an excess of 50%, the relative risk was reduced threefold. The absolute SAA concentration was strongly associated with a poor outcome. End-stage renal disease in older age was associated with a poor outcome. An underlying periodic fever syndrome was associated with a better outcome.

Familial Amyloidosis

Familial amyloidosis occurs when an inherited mutation leads to the production of proteins with an increased tendency to misfold into a β-pleated sheet conformation. A large number of proteins have been recognized to cause amyloidosis, including transthyretin, fibrinogen A α-chain, apolipoprotein AI, apolipoprotein AII, lysozyme, gelsolin, and cystatin C.[45] Worldwide, the most common protein associated with inherited amyloidosis is transthyretin. Clinically, the inherited amyloid syndromes are divided into those that are primarily associated with peripheral neuropathy and those that are primarily associated with cardiomyopathy. Before the recognition of the specific mutations, these were referred to as familial amyloid polyneuropathy and familial amyloid cardiomyopathy.[46] Therapeutic interventions to manage the aberrant protein that causes inherited amyloidosis depend on the source organ responsible for synthesis.

Transthyretin

Over 100 different mutations are associated with transthyretin amyloidosis. Transthyretin was previously known as thyroxin-binding prealbumin, because it migrates faster than albumin in the protein electrophoresis. The source of transthyretin is the liver and the choroid plexus.[47] TTR carries thyroxine and retinol. Worldwide, the most common mutation is VAL-30 MET, with large recognized pedigrees in Porto, Portugal; Sweden; and Nagano Prefecture, Japan. In northern Sweden, out of 500,000 people, 1.5% are known to carry the mutation. The penetrance is estimated to be 83%.[48] The clinical phenotype is highly variable among patients even with the same genetic mutation. Peripheral neuropathy is the most common initial clinical presentation in half of the known transthyretin mutations, followed by cardiac involvement. In patients who present with peripheral neuropathy, the late development of cardiac involvement is common. Rarer presentations include recurrent central nervous system hemorrhage, deposition of vitreous amyloid[49] causing visual loss, autonomic failure causing orthostatic syncope, bradycardia, intestinal pseudo-obstruction followed by incontinence, and urinary voiding difficulties.

The peripheral neuropathy of transthyretin amyloidosis begins with sensory loss in the lower extremities. Nonmyelinated fibers are affected early, so change in pain and temperature may be seen with normal electromyography and the absence of reflex loss. As the sensory neuropathy progresses, motor loss and muscular atrophy develop. Histologically, the amyloid is found in the nerve trunk, the vasa nervorum, the nerve plexus, and the nerve ganglia. Fiber degeneration is axonal.[50]

The prognosis of patients with TTR amyloidosis is superior to that of AL, with the median survival exceeding five years. Cardiac involvement, its severity, and the presence of nutritional failure related to peripheral neuropathy predict survival. The most common cause of death is cardiac failure or arrhythmias, or complications of peripheral and autonomic neuropathy.[51]

A particularly important form of inherited amyloidosis in the United States is the mutation Val122Ile. This mutation is found almost exclusively in Americans of African descent and causes late-onset heart failure passed down in an autosomal dominant fashion. These patients do not suffer from coronary artery disease and often have their cardiac amyloidosis misattributed as hypertensive heart disease with ventricular wall thickening or hypertrophic cardiomyopathy.[52] Recognition allows for appropriate screening and genetic counseling of first-degree relatives. Heart transplantation has been used for homozygous Val122Ile.[53]

Nontransthyretin Forms of Familial Amyloidosis

Fibrinogen A α-chain amyloidosis is the next most common form of amyloid, and it presents with very slowly progressive renal involvement. It will cause proteinuria followed by the development of renal insufficiency, often over the course of a decade. Patients frequently have associated hypertension.[54] The apolipoprotein A forms of amyloidosis also tend to be slowly progressive and can have both renal and peripheral nerve involvement. Slowly progressive glomerular and interstitial renal disease is seen. Apolipoprotein A amyloidosis was formerly known as familial amyloidosis of Ostertag.[55] Rare mutations of lysozyme, an enzyme that is produced by neurotrophils and macrophages, lead to widespread amyloid deposits in the kidney, liver, lung, and spleen.[56] Hepatic rupture has been reported with this form. Cystatin C amyloidosis produces amyloid angiopathy, and recurrent subcortical cerebral hemorrhages, and is seen primarily in Iceland.

Native Transthyretin Amyloidosis (Senile Cardiac Amyloidosis)

Senile systemic amyloidosis is caused by the deposition of native transthyretin in tissues, predominantly the heart, and can be responsible for nodular pulmonary amyloid deposits. Although associated with advanced age, we have seen it in patients as young as 55 years. Transthyretin is inherently amyloidogenic; and in patients with mutant transthyretin familial amyloidosis who undergo successful hepatic transplantation, amyloid progression can be seen by the deposition of native TTR[57] into pre-existent deposits in the heart and peripheral nerve.[58] Senile cardiac amyloidosis is found in 25% of autopsied patients over the age of 90. It is more common in men than in women.[59] Diagnosis usually requires endomyocardial biopsy, since peripheral deposits visible in the fat, bone marrow, and intestinal tract are less common. Peripheral neuropathy is rare in senile systemic amyloidosis. Nodular deposits in the lung can be mistaken for malignancy. The prognosis associated with senile cardiac amyloidosis is far better than in light chain amyloidosis. When matched for the anatomic extent of amyloid infiltration, patients regularly have a survival in excess of five years, often compared to less than one year in light chain amyloidosis. This, in part, may be related to direct toxic effects of the immunoglobulin light chain on cardiac tissue which is independent of the fixed deposits of amyloid seen in between myocardial fibers.[60]

Localized Amyloidosis

Localized amyloidosis is defined as deposition in a single organ without evidence of systemic disease in the absence of a systemic plasma cell proliferative disorder. In most instances, the amyloid deposits are composed of an immunoglobulin light chain, but a circulating monoclonal protein and clonal plasma cells in the bone marrow are lacking. Almost any anatomic location and tissue in the human body has been reported to present with amyloid deposition. The most important clinical presentations include the genitourinary tract, the tracheobronchial tree, and the skin. Patients presenting with localized amyloidosis do not have an increased risk of developing systemic AL later in the course of their disease. The treatment is usually local. The problem tends to be recurrent over time. Chemotherapy is contraindicated.[61]

Amyloidosis of the Renal Pelvis, Ureter, Bladder, and Urethra

Localized bladder amyloidosis presents with gross hematuria. Cystoscopic findings generally are consistent with bladder cancer but show no evidence of malignancy on biopsy. Ninety percent are shown to be of light chain etiology. The gross hematuria is usually treated by fulguration. When the deposits are too extensive and would require major bladder resection, intravesicle DMSO has been used successfully in management. Renal pelvis and ureteral amyloidosis generally present with obstruction. Most commonly, the preoperative diagnosis is transitional cell cancer. Nephrectomy often results for this nonmalignant condition. Recurrence in the contralateral ureter has been reported[62]

Tracheobronchial Tree and Laryngeal Amyloid

These patients tend to present with hoarseness, dysphonia, and cough. Deposits limited to the vocal cords and larynx are often treated with YAG laser therapy. Obstruction is uncommon, but recurrence often requires repeated laser therapies over time. In rare instances, profound narrowing of the distal bronchial tree has resulted in respiratory failure. Abnormalities of gas exchange are generally not found. In rare instances of diffuse bronchial amyloid, external beam radiation has been reported to be successful.[63]

SKIN

Cutaneous amyloidosis may be a reflection of localized disease such as macular and lichen-like amyloidosis, in which the deposits are actually degenerated keratin filaments. However, cutaneous deposits may sometimes be an early manifestation of systemic light chain amyloidosis and are a cutaneous manifestation of visceral involvement. These typically are nodular forms of cutaneous amyloidosis. The majority are associated with a clonal population of plasma cells in the bone marrow, with deposits appearing in the dermis. Nodules and plaques on the trunk or extremities are seen.[64]

The prognosis of those forms of amyloid that are related to keratin is "quite benign," with localized therapy used only if symptoms require it.[65] However, the nodular forms of amyloidosis are often associated with cardiac and hepatic amyloidosis and require therapeutic intervention associated with the specific viscera.

Localized amyloid can be found associated with joint, cartilage, valve, and the atria. Localized amyloidosis is frequently found as an incidental finding in resected specimens of degenerated hips.[66] It has also been found as a small component in resected valves, particularly the aortic valve, and has been identified in the atria as being composed of atrial natriuretic peptide.[67] Amyloid depositing in the media of the aorta, particularly in patients with aneurysms, has also been described. None of these forms of amyloid are systemic and they are not actively treated, although the atrial amyloid has been associated with supraventricular arrhythmias.

THERAPY

Treatment Overview

The treatment of amyloidosis depends on the specific type, which organs are involved, and the extent of organ dysfunction at the time of diagnosis.

Ideally, treatments designed to interfere with protein folding and the conformational change leading to fibril formation are explored. The use of a nonsteroidal anti-inflammatory agent to interfere with the folding of the transthyretin tetramer is being studied. Efforts to interfere with the binding of the amyloid P-component to the amyloid fibril have also been made to destabilize the protein.[68] Moreover, specific antibody therapy designed to disrupt the fibrillar confirmation has been reported in animal models of light chain amyloidosis.[69] None of these techniques, however, are available and the primary treatment strategies are aimed at eradicating the production of the amyloidogenic protein.

In inherited amyloidosis, liver transplantation has been used when the source of the mutated protein is primarily hepatic, as is the case in transthyretin, the fibrinogen A α-chain, and apolipoprotein forms of amyloidosis.[70] In these situations, hepatic transplantation can be curative. In AA amyloidosis, efforts to reduce the inflammatory process that leads to the increased production of serum amyloid A protein have been successful. Colchicine is highly specific for the prevention of amyloidosis in familial Mediterranean fever. The eradication of leprosy and tuberculosis with the use of specific antibiotics has also been reported to lead to regression of amyloidosis. Since the majority of secondary amyloidosis in the Western world is related to inflammatory arthropathies, immunosuppressive agents such as chlorambucil and, more recently, inhibitors of tumor necrosis factors such as infliximab and etanercept have been successfully used.[71] Anakinra, a powerful inhibitor of interleukin-6, has also been used in the management of secondary amyloidosis.[72]

In immunoglobulin light chain amyloidosis, the most common intervention involves cytotoxic chemotherapy designed to eradicate the clonal plasma cell population that produces the immunoglobulin light chain in the bone marrow. The usual fraction of plasma cells is small, and they tend to be nonproliferative over time, which often frustrates the attempt to eradicate the clone since systemic chemotherapy tends to be most effective against proliferative malignant cells.

Alkylator-based Therapy

Historically, the standard of care for the treatment of amyloidosis was melphalan and prednisone given orally for 4–7 days every 4–6 weeks. This technique is used infrequently today, because of low response rates and lengthy time to response. In light chain amyloidosis, rapid response is preferable since organ damage may be ongoing while one is waiting for systemic cytotoxic chemotherapy to reduce the production of light chains that create the amyloid. Moreover, long-term melphalan exposure can produce a significant risk of late myelodysplasia. The major advantage of melphalan and prednisone is the fact that they can be administered to almost any patient without regard to age, performance status, serum creatinine, and cardiac functional status.[73]

Currently, the standard of care for non-stem-cell transplantation patients is considered to be melphalan with high-dose dexamethasone instead of prednisone. When studied, patients were selected on the basis of their ineligibility for high-dose melphalan and transplant therapy. Among 46 patients, the hematologic response was seen in 31 and a hematologic complete response was seen in 15 (33%).[74] Improvement in organ dysfunction was seen in 22 patients. The day-100 mortality was 4%. Resolution of cardiac failure was seen in 6 of 32 patients with a median time to response of 4.5 months and 11% adverse effects. At a 5-year followup, the actuarial survival was 50%, and the progression-free survival was 40%. Patients who relapsed could successfully be reinduced with melphalan and dexamethasone.[75] Thirty of 41 evaluable patients were responders to chemotherapy, with the median survival not reached by 4 responders. Others, however, have not been as successful in the treatment of amyloidosis with melphalan and dexamethasone. A study of 61 patients treated with melphalan parenterally with dexamethasone reported a median survival of only 17.5 months, with a 3-month all-cause mortality of 28%.[76] Using a similar melphalan and dexamethasone-based regimen, a median survival of 10.5 months has been reported in a group of light chain amyloidosis patients ineligible for stem cell transplantation. In a phase III trial comparing melphalan–dexamethasone with transplant, the melphalan–dexamethasone group had a median survival of 56.9 months. Caution is therefore required in interpreting the data when four reports have a median survival ranging from 10.5 to 61 months.[77] The proportion of patients with cardiac amyloidosis in each cohort is likely playing an important role in the ultimate prediction of survival. Most studies do not report the cardiac stage as measured by serum troponin and brain natriuretic peptide. As a consequence, it is difficult to determine the exact effectiveness of chemotherapy and compare it directly with the fraction of patients destined to do poorly due to advanced cardiac involvement.

The Boston University Medical Center has used low-dose melphalan in patients ineligible for stem cell transplantation; 15 patients (median age 55) received a median of 3 cycles. The median survival was 2 months; 10 patients died within 5 weeks of starting therapy, a reflection of the ineffectiveness of gentle chemotherapy in those patients with advanced cardiac involvement.[78] Melphalan and dexamethasone have replaced melphalan and prednisone as the standard of care in patients who are not transplant-eligible, but the overall benefit of these regimens is highly dependent upon the patient population exposed to therapy. Figure 1 gives an algorithm on the therapy of newly diagnosed AL amyloidosis.

Stem Cell Transplantation

In the first phase II trial exploring dose-intensive melphalan in autologous stem cell transplantation, hematologic responses were seen in nearly two-thirds of patients. Validation of the organ responses following stem cell

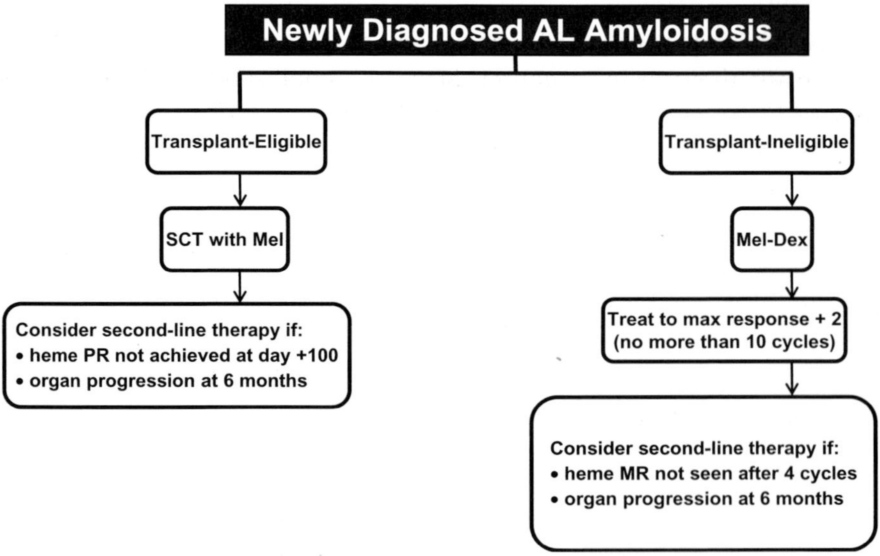

Fig. 1 Algorithm to approach treatment decisions in newly diagnosed AL amyloidosis.

transplantation using amyloid P-component scanning has been performed. Quality of life following stem cell transplantation is improved. Patients up to the age of 70 can be transplanted.[79] It is clear that the best outcomes are achieved in those patients who have a complete hematologic response or a >90% reduction in the involved light chain. Amyloid-related organ dysfunction improves in the majority of patients who respond and virtually all who achieve a complete hematologic response.[80] The determinants of outcome following stem cell transplantation include the post-transplantation development of acute renal insufficiency, the recovery of the absolute lymphocyte count, the development of excessive fluid accumulation during mobilization, as well as the pretransplant value of the immunoglobulin light chain. In a recent report, among more than 400 patients receiving high-dose chemotherapy and stem cell transplantation, the frequency of renal involvement was 70%, echocardiographic cardiac involvement 49%, peripheral nerve involvement 12%, and hepatic involvement 14%; 47% had single-organ involvement, 39% two-organ involvement, and 14% greater than two-organ involvement. The median 24 h urine protein excretion for the group was 3.68 g for the 24 h period. The median number of plasma cells was 7%, with an interquartile range of 3%–13%. The median age of patients transplanted was 57, with an interquartile range of 51–63. The median time from diagnosis to stem cell transplantation was four months. Stem cell mobilization is done without cytotoxic chemotherapy, using growth factors alone. The median number of stem cell collections required for safe transplantation is 2. The median time from the completion of stem cell collection to stem cell transplantation was nine days. Approximately 10% of patients need to have delay due to marked fluid retention during the collection procedure. Melphalan 200 mg/m^2 was given to 62%. Melphalan 140 mg/m^2 was given to 28%. The median length of the hospital stay for patients was eight days, and 19% completed the entire transplant procedure as an outpatient. Posttransplantation growth factor is not used to accelerate neutrophil recovery, and 500 neutrophils/μL is achieved at a median of 13.5 days. The median length of time to engraftment of 50,000 platelets is 17.5 days. Patients with amyloidosis stage I were 37%, stage II 38%, and stage III 25%. The pretransplant involved immunoglobulin-free light chain level was 15.3 mg/dL — five times the upper limit of normal.[81]

A hematologic response was seen in 75.8% of patients. Complete hematologic response was seen in 168, or 38.7% of patients. Hematologic relapses have been recorded in 119 patients (27.4%). The median time to progression for those patients was 24.5 months. Organ responses have been recorded in 203 patients (46.8%).

The major disadvantage of stem cell transplantation in the management of amyloidosis is treatment-related mortality. Day-100 all-cause mortality at Mayo Clinic is 10.1%, with most patients succumbing to multiorgan failure, respiratory distress syndrome, and ventilatory failure. The most important determinant of treatment-related mortality is the amyloidosis stage as measured by cardiac biomarkers.[82] The median survival for amyloid stage III is 58 months and has not been reached for stages I and II. The pretreatment free light chain level is also predictive of outcome. Patients whose serum-free light chain is >13.5 had a median survival of 87.6 months and this has not been reached for those patients with a light chain level lower than 13.5 at the time of transplant. Response was the strongest predictor of outcome in patients. The median survival has not been reached for complete responders, is 107 months for partial responders, and 32 months for nonresponders.[83]

Stem cell transplantation patients are highly selected and represent 20% of all patients seen at Mayo Clinic. In a proportional hazards model for variables that affect survival, the only relevant predictor of outcome was the stage. When best response was incorporated into a landmark model, the response was the strongest predictor of survival. A meta-analysis of studies has been published and it concludes that stem cell transplantation remains an unproven technique for the treatment of amyloidosis, and treatment-related mortality remains a major drawback of wider application of this technique.[84] Ten-year survivals have now been reported and are 53% for those patients who achieve complete hematologic remission. Transplant-related mortality is falling as well and is now at 7% all-cause, day-100 at Mayo Clinic. Eligible patients who can safely undergo stem cell transplantation should receive this therapy. Figure 2 gives Mayo guidelines for assessment of stem cell transplantation eligibility. It is unclear whether induction chemotherapy is required for patients with amyloidosis, because of the small number of plasma cells generally seen in the bone marrow pretransplant. In a phase II study

Transplant Eligibility Criteria

- **"Physiologic" age ≤ 70 years**
- **Performance score ≤ 2**
- **TnT < 0.06 ng/ml**
- **CrCl ≥ 30 ml/min***
  (unless on chronic dialysis)
- **NYHA class I/II***
- **No more than 2 organs *significantly* involved**

*Selected patients may become eligible for PBSCT with cardiac and renal transplantation.

Fig. 2 Mayo Clinic eligibility criteria for stem cell transplantation.

using dexamethasone induction followed by high-dose melphalan of 30 registered, 22 have been transplanted. The hematologic response rate was 52% and the complete response rate was 52%. An organ response rate of 57% was seen. The median overall survival has not been reached.[85]

Novel-Agent-Containing Regimens

Thalidomide

The first reports of benefit with the use of thalidomide were single case reports. Outcomes will vary based on the specific organ involved, specifically the severity of cardiac involvement and the number of organs involved. A single patient with cardiac amyloidosis was treated with thalidomide and showed improvement in functional capacity, cardiac function, and laboratory parameters without toxicity.[86] Hematologic response was complete at eight weeks and improvement in cardiac failure to class II 2.5 years after the start of therapy. A second reported patient was a 62-year-old male with a septal thickness of 19 mm, and an endomyocardially proven amyloid with a G λ monoclonal protein. With the use of thalidomide, beginning at 100 mg daily, increasing to 200 mg daily with dexamethasone, a complete hematologic response was obtained, and the patient was alive after 18 months on continued maintenance with alternate-day prednisone, with improvement in cardiac failure from class IV to class III and occasionally class II, and a 6 mm reduction in the septal thickness from its peak.[87] Two published series on the use of thalidomide as a single agent with steroids in amyloidosis have been published. In a trial from Boston University,[88] 16 patients were enrolled (median age 62) — 14 renal, 4 cardiac, 4 liver, and 2 soft tissue amyloidosis. The maximum tolerated dose was 300 mg. Fatigue was the major dose-limiting toxicity. Toxicity in patients with amyloid was far greater than in patients with multiple myeloma, which included exacerbation of both peripheral and pulmonary edema and worsening azotemia. Grade 3–4 toxicity was seen in 50% of patients, and 25% had to discontinue the use of thalidomide. A study from Mayo Clinic[89] enrolled 12 patients. Progressive edema, cognitive difficulties, and constipation occurred in 75%, with dyspnea, dizziness, and rash in 50%. Five developed progressive renal insufficiency. Deep vein thrombosis and syncope each occurred in two. The median time in the study was only 72 days. Patients were found to be intolerant of thalidomide. The Center for Amyloidosis in Pavia, Italy, reported on 31 patients receiving thalidomide beginning at 100 mg with 100 mg increments every two weeks up to 400 mg and dexamethasone 20 mg, 4 days every 21. Thirty-one patients were treated and 35% tolerated thalidomide at 400 mg/day. The hematologic response rate was 48%, with 19% complete responses and 26% organ responses. The median time to response was only 3.6 months. Treatment-related toxicity was frequent. Symptomatic bradycardia occurred in 26%. No patients with cardiac amyloid had an organ response, and there were no treatment-related deaths.[90] Thalidomide is active in amyloidosis, but the toxicity is substantial. The thalidomide dose should not exceed 50 mg initially and probably should never go above 100 mg per day.

Thalidomide has been combined with melphalan and dexamethasone to treat amyloidosis.[75] Twenty-two patients with advanced cardiac amyloidosis received a combination of the three agents. Six patients died from cardiac amyloid before cycle 3. Early death was associated with a reduced ejection fraction. Eight patients achieved a hematologic response, and 4 achieved a durable improvement in cardiac function. Treatment was feasible, but effective only in those with preserved cardiac systolic function. Thalidomide has been combined with cyclophosphamide and dexamethasone in a risk-adapted approach, with the dosing being determined by patient tolerance.[91] Seventy-five patients with advanced amyloidosis were treated. Fifty-one received Cytoxan, thalidomide, and dexamethasone, and 24 received the same combination with attenuated doses adjusted by risk. A hematologic response was seen in 74% of 65 evaluable patients, with complete responses in 21%. The median overall survival from the onset of therapy was 41 months. Toxicity necessitating cessation of therapy occurred in 8% and was grade 2 in 52%. Treatment-related mortality was 4%. The combination of thalidomide with an alkylator and a steroid is an option in poor-risk patients with amyloidosis.

Lenalidomide

Lenalidomide and dexamethasone have been used in the treatment of amyloidosis.[92] Thirty-four patients were enrolled. The initial dose of lenalidomide was 25 mg but was poorly tolerated, and a reduced dose of 15 mg was well tolerated. Of 24 evaluable patients, 7 (29%) achieved a hematologic complete response, and an additional 9 (38%) achieved a partial hematologic response, for an overall hematologic response rate of 67%. Fatigue and myelosuppression were the most common treatment-related adverse effects. Thromboembolic complications were seen in 9%. A Mayo Clinic study enrolled 23 patients, 13 of whom were previously treated. With the first three cycles of therapy, 10 patients discontinued treatment; there were four early deaths, three adverse events, and three other causes. At a median followup of 17 months, 10 patients responded to treatment, including nine hematologic, four renal, two cardiac, and two hepatic responses. All but one required dexamethasone to achieve the response. The most common adverse events were neutropenia, thrombocytopenia, rash, and fatigue. Significant activity of lenalidomide with dexamethasone was seen.[93]

In a secondary analysis of an ongoing clinical trial, 41 patients with AL amyloidosis received lenalidomide with or without dexamethasone. Twenty-seven of 41 patients (66%) developed kidney dysfunction during the lenalidomide treatment. The dysfunction was severe in 13 (32%). Four of them required initiation of dialysis (10%). The median time to kidney dysfunction after the start of the lenalidomide treatment was 44 days. Four of eight patients without underlying renal amyloidosis developed renal dysfunction. Patients with severe kidney dysfunction had a higher frequency of underlying renal amyloidosis. Recovery of renal function occurred in 12 (44%). Among patients with AL, worsening of the kidney occurs frequently during lenalidomide treatment, and kidney function needs to be closely monitored during therapy.[94]

A rise in the serum NT-proBNP has also been observed during lenalidomide therapy.[95] The recognition of potential drug-induced cardiac toxicities is important, so increased cardiac surveillance while the patient is on lenalidomide therapy is necessary.[95] In a followup of the original study in Boston,[96] a total of 69 patients were treated with lenalidomide and dexamethasone. Hematologic complete responses were seen in 16% of the patients. Most occurred by six months of treatment, and 60% of the complete responses were durable. The median progression-free survival in this study was 49.8 months. A French phase I–II study on the use of lenalidomide with melphalan and dexamethasone has also been reported. The maximum tolerated dose of lenalidomide was 15 mg. A complete hematologic response was achieved in 42% at a dose of 15 mg of lenalidomide per day. After a median followup of 19 months, the two-year overall and event-free survivals were 80.8% and 53.8%, respectively. Hematologic and organ response rates were both associated with superior event-free survival. An improved event-free survival was seen in patients whose free light chains decreased by more than 50% during therapy. Melphalan–dexamethasone with lenalidomide 15 mg per day is an effective combination therapy in newly diagnosed AL.[97]

Bortezomib

By inhibiting proteasome function in plasma cells, bortezomib activates stress-activated protein kinase in mitochondrial apoptotic signaling.[98] Amyloid-forming light chains can produce a load for the endoplasmic reticulum that makes these cells more sensitive to proteasome inhibition.[99] Eighteen patients, including seven who had relapsed or progressed after previous therapy, were treated with bortezomib. Sixty-one percent had two or more organs involved; the kidneys and the heart were affected in 14 and 15 patients, respectively. Ninety-four percent had a hematologic response and 44% a complete hematologic response; five patients (28%) had a response in at least one organ. The median time to response was 0.9 months. The median time to organ response was four months. Neurotoxicity, fatigue, peripheral edema, constipation, and postural hypotension were significant toxicities.[100] Bortezomib was administered to 20 patients with AL who had a median of three lines of prior chemotherapy including thalidomide in all. A median of three cycles of bortezomib was administered, and 45% received dexamethasone. Fifteen percent of the patients achieved a complete hematologic response, with 65% achieving a partial hematologic response. Toxicity was seen in 75% of the patients and was severe enough to require discontinuation of therapy in 40% of the patients.[101]

In a phase I dose-escalation study, bortezomib was given in weekly and twice-weekly schedules.[102] Thirty-one patients were enrolled across seven cohorts. No corticosteroids were given. Dose-limiting toxicity included grade three congestive heart failure in two. The most commonly reported toxicities were gastrointestinal fatigue and peripheral neuropathy. Discontinuation and dose reduction for toxicity were reported in 12 and four patients, respectively. Hematologic response was seen in 50% of 30 evaluable patients, including 20% complete responses and a median time to first response of 1.2 months. There has been a shift from twice-weekly to once-weekly bortezomib therapy in the amyloidosis community to reduce the toxicity in this fragile population. Amyloidosis relapsing after autologous stem cell transplantation has been treated with bortezomib, resulting in the normalization of detectable serum-free light chains and reversing tissue damage, suggesting that this is an effective salvage therapy and may actually render patients who had previously been ineligible for high-dose therapy capable of receiving this effective management strategy.[103] A multicenter survey of bortezomib with or without dexamethasone was performed across three amyloidosis treatment centers. Ninety-four patients were analyzed. Hematologic response was achieved in 71% at a median time of 52 days, with 25% complete responses. In previously untreated patients, the complete response rate was 47%. Twenty-nine percent of patients had organ progression at 12 months. Twenty-seven percent had hematologic progression. The one-year survival was 76%. NT-proBNP was independently associated with survival. Toxicity included neuropathy, orthostasis, peripheral edema, constipation, and diarrhea. Bortezomib, with or without dexamethasone, is active in AL and induces rapid responses and higher rates of hematologic and organ responses. There is now a United States and European study that is randomizing patients between melphalan–dexamethasone–bortezomib and melphalan–dexamethasone to determine the optimal therapy for non-transplant-related patients.

An Austrian group reported a retrospective evaluation of the efficacy of bortezomib and dexamethasone in 26 patients with AL, 18 of whom received the combination as first-line therapy. All had renal involvement, and 35% cardiac involvement. The overall response rate was 54%, with eight patients achieving a hematologic complete response (31%). The median time to response was 7.5 weeks. Improvement in organ function was noticed in three patients (12%). The median progression-free and overall survivals were five and 18.7 months, respectively. Median progression-free and overall survivals have not been reached by complete responders. Hematologic side effects were the most common.[104] Bortezomib is a highly active management tool for amyloidosis. Patients achieving a complete response have a marked benefit from therapy.

Pomalidomide

Pomalidomide, a derivative of thalidomide that is structurally similar to lenalidomide and thalidomide, was administered to 20 patients, with 17 evaluable for toxicity. All had been previously treated. All had received prior alkylator, 44% prior lenalidomide, and 37% prior bortezomib. The most common adverse event was myelosuppression. The most common nonhematologic adverse event was fatigue. Three patients discontinued therapy, two due to progression, and there was one death five days into therapy. However, responses were reported and the potential of pomalidomide to be an effective agent in the treatment of amyloidosis remains to be defined and will require further followup.[105]

THERAPY OF SECONDARY AMYLOIDOSIS

The treatment of secondary amyloidosis, when due to an infectious cause, is related to the treatment of the underlying infectious disease. However, since the majority of patients now seen have problems associated with sustained inflammatory response and cytokine-mediated inflammation, agents that suppress tumor necrosis factor have recently been introduced. Etanercept was tested in patients with rheumatoid arthritis and AA amyloidosis, and was found to lead to significant improvement in SAA levels, creatinine clearance, and proteinuria. Previously, chlorambucil had been used for the treatment of secondary amyloidosis, but it will produce azospermia and can produce myelodysplasia, and is used infrequently today. Anti-TNFα therapy can induce rapid resolution of amyloid deposits associated with rheumatoid arthritis.[106] Infliximab has been reported to improve secondary amyloidosis from chronic inflammatory arthritis.[107] Even when chronic infection is present, such as hidradenitis suppurativa, infliximab has produced responses.[108] In a patient with renal amyloidosis secondary to inflammatory bowel disease, infliximab therapy led to a response.[109] Etanercept has been reported to induce resolution of renal deterioration in patients with rheumatoid arthritis.[110] Demonstrated suppression of SAA levels in patients with inflammatory arthritis and secondary amyloidosis has been demonstrated.[111] There have been no comparative trials comparing infliximab and etanercept.

TRANSPLANTATION FOR FAMILIAL AMYLOIDOSIS

When familial amyloidosis is produced by mutations in transthyretin, liver transplantation can remove the primary source of the amyloidogenic protein. However, treatment-related mortality associated with liver transplantation is strongly related to the severity of the patient's neuropathy and their modified body mass index at the time of transplantation.[112,113] Progression in the myocardium has been reported to occur despite liver transplantation, presumably owing to deposition of native transthyretin on a nidus of pre-existent mutant transthyretin.[58] A report from a registry in 16 countries involving 539 patients gave a five-year median survival after transplantation of 77%.[114] Combined liver and kidney transplantation and six heart and liver transplants were reported. One hundred and forty-nine evaluable patients showed improvement in neurological, musculoskeletal, and gastrointestinal involvement in half. Only 20% of patients with cardiac symptoms improved. Domino liver transplantation has been used where the amyloid-producing liver is given to someone with native liver disease, and case reports now exist in the literature of the development of amyloidosis in the recipients of these livers at approximately seven years.[115] Surveillance biopsies in five patients revealed amyloid deposits in two of the five 47 months following the transplant. Other forms of amyloidosis that are inherited are less amenable to liver transplantation. Apolipoprotein A1 is 50% produced by the gastrointestinal tract. Despite the fact that transplantation only removes part of the amyloid source, clinical improvement has been reported.[116] The fibrinogen A α-chain is produced exclusively in the liver, and hepatic transplantation appears to be curative, where renal transplantation alone results in amyloid deposition and dysfunction of the transplanted kidney within 1–10 years.[117] Lysozyme amyloidosis is produced by neutrophils, and no interventions are known to be effective for this form of amyloidosis.[56]

AGENTS THAT DESTABILIZE THE AMYLOID PROTEIN STRUCTURE

Chemical Agents Disrupting the Serum Amyloid P Component

CPHPC was identified by screening a compound library for inhibitors of SAP binding to amyloid fibrils and then chemically modifying the resulting compounds to enhance the avidity of the protein–ligand interaction. It was tested by a 48 h intravenous infusion in seven patients with systemic amyloidosis and resulted in rapid and consistent depletion of SAP by the end of the infusion. Sustained pharmacologic depletion of the serum amyloid P component has been reported.[118] Recently, the combination of CPHPC with an antibody to SAP has been investigated using CPHPC to deplete circulating human SAP, enabling injected anti-SAP antibodies to reach deposits in tissues. Studies on mice have demonstrated the ability of this combined therapy to eliminate amyloid deposition and it could be applicable to all forms of systemic and localized amyloid.[68]

Monoclonal Antibody Treatment of Amyloidosis

Antibodies have been developed against human immunoglobulin light chains and specifically epitopes of the β-pleated sheet structure of these proteins. This hypothesis has been tested on mice exposed to the monoclonal antibody, and tissue biopsies have shown inflammation within the amyloid deposits and opsonization of the amyloid fibrils.[119] An epitope has been identified that is common to both nonnative and fibrillary immunoglobulin light chains and has been proposed to be used for the rational design of amyloid reactive antibodies.[120] Recently, the use of these antibodies for the radio-immunodetection of amyloid deposits has been demonstrated.[69] The M11–1F4 antibody can be employed to identify candidates for passive immunotherapy using a chimeric form of the antibody. Inhibition of free light chain production, the precursor to amyloid deposition, by using interfering RNA molecules has also been investigated by transfecting cells with the interfering RNAs, leading to a 40% reduction in messenger RNA production.[121] These antibodies offer the opportunity for an immunologic approach to the management of amyloidosis.

CONCLUSION

Amyloidosis is a complex, diverse group of diseases, all of which have protein misfolding in common but have diverse protein precursors. The diagnosis is difficult and is often missed if not suspected. Accurate identification of the precursor protein followed by directive therapy is critical for a positive outcome.

REFERENCES

1. Sipe JD, Cohen AS. (2000) Review: history of the amyloid fibril. *J Struct Biol* **130**: 88–98.
2. Reznik M. (1989) Paul Divry: the discovery of cerebral amyloidosis. *Acta Neurol Belg* **89**: 168–178.
3. Cohen AS, Calkins E. (1959) Electron microscopic observations on a fibrous component in amyloid of diverse origins. *Nature* **183**: 1202–1203.
4. Relini A, Torrassa S, Ferrando R, *et al.* (2010) Detection of populations of amyloid-like protofibrils with different physical properties. *Biophys J* **98**: 1277–1284.
5. Sunde M, Serpell LC, Bartlam M, *et al.* (1997) Common core structure of amyloid fibrils by synchrotron X-ray diffraction. *J Mol Biol* **273**: 729–739.
6. Baden EM, Sikkink LA, Ramirez-Alvarado M. (2009) Light chain amyloidosis — current findings and future prospects. *Curr Protein Pept Sci* **10**: 500–508.

7. Lindgren M, Hammarstrom P. (2010) Amyloid oligomers: spectroscopic characterization of amyloidogenic protein states. *Febs J* **277**: 1380–1388.
8. Ancsin JB. (2003) Amyloidogenesis: historical and modern observations point to heparan sulfate proteoglycans as a major culprit. *Amyloid* **10**: 67–79.
9. Enqvist S, Sletten K, Westermark P. (2009) Fibril protein fragmentation pattern in systemic AL-amyloidosis. *J Pathol* **219**: 473–480.
10. Bellotti V, Nuvolone M, Giorgetti S, *et al.* (2007) The workings of the amyloid diseases. *Ann Med* **39**: 200–207.
11. Comenzo RL, Zhang Y, Martinez C, *et al.* (2001) The tropism of organ involvement in primary systemic amyloidosis: contributions of Ig V(L) germ line gene use and clonal plasma cell burden. *Blood* **98**: 714–720.
12. Reixach N, Deechongkit S, Jiang X, *et al.* (2004) Tissue damage in the amyloidoses: transthyretin monomers and nonnative oligomers are the major cytotoxic species in tissue culture. *Proc Natl Acad Sci USA* **101**: 2817–2822.
13. Hartley DM, Walsh DM, Ye CP, *et al.* (1999) Protofibrillar intermediates of amyloid beta-protein induce acute electrophysiological changes and progressive neurotoxicity in cortical neurons. *J Neurosci* **19**: 8876–8884.
14. Andersson K, Olofsson A, Nielsen EH, *et al.* (2002) Only amyloidogenic intermediates of transthyretin induce apoptosis. *Biochem Biophys Res Commun* **294**: 309–314.
15. Perfetto F, Moggi-Pignone A, Livi R, *et al.* (2010) Systemic amyloidosis: a challenge for the rheumatologist. *Nat Rev Rheumatol* **6**: 417–429.
16. Gertz MA, Lacy MQ, Dispenzieri A, Hayman SR. (2005) Amyloidosis: diagnosis and management. *Clin Lymphoma Myeloma* **6**: 208–219.
17. Haas M, Meehan SM, Karrison TG, Spargo BH. (1997) Changing etiologies of unexplained adult nephrotic syndrome: a comparison of renal biopsy findings from 1976–1979 and 1995–1997. *Am J Kidney Dis* **30**: 621–631.
18. Olivieri B, Rai AJ. (2010) A primer on clinical applications and assays using urine: focus on analysis of plasma cell dyscrasias using automated electrophoresis and immunofixation. *Methods Mol Biol* **641**: 13–26.
19. Grasberger H, Golcher HM, Fingerhut A, Janssen OE. (2002) Loop variants of the serpin thyroxine-binding globulin: implications for hormone release upon limited proteolysis. *Biochem J* **365**: 311–316.
20. Mollee P. (2009) Current trends in the diagnosis, therapy and monitoring of the monoclonal gammopathies. *Clin Biochem Rev* **30**: 93–103.
21. Bely M, Makovitzky J. (2006) Sensitivity and specificity of Congo red staining according to Romhanyi. Comparison with Puchtler's or Bennhold's methods. *Acta Histochem* **108**: 175–180.

22. Solomon A, Murphy CL, Westermark P. (2008) Unreliability of immunohistochemistry for typing amyloid deposits. *Arch Pathol Lab Med* **132**: 14; author reply 14–15.
23. Plante-Bordeneuve V, Ferreira A, Lalu T, *et al.* (2007) Diagnostic pitfalls in sporadic transthyretin familial amyloid polyneuropathy (TTR-FAP). *Neurology* **69**: 693–698.
24. Vrana JA, Gamez JD, Madden BJ, *et al.* (2009) Classification of amyloidosis by laser microdissection and mass spectrometry-based proteomic analysis in clinical biopsy specimens. *Blood* **114**: 4957–4959.
25. Roden AC, Aubry MC, Zhang K, *et al.* (2010) Nodular senile pulmonary amyloidosis: a unique case confirmed by immunohistochemistry, mass spectrometry, and genetic study. *Hum Pathol* **4**(7): 1040–1045.
26. Gono T, Matsuda M, Dohi N, *et al.* (2002) Gastroduodenal lesions in primary AL amyloidosis. *Gastrointest Endosc* **56**: 563.
27. Belkin RN, Kupersmith AC, Khalique O, *et al.* (2010) A novel two-dimensional echocardiographic finding in cardiac amyloidosis. *Echocardiography* **27**: 1171–1176.
28. Piper C, Butz T, Farr M, *et al.*(2010) How to diagnose cardiac amyloidosis early: impact of ECG, tissue Doppler echocardiography, and myocardial biopsy. *Amyloid* **17**: 1–9.
29. Syed IS, Glockner JF, Feng D, *et al.* (2010) Role of cardiac magnetic resonance imaging in the detection of cardiac amyloidosis. *JACC Cardiovasc Imaging* **3**: 155–164.
30. Hazenberg BP, van Rijswijk MH, Piers DA, *et al.* (2006) Diagnostic performance of 123I-labeled serum amyloid P component scintigraphy in patients with amyloidosis. *Am J Med* **119**: 355, e315–e324.
31. Rajkumar SV, Gertz MA, Kyle RA. (1998) Primary systemic amyloidosis with delayed progression to multiple myeloma. *Cancer* **82**: 1501–1505.
32. Perfetti V, Casarini S, Palladini G, *et al.* (2002) Analysis of V(lambda)–J(lambda) expression in plasma cells from primary (AL) amyloidosis and normal bone marrow identifies 3r (lambdaIII) as a new amyloid-associated germline gene segment. *Blood* **100**: 948–953.
33. Park MA, Mueller PS, Kyle RA, *et al.* (2003) Primary (AL) hepatic amyloidosis: clinical features and natural history in 98 patients. *Medicine (Baltimore)* **82**: 291–298.
34. Dispenzieri A, Gertz MA, Kyle RA, *et al.* (2004) Serum cardiac troponins and N-terminal pro-brain natriuretic peptide: a staging system for primary systemic amyloidosis. *J Clin Oncol* **22**: 3751–3757.

35. Dispenzieri A, Lacy MQ, Katzmann JA, *et al.* (2006) Absolute values of immunoglobulin free light chains are prognostic in patients with primary systemic amyloidosis undergoing peripheral blood stem cell transplantation. *Blood* **107**: 3378–3383.
36. Nakamura T. (2008) Clinical strategies for amyloid A amyloidosis secondary to rheumatoid arthritis. *Mod Rheumatol* **18**: 109–118.
37. Marhaug G, Dowton SB. (1994) Serum amyloid A: an acute phase apolipoprotein and precursor of AA amyloid. *Baillieres Clin Rheumatol* **8**: 553–573.
38. Rocken C, Shakespeare A. (2002) Pathology, diagnosis and pathogenesis of AA amyloidosis. *Virchows Arch* **440**: 111–122.
39. McMahan ZH, Sailors JL, Toto R, Olsen NJ. (2010) Systemic amyloidosis presenting as chronic diarrhea in a patient with ankylosing spondylitis. *J Clin Rheumatol* **16**: 22–25.
40. Skinner M, Pinnette A, Travis WD, *et al.* (1988) Isolation and sequence analysis of amyloid protein AA from a patient with cystic fibrosis. *J Lab Clin Med* **112**: 413–417.
41. Simsek I, Kaya A, Erdem H, *et al.* (2010) No regression of renal amyloid mass despite remission of nephrotic syndrome in a patient with TRAPS following etanercept therapy. *J Nephrol* **23**: 119–123.
42. Hazenberg BP, van Rijswijk MH. (2000) Where has secondary amyloid gone? *Ann Rheum Dis* **59**: 577–579.
43. Yonem O, Bayraktar Y. (2007) Secondary amyloidosis due to FMF. *Hepatogastroenterology* **54**: 1061–1065.
44. Gillmore JD, Lovat LB, Persey MR, *et al.* (2001) Amyloid load and clinical outcome in AA amyloidosis in relation to circulating concentration of serum amyloid A protein. *Lancet* **358**: 24–29.
45. Bhat A, Selmi C, Naguwa SM, *et al.* (2010) Current concepts on the immunopathology of amyloidosis. *Clin Rev Allergy Immunol* **38**: 97–106.
46. Rapezzi C, Quarta CC, Riva L, *et al.* (2010) Transthyretin-related amyloidoses and the heart: a clinical overview. *Nat Rev Cardiol* **7**: 398–408.
47. Benson MD, Kincaid JC. (2007) The molecular biology and clinical features of amyloid neuropathy. *Muscle Nerve* **36**: 411–423.
48. Saporta MA, Zaros C, Cruz MW, *et al.* (2009) Penetrance estimation of TTR familial amyloid polyneuropathy (type I) in Brazilian families. *Eur J Neurol* **16**: 337–341.
49. Kawaji T, Ando Y, Ando E, *et al.* (2010) Transthyretin-related vitreous amyloidosis in different endemic areas. *Amyloid* **17**: 105–108.
50. Said G, Plante-Bordeneuve V. (2009) Familial amyloid polyneuropathy: a clinico-pathologic study. *J Neurol Sci* **284**: 149–154.

51. Rapezzi C, Merlini G, Quarta CC, *et al.* (2009) Systemic cardiac amyloidoses: disease profiles and clinical courses of the 3 main types. *Circulation* **120**: 1203–1212.
52. Connors LH, Prokaeva T, Lim A, *et al.* (2009) Cardiac amyloidosis in African Americans: comparison of clinical and laboratory features of transthyretin V122I amyloidosis and immunoglobulin light chain amyloidosis. *Am Heart J* **158**: 607–614.
53. Hamour IM, Lachmann HJ, Goodman HJ, *et al.* (2008) Heart transplantation for homozygous familial transthyretin (TTR) V122I cardiac amyloidosis. *Am J Transplant* **8**: 1056–1059.
54. Gillmore JD, Lachmann HJ, Rowczenio D, *et al.* (2009) Diagnosis, pathogenesis, treatment, and prognosis of hereditary fibrinogen A alpha-chain amyloidosis. *J Am Soc Nephrol* **20**: 444–451.
55. Benson MD. (2005) Ostertag revisited: the inherited systemic amyloidoses without neuropathy. *Amyloid* **12**: 75–87.
56. Granel B, Valleix S, Serratrice J, *et al.* (2006) Lysozyme amyloidosis: report of 4 cases and a review of the literature. *Medicine (Baltimore)* **85**: 66–73.
57. Liepnieks JJ, Zhang LQ, Benson MD. (2010) Progression of transthyretin amyloid neuropathy after liver transplantation. *Neurology* **75**: 324–327.
58. Ihse E, Suhr OB, Hellman U, Westermark P. (2010) Variation in amount of wild-type transthyretin in different fibril and tissue types in ATTR amyloidosis. *J Mol Med* **89**(2): 171–180.
59. Chee CE, Lacy MQ, Dogan A, *et al.* (2010) Pitfalls in the diagnosis of primary amyloidosis. *Clin Lymphoma Myeloma Leuk* **10**: 177–180.
60. Ng B, Connors LH, Davidoff R, *et al.* (2005) Senile systemic amyloidosis presenting with heart failure: a comparison with light chain–associated amyloidosis. *Arch Intern Med* **165**: 1425–1429.
61. Biewend ML, Menke DM, Calamia KT. (2006) The spectrum of localized amyloidosis: a case series of 20 patients and review of the literature. *Amyloid* **13**: 135–142.
62. Ng CS, Wan S, Yim AP, Vale J. (2002) Idiopathic localised bladder amyloidosis: rare cause of haematuria. *Int Urol Nephrol* **34**: 55–58.
63. Utz JP, Swensen SJ, Gertz MA. (1996) Pulmonary amyloidosis: the Mayo Clinic experience from 1980 to 1993. *Ann Intern Med* **124**: 407–413.
64. Schreml S, Szeimies RM, Vogt T, *et al.* (2010) Cutaneous amyloidoses and systemic amyloidoses with cutaneous involvement. *Eur J Dermatol* **20**: 152–160.
65. Hashimoto K, Ito K, Taniguchi Y, *et al.* (1990) Keratin in cutaneous amyloidoses. *Clin Dermatol* **8**: 55–65.

66. Goffin YA, McCrickard EL, Ameryckx JP, *et al.* (1985) Amyloidosis of the joints: evidence that human hip capsules have a unique predisposition for amyloid of the senile systemic type. *Appl Pathol* **3**: 88–95.
67. Pucci A, Wharton J, Arbustini E, *et al.* (1991) Atrial amyloid deposits in the failing human heart display both atrial and brain natriuretic peptide-like immunoreactivity. *J Pathol* **165**: 235–241.
68. Bodin K, Ellmerich S, Kahan MC, *et al.* (2010) Antibodies to human serum amyloid P component eliminate visceral amyloid deposits. *Nature* **468**: 93–97.
69. Wall JS, Kennel SJ, Stuckey AC, *et al.* (2010) Radioimmunodetection of amyloid deposits in patients with AL amyloidosis. *Blood* **116**: 2241–2244.
70. Wilczek HE, Larsson M, Yamamoto S, Ericzon BG. (2008) Domino liver transplantation. *J Hepatobil Pancreat Surg* **15**: 139–148.
71. Yuksel S, Yalcinkaya F, Acar B, *et al.* (2006) Clinical improvement with infliximab in a child with amyloidosis secondary to familial Mediterranean fever. *Rheumatology (Oxford)* **45**: 1307–1308.
72. Leslie KS, Lachmann HJ, Bruning E, *et al.* (2006) Phenotype, genotype, and sustained response to anakinra in 22 patients with autoinflammatory disease associated with CIAS-1/NALP3 mutations. *Arch Dermatol* **142**: 1591–1597.
73. Gertz MA, Lacy MQ, Dispenzieri A. (2004) Therapy for immunoglobulin light chain amyloidosis: the new and the old. *Blood Rev* **18**: 17–37.
74. Palladini G, Russo P, Nuvolone M, *et al.* (2007) Treatment with oral melphalan plus dexamethasone produces long-term remissions in AL amyloidosis. *Blood* **110**: 787–788.
75. Palladini G, Russo P, Lavatelli F, *et al.* (2009) Treatment of patients with advanced cardiac AL amyloidosis with oral melphalan, dexamethasone, and thalidomide. *Ann Hematol* **88**: 347–350.
76. Dietrich S, Schonland SO, Benner A, *et al.* (2010) Treatment with intravenous melphalan and dexamethasone is not able to overcome the poor prognosis of patients with newly diagnosed systemic light chain amyloidosis and severe cardiac involvement. *Blood* **116**: 522–528.
77. Gertz MA. (2010) I don't know how to treat amyloidosis. *Blood* **116**: 507–508.
78. Sanchorawala V, Wright DG, Seldin DC, *et al.* (2002) Low-dose continuous oral melphalan for the treatment of primary systemic (AL) amyloidosis. *Br J Haematol* **117**: 886–889.
79. Seldin DC, Anderson JJ, Sanchorawala V, *et al.* (2004) Improvement in quality of life of patients with AL amyloidosis treated with high-dose melphalan and autologous stem cell transplantation. *Blood* **104**: 1888–1893.

80. Gertz MA, Lacy MQ, Dispenzieri A, *et al.* (2007) Effect of hematologic response on outcome of patients undergoing transplantation for primary amyloidosis: importance of achieving a complete response. *Haematologica* **92**: 1415–1418.
81. Gertz MA, Lacy MQ, Dispenzieri A, *et al.* (2010) Autologous stem cell transplant for immunoglobulin light chain amyloidosis: a status report. *Leuk Lymphoma* **51**: 2181–2187.
82. Gertz MA, Lacy MQ, Dispenzieri A, *et al.* (2010) Trends in day 100 and 2-year survival after auto-SCT for AL amyloidosis: outcomes before and after 2006. *Bone Marrow Transplant* **46**(7): 970–975.
83. Kumar S, Dispenzieri A, Katzmann JA, *et al.* (2010) Serum immunoglobulin free light-chain measurement in primary amyloidosis: prognostic value and correlations with clinical features. *Blood* **116**: 5126–5129.
84. Mhaskar R, Kumar A, Behera M, *et al.* (2009) Role of high-dose chemotherapy and autologous hematopoietic cell transplantation in primary systemic amyloidosis: a systematic review. *Biol Blood Marrow Transplant* **15**: 893–902.
85. Cohen AD, Zhou P, Chou J, *et al.* (2007) Risk-adapted autologous stem cell transplantation with adjuvant dexamethasone +/– thalidomide for systemic light-chain amyloidosis: results of a phase II trial. *Br J Haematol* **139**: 224–233.
86. Oh IY, Kim HK, Kim YJ, *et al.* (2006) An intriguing case of primary amyloidosis with cardiac involvement: symptomatic and echocardiographic improvement with thalidomide treatment. *Int J Cardiol* **113**: 141–143.
87. Campbell P, Murdock C. (2006) Cardiac amyloidosis — sustained clinical and free light chain response to low dose thalidomide and corticosteroids. *Intern Med J* **36**: 137–139.
88. Seldin DC, Choufani EB, Dember LM, *et al.* (2003) Tolerability and efficacy of thalidomide for the treatment of patients with light chain–associated (AL) amyloidosis. *Clin Lymphoma* **3**: 241–246.
89. Dispenzieri A, Lacy MQ, Rajkumar SV, *et al.* (2003) Poor tolerance to high doses of thalidomide in patients with primary systemic amyloidosis. *Amyloid* **10**: 257–261.
90. Palladini G, Perfetti V, Perlini S, *et al.* (2005) The combination of thalidomide and intermediate-dose dexamethasone is an effective but toxic treatment for patients with primary amyloidosis (AL). *Blood* **105**: 2949–2951.
91. Wechalekar AD, Goodman HJ, Lachmann HJ, *et al.* (2007) Safety and efficacy of risk-adapted cyclophosphamide, thalidomide, and dexamethasone in systemic AL amyloidosis. *Blood* **109**: 457–464.

92. Sanchorawala V, Wright DG, Rosenzweig M, *et al.* (2007) Lenalidomide and dexamethasone in the treatment of AL amyloidosis: results of a phase 2 trial. *Blood* **109**: 492–496.
93. Dispenzieri A, Lacy MQ, Zeldenrust SR, *et al.* (2007) The activity of lenalidomide with or without dexamethasone in patients with primary systemic amyloidosis. *Blood* **109**: 465–470.
94. Specter R, Sanchorawala V, Seldin DC, *et al.* (2010) Kidney dysfunction during lenalidomide treatment for AL amyloidosis. *Nephrol Dial Transplant* **26**(3): 881–886.
95. Dispenzieri A, Dingli D, Kumar SK, *et al.* (2010) Discordance between serum cardiac biomarker and immunoglobulin-free light-chain response in patients with immunoglobulin light-chain amyloidosis treated with immune modulatory drugs. *Am J Hematol* **85**: 757–759.
96. Sanchorawala V, Finn KT, Fennessey S, *et al.* (2010) Durable hematologic complete responses can be achieved with lenalidomide in AL amyloidosis. *Blood* **116**: 1990–1991.
97. Moreau P, Jaccard A, Benboubker L, *et al.* (2010) Lenalidomide in combination with melphalan and dexamethasone in patients with newly-diagnosed AL amyloidosis: a multicenter phase 1/2 dose escalation study. *Blood* **116**(23): 4777–4782.
98. Sterz J, von Metzler I, Hahne JC, *et al.* (2008) The potential of proteasome inhibitors in cancer therapy. *Expert Opin Investig Drugs* **17**: 879–895.
99. Sitia R, Palladini G, Merlini G. (2007) Bortezomib in the treatment of AL amyloidosis: targeted therapy? *Haematologica* **92**: 1302–1307.
100. Kastritis E, Anagnostopoulos A, Roussou M, *et al.* (2007) Treatment of light chain (AL) amyloidosis with the combination of bortezomib and dexamethasone. *Haematologica* **92**: 1351–1358.
101. Wechalekar AD, Lachmann HJ, Offer M, *et al.* (2008) Efficacy of bortezomib in systemic AL amyloidosis with relapsed/refractory clonal disease. *Haematologica* **93**: 295–298.
102. Reece DE, Sanchorawala V, Hegenbart U, *et al.* (2009) Weekly and twice-weekly bortezomib in patients with systemic AL amyloidosis: results of a phase 1 dose-escalation study. *Blood* **114**: 1489–1497.
103. Brunvand MW, Bitter M. (2010) Amyloidosis relapsing after autologous stem cell transplantation treated with bortezomib: normalization of detectable serum-free light chains and reversal of tissue damage with improved suitability for transplant. *Haematologica* **95**: 519–521.
104. Lamm W, Willenbacher W, Lang A, *et al.* (2010) Efficacy of the combination of bortezomib and dexamethasone in systemic AL amyloidosis. *Ann Hematol* **90**(2): 201–206.

105. Dispenzieri A, Gertz MA, Hayman SR, *et al.* (2010) A phase-2 study of pomalidomide and dexamethasone In previously-treated light-chain (AL) amyloidosis. *ASH Annual Meeting Abstracts* **116**: 987.
106. Kuroda T, Wada Y, Kobayashi D, *et al.* (2009) Effective anti-TNF-alpha therapy can induce rapid resolution and sustained decrease of gastroduodenal mucosal amyloid deposits in reactive amyloidosis associated with rheumatoid arthritis. *J Rheumatol* **36**: 2409–2415.
107. Keersmaekers T, Claes K, Kuypers DR, *et al.* (2009) Long-term efficacy of infliximab treatment for AA-amyloidosis secondary to chronic inflammatory arthritis. *Ann Rheum Dis* **68**: 759–761.
108. Montes-Romero JA, Callejas-Rubio JL, Sanchez-Cano D, *et al.* (2008) Amyloidosis secondary to hidradenitis suppurativa. Exceptional response to infliximab. *Eur J Intern Med* **19**: e32–e33.
109. Hatakeyama T, Komatsuda A, Matsuda A, *et al.* (2008) Renal amyloidosis associated with extracapillary glomerulonephritis and vasculitis in a patient with inflammatory bowel disease treated with infliximab. *Clin Nephrol* **70**: 240–244.
110. Nakamura T, Higashi S, Tomoda K, *et al.* (2010) Etanercept can induce resolution of renal deterioration in patients with amyloid A amyloidosis secondary to rheumatoid arthritis. *Clin Rheumatol* **29**: 1395–1401.
111. Perry ME, Stirling A, Hunter JA. (2008) Effect of etanercept on serum amyloid A protein (SAA) levels in patients with AA amyloidosis complicating inflammatory arthritis. *Clin Rheumatol* **27**: 923–925.
112. Okamoto S, Wixner J, Obayashi K, *et al.* (2009) Liver transplantation for familial amyloidotic polyneuropathy: impact on Swedish patients' survival. *Liver Transpl* **15**: 1229–1235.
113. Yamamoto S, Wilczek HE, Nowak G, *et al.* (2007) Liver transplantation for familial amyloidotic polyneuropathy (FAP): a single-center experience over 16 years. *Am J Transplant* **7**: 2597–2604.
114. Herlenius G, Wilczek HE, Larsson M, Ericzon BG. (2004) Ten years of international experience with liver transplantation for familial amyloidotic polyneuropathy: results from the Familial Amyloidotic Polyneuropathy World Transplant Registry. *Transplantation* **77**: 64–71.
115. Llado L, Baliellas C, Casasnovas C, *et al.* (2010) Risk of transmission of systemic transthyretin amyloidosis after domino liver transplantation. *Liver Transpl* **16**: 1386–1392.
116. Testro AG, Brennan SO, Macdonell RA, *et al.* (2007) Hereditary amyloidosis with progressive peripheral neuropathy associated with apolipoprotein AI Gly26Arg: outcome of hepatorenal transplantation. *Liver Transpl* **13**: 1028–1031.

117. Stangou AJ, Banner NR, Hendry BM, *et al.* (2010) Hereditary fibrinogen A alpha-chain amyloidosis: phenotypic characterization of a systemic disease and the role of liver transplantation. *Blood* **115**: 2998–3007.
118. Gillmore JD, Tennent GA, Hutchinson WL, *et al.* (2010) Sustained pharmacological depletion of serum amyloid P component in patients with systemic amyloidosis. *Br J Haematol* **148**: 760–767.
119. Hrncic R, Wall J, Wolfenbarger DA, *et al.* (2000) Antibody-mediated resolution of light chain–associated amyloid deposits. *Am J Pathol* **157**: 1239–1246.
120. O'Nuallain B, Allen A, Kennel SJ, *et al.* (2007) Localization of a conformational epitope common to non-native and fibrillar immunoglobulin light chains. *Biochemistry* **46**: 1240–1247.
121. Phipps JE, Kestler DP, Foster JS, *et al.* (2010) Inhibition of pathologic immunoglobulin-free light chain production by small interfering RNA molecules. *Exp Hematol* **38**: 1006–1013.

CHAPTER 4

Adiposopathy

Harold E. Bays and J. Michael Gonzalez-Campoy†*

INTRODUCTION

Individuals with excessive body fat (adiposity) are often categorized as being overweight or obese. In many developed countries, obesity and its comorbidities are epidemics.[1] Adiposity alone causes fat-mass-related disorders of the cardiovascular, neurological, respiratory, musculoskeletal, integumentary, gastrointestinal, and genitourinary systems, as well as psychological disorders. It can also cause adipocyte and adipose tissue dysfunction which contributes to, or worsens metabolic disease. For decades, scientific evidence has supported the concept that adipose tissue is an active endocrine and immune organ. However, many clinicians and medical organizations have been slow to recognize the science behind the full pathogenic potential of adipose tissue. This has resulted in the curious notion that excessive body fat is not necessarily a "disease," even when associated with adverse clinical consequences. It is the purpose of this review to describe the pathogenic potential of adipose tissue.

*Louisville Metabolic and Atherosclerosis Research Center, 3288 Illinois Avenue, Louisville KY 40213, USA. www.lmarc.com; hbaysmd@aol.com.

†Minnesota Center for Obesity, Metabolism and Endocrinology, PA, 1185 Town Centre Drive, Suite 220, Eagan, MN 55123, USA. www.mncome.com; drmike@mncome.com.

DEFINITION

Adiposopathy is defined as anatomic, structural, and functional abnormalities of adipocytes and adipose tissue leading to metabolic disease. It is promoted by positive caloric balance, unhealthy nutritional intake, and sedentary lifestyle in genetically and environmentally susceptible individuals. It is exacerbated by limited or impaired adipogenesis in peripheral subcutaneous adipose tissue during positive caloric balance, leading to increased fat deposition in other adipose tissue depots and other body organs. Adiposopathy is anatomically manifested by adipocyte hypertrophy and visceral fat accumulation; structurally manifested by extracellular matrix abnormalities and impaired vascular delivery to adipose tissue (whose expansion may exceed vascular supply); and functionally manifested by adipocyte and adipose tissue hypoxia, mitochondrial and endoplasmic reticulum dysfunction, and other cellular abnormalities. These anatomic, structural, and functional abnormalities all lead to pathophysiologic adipocyte and adipose tissue endocrine and immune responses that result in metabolic disease. The degree to which adiposopathy causes or worsens metabolic disease is dependent upon the interactions and crosstalk with other body tissues and other body organ systems. In summary, adiposopathy is a "disease" of adipocytes and adipose tissue that contributes to some of the most common disorders encountered in clinical practice, including diabetes mellitus, high blood pressure, dyslipidemia, hepatosteatosis, hyperandrogenemia in women, hypoandrogenemia in men, some cancers, and other metabolic abnormalities. Because the onset or worsening of these disorders occurs with varying degrees of adiposity, adiposopathy better defines the potential pathogenic relationship between adipose tissue and metabolic disorders compared to measures of fat mass alone.

IMPAIRED ADIPOGENESIS AS AN "ACQUIRED LIPODYSTROPHY"

Adipose tissue provides important mechanical and physiological functions. Physically, adipose tissue that surrounds visceral organs provides protective padding from body trauma and/or during ambulation. Physiologically, it is also an active endocrine and immune organ integral to glucose and lipid metabolism, as well as other metabolic body processes. The major physiologic function of adipose tissue is to store energy in the form of intra-adipocyte triglycerides, for use by body tissues in postabsorptive states.

During positive caloric balance, the increased amount of triglycerides stored in fat cells increases the size of adipocytes (hypertrophy). Adipocyte hypertrophy normally triggers adipogenic signaling for the recruitment, proliferation, and differentiation of new fat cells (hyperplasia). If any stage of adipogenesis is impaired in peripheral subcutaneous adipose tissue, then this may result in the spillover or overflow of fat deposition into other fat depots (e.g. visceral, periorgan, and perivascular fat) and other body organs (e.g. the liver, muscle, and pancreas). Furthermore, if the recruitment, proliferation, and/or differentiation of new fat cells are impaired or inadequate, then existing fat cells may become excessively enlarged, causing them to be dysfunctional. Thus, adipogenesis plays a critical role in determining the metabolic consequences of positive caloric balance.

ADIPOCYTES AND ADIPOSE TISSUE: ANATOMIC ABNORMALITIES

Two of the most-described anatomic abnormalities characterizing adiposopathy are adipocyte hypertrophy and visceral fat accumulation.[2] Other adiposopathic anatomic findings include increased periorgan fat accumulation (e.g. pericardial and perivascular fat), increased abdominal (truncal) subcutaneous fat accumulation, hepatosteatosis ("fatty liver"), and increased fatty infiltration of muscle. If peripheral, subcutaneous adipose tissue expansion is limited (i.e. due to impaired adipogenesis, dysfunctional extracellular matrix dissolution and reformation, lack of sufficient oxygenation/angiogenesis, etc.), then the energy overflow from positive caloric balance may cause these anatomic manifestations of adiposopathy. From a practical standpoint, an increase in total body fatness [as reflected by an increase in the body mass index (BMI)] can also be an anatomic manifestation of adiposopathy. This is because fat accumulation rarely occurs in peripheral subcutaneous fat depots alone. Instead, during fat weight gain, excessive fat is usually deposited in multiple fat depots and body organs,[3] with varying degrees of pathogenic potential, and with the proportional locale of fat deposition being dependent upon the individual's genetic and environmental predisposition.

The amount of visceral adipose tissue deposition varies, but generally represents about 20% of total body fat. Examples of visceral fat are intraperitoneal (omental, mesenteric, and umbilical), extraperitoneal (peripancreatic and perirenal), and intrapelvic (gonadal/epidydimal and urogenital) adipose tissue. Visceral adipose tissue is often considered to be the most metabolically active fat depot (i.e. lipolysis, as well as other endocrine and immune responses), followed by subcutaneous, abdominal adipose tissue. Subcutaneous, peripheral adipose tissue (e.g. truncal, gluteofemoral, mammary, and inguinal) is generally considered the least metabolically active per gram of tissue. However, even subcutaneous adipose tissue (which constitutes approximately 80% of body fat) has pathogenic potential for causing or promoting metabolic disease.[4]

ADIPOCYTES AND ADIPOSE TISSUE: STRUCTURAL ABNORMALITIES

The majority of cells in adipose tissue are adipocytes. By far, the greatest volume component of adipose tissue is the fat (triglycerides) contained within adipocytes. Stromal vascular fraction cells bordering adipocytes include mesenchymal cells, fibroblasts, preadipocytes, endothelial precursor cells, smooth muscle cells, blood cells, and immune cells (e.g. monocytes and macrophages), which are located in a meshwork of fibrous, loose connective tissue, collagen, nerves, and blood vessels. Both adipocytes and stromal vascular fraction cells contribute to and are in a state of dynamic equilibrium with an extracellular matrix (ECM) that surrounds adipocytes. The ECM provides a structural scaffold on which cells reside. It also offers structural protection against disruption of fat cells that otherwise would only have a lipid monolayer to hold fat and cytosol during mechanical stress. Adipose tissue ECM is in constant flux (being broken down, and then rebuilt), during times of fat gain or fat loss.

On a cellular level, adipocyte hypertrophy may result in fat cell hypoxia. On an organ level, if fat accumulation outpaces angiogenesis, then this may contribute to adipose tissue hypoxia. Adipose cellular or organ hypoxia may contribute not only to cellular and organ endocrine dysfunction, but also to proinflammatory responses, such as an increased release of proinflammatory factors, and a decrease in release of anti-inflammatory factors (e.g. adiponectin). Adipose tissue macrophage infiltration may also occur, resulting in a net increase in adipose tissue inflammatory response.

From a structural standpoint, optimal creation and maintenance of the ECM is dependent upon normal adipocyte function. If adiposity results in dysfunctional adipocyte secretions, and increased inflammation, then periadipocyte ECM remodeling may be impaired, physically limiting an increase in adipocyte size, and adipose tissue accumulation, and thus limiting further fat storage. The resultant increase in nonstored excess energy may be manifested by increased circulating free fatty acids and fat accumulation in other fat depots, and in other body organs. Additionally, hypoxia-driven inflammation may promote ECM instability with excessive synthesis of ECM components that, in the long term, may interfere with cell-to-cell contact and adipogenic signaling, which in turn may result in adiposopathic metabolic disturbances even after weight loss.

INTRA-ADIPOCYTE ORGANELLE-OPATHY

The intracellular organelles most commonly described to undergo dysfunction in excessively hypertrophied adipocytes are the endoplasmic reticulum and mitochondria. The endoplasmic reticulum (ER) is a network of interconnected tubules, vesicles, and cisternae that (among other functions) produce proteins and lipids, and transport protein and carbohydrates necessary for normal cellular function. Mitochondria are membrane-enclosed organelles containing enzymes responsible for transforming nutrients into cellular energy via the production of adenosine triphosphate (ATP). Adipocyte hypertrophy can result in increased markers of adipocyte ER stress, which is associated with inflammation, cellular dysfunction, and metabolic diseases. Similarly, it can lead to increased markers of adipocyte mitochondrial stress, which is associated with obesity, insulin resistance, and metabolic disease such as diabetes mellitus. ER stress and mitochondrial stress are intracellular manifestations of "sick fat" and help explain the pathogenesis of the endocrine and immune abnormalities associated with adiposopathy.

ENDOCRINOPATHY

Adipose tissue is an active endocrine organ. Adipocytes and adipose tissue produce biologic factors necessary for metabolic health. These include factors applicable to angiogenesis, adipogenesis, extracellular matrix dissolution and reformation, lipogenesis, growth factor production, glucose metabolism, production of factors associated with the renin angiotensin system, lipid metabolism, enzyme production, hormone production, steroid metabolism, immune response, hemostasis, and element binding. Adipocytes and adipose tissue also have receptors for traditional peptides and glycoprotein hormones, receptors for nuclear hormones, other nuclear receptors, receptors for cytokines or a different kind of cytokine-like activity, receptors for growth factors, catecholamine receptors, and other receptors. In addition to increased circulating free fatty acids, the adipocyte and adipose tissue endocrine factors most commonly associated with metabolic disease are leptin, loss of adiponectin, hormones associated with the Renin–angiotensin aldosterone system, androgens, and estrogens.

IMMUNOPATHY

Adipose tissue is an active immune organ. The net inflammatory responses from adipocytes and adipose tissue depend upon the balance of adipocyte and adipose tissue production of proinflammatory and anti-inflammatory factors. Examples of proinflammatory factors produced by adipocytes and adipose tissue are: (1) *factors with cytokine activity*, such as adipsin, interleukin-1B, -6, -8, -17D, -18, leptin, macrophage colony-stimulating factor-1, monocyte chemoattractant protein-1, macrophage migration inhibitory factor, resistin, tumor necrosis factor alpha, regulated on activation, normal T cell expressed and secreted, visceral adipose tissue derived serpin; (2) *acute phase response proteins*, including alpha-1 acid glycoprotein, ceruloplasmin, C-reactive protein, haptoglobin, interleukin-1RA, lipocalins, metallothionein, pentraxin-3, plasminogen activator inhibitor-1, and serum amyloid A; (3) *proteins of the alternative complement system*, including adipsin, acylation-stimulating protein, complement C3 and B; (4) *chemotactic/chemoattractants*, which include eotaxin, interferon inducible protein, macrophage colony-stimulating factor-1, monocyte chemoattractant protein-1, macrophage migration inhibitory factor, regulated on activation normal T cell expressed and secreted, resistin, stromal-derived factor-1, vascular lesion protein-1, and vascular adhesion molecule-1; and (5) *eicosanoids/prostaglandins*, which include prostaglandin E2. Examples of anti-inflammatory adipocyte and adipose tissue factors are adiponectin, annexin-1, interleukin-6 and -10, transforming growth factor beta, bone morphogenic factor, nitric oxide, and interleukin-1 receptor antagonist. The adipocyte and adipose tissue immune factors most commonly associated with metabolic disease are leptin, adiponectin, and tumor necrosis alpha.

COMMUNICOPATHY AND EXTRA-ADIPOSE ORGANOPATHY

Adiposopathy alone does not cause or worsen metabolic disease. The ultimate determinant as to whether adiposopathic pathogenic endocrine and immune responses result in clinical metabolic disease is the interaction or crosstalk with other body organs or body organ systems (e.g. the liver, muscle, and pancreas, as well as organs of the cardiovascular, endocrine, immune, nervous, genitourinary, gastrointestinal, integument, and other body organ systems). If adiposopathic endocrine and immune responses are adequately mitigated by other body organs, then adiposity may not result in clinical metabolic disease. Conversely, if adiposopathic responses are not mitigated by other body organs, as occurs when other body organs or body organ systems are limited or impaired in their communication or interaction with adipose tissue, then even mild adiposity-induced adiposopathic responses may result in adverse metabolic consequences. Specifically, if communications between adipocytes/adipose tissue and other body organs are pathologically disrupted (i.e. communicopathy), or if other body organs are limited in their ability to mitigate adiposopathic responses (i.e. organopathy, as might occur with the genetic or environmental impairment of body organs or body organ systems), then this often results in adiposopathy, or adiposity-induced metabolic disease.

FREE FATTY ACIDS AND LIPOTOXICITY

One of the more commonly described adverse consequences of impaired adipocyte and adipose tissue function is the net increased release of free fatty acids resulting in increased circulating free fatty acids. When positive caloric balance is accompanied by excessive fat storage in adipocytes, then this may impair the functionality of lipid-acting enzymes. Excessive adipocyte enlargement may also cause physical and other limitations in free fatty acid storage, which increase circulating fatty acid blood levels and promote "lipotoxicity," and thus contribute to metabolic disease.

The vast majority of adipocyte content is triglycerides (e.g. triacylglycerides). Triglycerides are composed of three fatty acids of variable lengths and variable single and double bonds attached to a three-carbon glycerol structural backbone. Hormone-sensitive lipase is a key enzyme that catalyzes the hydrolysis of intra-adipocyte triglycerides, resulting in the production of nonesterified fatty acids ("free fatty acids") and glycerol. Quantitatively, free fatty acids are the major secretory product derived from adipose tissue. Circulating free fatty acid levels are determined by: (1) whether the individual is immediately postprandial or postabsorptive, (2) the storage capacity of adipose tissue, (3) the degree to which other body organs store free fatty acids as triglycerides, and (4) the degree to which body organs metabolize free fatty acids. The type of circulating fatty acids (e.g. saturated or unsaturated) is largely dependent upon the specific fatty acids in consumed fats, which can be assessed by fat tissue biopsy or red blood cell phospholipid analyses.

Two important adipocyte functional abnormalities associated with adiposopathy are the decreased production of hormone-sensitive lipase and of lipoprotein lipase. Adiposity is associated with a decrease in intracellular hormone-sensitive lipase activity. Impaired hormone-sensitive lipase function reduces intra-adipocyte lipolysis of stored triglycerides. When coupled with continued triglyceride formation, decreased lipolysis and continued lipogenesis increase adipocyte lipid storage, which may cause excessive adipocyte hypertrophy and dysfunction. Adiposity is also associated with a decrease in extracellular lipoprotein lipase activity, which decreases lipolysis of circulating lipoproteins, and contributes to increased triglyceride blood levels.

More specifically, hormone-sensitive lipase is an intracellular, rate-limiting enzyme produced by adipocytes that hydrolyzes triglyceride esters into a free fatty acid and a diglyceride. The diglyceride is rapidly cleaved to a monoglyceride by diglyceride lipase, and then a remaining fatty acid is cleaved by monoglyceride lipase. Hormone-sensitive lipase is the only one of these lipolytic enzymes affected by hormones, and is upregulated by catecholamines (i.e. beta-adrenergic stimulation) and adrenocorticotropic hormone. Hormone-sensitive lipase is downregulated by insulin. Through this mechanism, increased sympathetic nervous system activity and increased cortisol secretion are catabolic, while hyperinsulinemia is anabolic with regard to triglyceride storage in adipocytes. The fatty acids most easily released from adipose tissue by hormone-sensitive lipase are shorter (16–20 carbons) and more unsaturated.[5]

Lipoprotein lipase is an enzyme produced by adipocytes and secreted into extracellular surroundings, which hydrolyzes triglycerides into glycerol and free fatty acids. Human adipocytes rely upon lipoprotein lipase to hydrolyze circulating triglycerides to provide free fatty acids for lipogenesis, because adipocytes do not synthesize lipids. In the postprandial state, lipoprotein lipase interacts with chylomicrons and other triglyceride-rich lipoproteins. In the fasting state, lipoprotein lipase mainly interacts with very low density lipoproteins and other triglyceride-rich lipoproteins.

After the extracellular hydrolysis of triglycerides by lipoprotein lipase, free fatty acids undergo uptake by adipocytes, which are activated by acetyl CoA. Glycerol originates from glucose or pyruvate, which means that adipocytes must have access to glucose in order to store fatty acids as triglycerides. When glycerol is phosphorylated by glycerol kinase, it is acylated to a fatty acid-CoA by glycerol-3-phosphate acyltransferase (GPAT) to form lysophosphatidic acid, which is acylated to another fatty acid-CoA by acylglycerophosphate acyltransferase (AGPAT) to form phosphatidic acid. Phosphatidic acid is dephosphorylated by phosphatidic acid phosphohyrolase (PAP) to form diacylglycerol, and then acylated to a final fatty acid-CoA by diacylglycerol acyltransferase (DGAT) to form triacylglycerol. By these mechanisms, glycerol is esterified to intra-adipocyte free fatty acids to form and store intra-adipocyte triglycerides.

As with hormone-sensitive lipase, lipoprotein lipase is affected by hormones. Insulin increases lipoprotein lipase activity (and decreases hormone-sensitive lipase activity), which helps explain why insulin therapy can rapidly reduce very high triglyceride blood levels in poorly controlled hyperglycemic patients with diabetes mellitus. Increasing lipoprotein lipase activity is a major mechanism as to how therapeutic agents such as fibrates and omega-3 fatty acids lower triglyceride levels.

Largely because of nutrient-induced increases in insulin levels which downregulate hormone-sensitive lipase and upregulate lipoprotein lipase, free fatty acids are dramatically decreased (by ~70% or more) after meals. This helps create and "trap" circulating free fatty acids into adipocytes. Conversely, during fasting, decreased insulin levels upregulate hormone-sensitive lipase and downregulate lipoprotein lipase. This increases free fatty acid release from fat, increases circulating free fatty acids, increases the free fatty acids available to muscle for beta oxidation (free fatty acids are a primary source of energy in muscle), and increases the free fatty acid uptake in the liver (used for hepatic beta oxidation, ketogeneis, lipogenesis, and gluconeogenesis). This helps explain why prolonged fasting (greater than seven days) may markedly increase circulating free fatty acids, increase hepatosteatosis, increase ketosis, potentially result in a transient rise in triglyceride (and cholesterol) levels[6] and, through "lipotoxicity," worsen insulin sensitivity (even as glucose blood levels are reduced).[7]

If adiposopathic processes prevent adequate storage of excessive free fatty acids in adipose tissue (as might occur with insufficient adipogenesis and impaired adipocyte function), then increased free fatty acids are shunted to other body tissues, such as the liver, muscle, and pancreas. This dysfunctional deposition of free fatty acids may have pathophysiologic effects, often termed "lipotoxicity." Lipotoxicity is the dysfunction of body organs induced by the adverse effects of excessive free fatty acids and their products (e.g. ceramides and diacylglycerols) which may result in hepatic and muscle insulin resistance, potential insulinopenia from the pancreas, and dysfunction of other body organs (the heart, vasculature, kidney, etc.). Lipotoxicity helps explain why those with type 2 diabetes mellitus have increased circulating fasting and postprandial circulating free fatty acids compared to those without type 2 diabetes mellitus, increased insulin resistance, and decreased pancreatic insulin release (at least relative to ambient glucose levels).

NUTRITION AND ADIPOSOPATHY

The effect of consumed nutrients upon adipocyte and adipose tissue function is related not only to food quantity (caloric content), but also to the quality or nutritional content of food. Thus, depending upon their quantity and/or quality, nutritional intake may promote pathologic processes that contribute to metabolic disease. Foods — broadly categorized as carbohydrates (Fig. 1), fats (Fig. 2), and proteins (Fig. 3) containing minerals, vitamins, and other chemical compounds — may either have unfavorable health effects and be pathogenic, or have favorable health

Carbohydrates

Simple carbohydrates [monosaccharides and disaccharides]

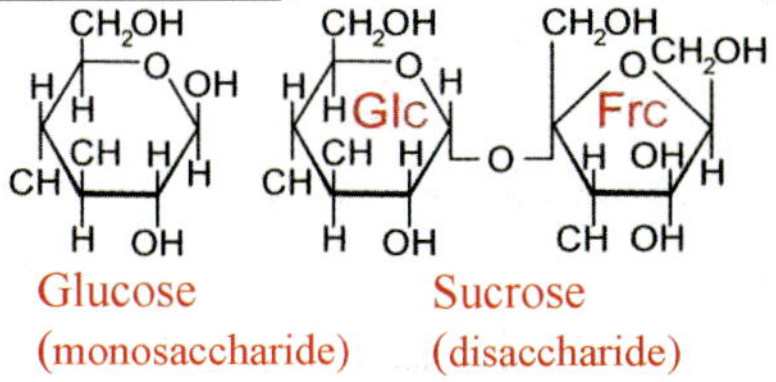

- Glucose (a monosaccharide found in corn syrup)
- Fructose (a monosaccharide found in fruits, high fructose corn syrup)
- Fructose and glucose (honey, molasses)
- Galactose (a monosaccharide component of lactose)
- Lactose [a disaccharide (glucose and lactose) found in milk]
- Sucrose [a disaccharide (glucose and fructose) found in beets, maple, brown, molasses, white, table, powdered sugar]
- Lactose

Complex carbohydrates (polysaccharides or "glycans")

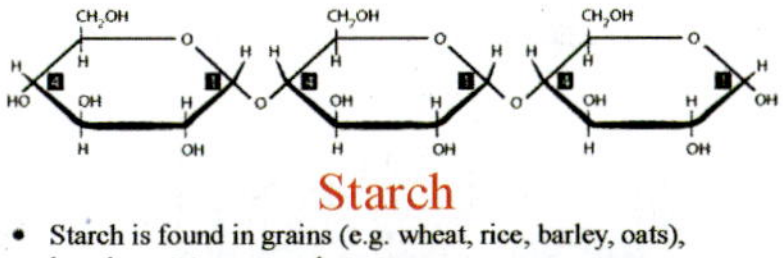

- Starch is found in grains (e.g. wheat, rice, barley, oats), breads, potatoes, corn, beans
- Glycogen is generated from glucose by the liver and muscle for energy storage, with small amounts consumed through intake of liver; minimal amount of glycogen in consumed muscle/meats
- Cellulose/hemicellulose is found in plant-derived insoluble fiber foodstuffs, grains, bran, vegetables, and fruits
- Gum is found in plant-derived soluble fibers from oats, legumes, guar, and barley
- Pectin is found in plant-derived soluble fibers from fruits
- Dextrin is found in fiber supplements
- Mucilage is found in psyllium, flax seed, marsh mallows

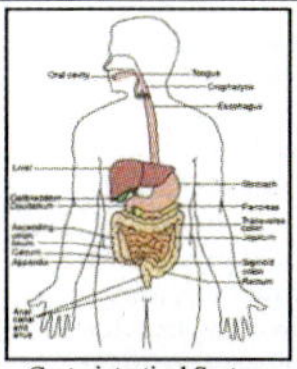

Gastrointestinal System

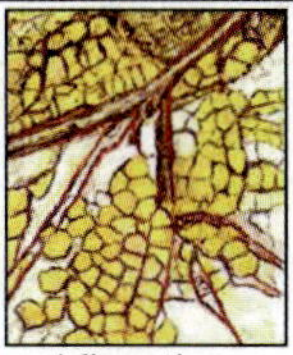

Adipose tissue

- Postprandially, in the mouth, amylase from salivary glands (parotid, submandibular, and buccal) catalyzes hydrolysis of starch to more simple sugars
- In the small intestine, amylase from the pancreas (through chyme-induced release of cholecystokinin) further catalyzes the hydrolysis of starch to more simple sugars
- In the small intestine, lactase, maltase, and sucrase from the brush border break down disaccharides to monosaccharides
- Monosaccharides (glucose, fructose, galactose) are absorbed by the intestine, and transported to peripheral tissues by GLUT hexose transporters
- In the liver and muscle, fructose and galactose are converted to glucose, and glucose phosphorylation to glucose 6-phosphate allows for glycogenesis towards formation of storage polysaccharides known as glycogen
- Lipogenesis is often thought to be virtually absent in muscle. In the liver, if glycogen stores are replete, then additional dietary carbohydrates are converted into glucose 6-phophate, which then through the glycolytic process, generate acetyl-CoAs which are used to form fatty acids that combine with glycerol to form hepatic triglycerides (lipogenesis). Increased liver lipids are then transported from the liver through the blood to peripheral tissues (e.g. adipocytes and muscle) via very low density lipoproteins. If the carbohydrate load is excessive, then lipids may be stored in the liver (hepatosteatosis); decreased carbohydrate intake (especially decreased simple sugars or carbohydrates high in glycemic index) may reduce hepatosteatosis.
- During fasting (postabsorption state), hepatic and muscle stored glycogen is broken down into glucose through glycogenolysis to generate glucose-6-phosphate
- Glucose-6-phophate then may undergo metabolism through glycolysis, which is a sequence of enzymatic reactions that generate energy. The glycolytic end-product of pyruvate can undergo an anaerobic pathway to create lactic acid (when energy demands exceed oxygen supply), orundergo an aerobic pathway (Kreb's citric acid cycle) to create acetyl-CoA, then citrate, and then through oxygen-mediated mitochondrial respiration, the generation of ATP, CO2, and water.

Fig. 1 The contribution of carbohydrate intake to adiposopathy.

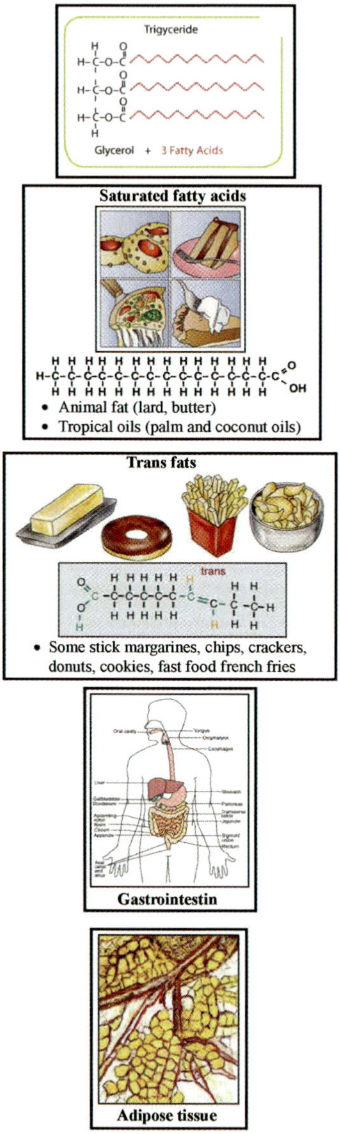

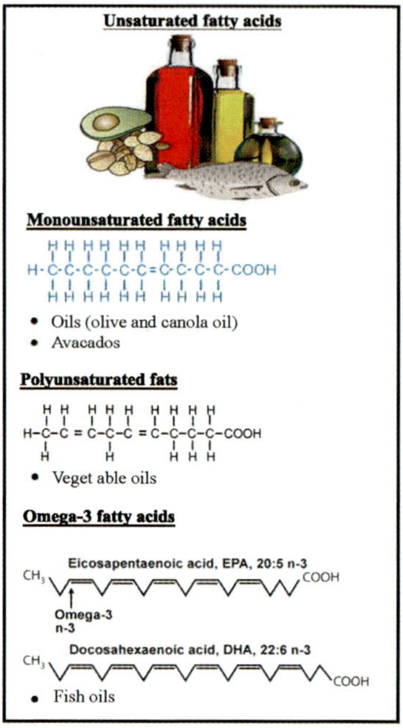

- Postprandially, over 90% of consumed fats are triglycerides (TG), which upon entering the small intestine, stimulate cholecystokinin release which in turn, stimulates bile release from the gallbladder and liver (wherein bile acids emulsify fats into micelles) and lipase release from the pancreas.
- Both pancreatic and intestinal lipases hydrolyze TG to di- and monoglycerides, glycerol, and free fatty acids which enter into intestinal cells, are re-esterified into TG, and then packaged along with re-esterified cholesterol into apo B48-containing chylomicrons.
- Chylomicrons are secreted into mesenteric lymph vessels and eventually into the circulation for delivery of lipids to peripheral tissues.
- The vast majority of body fat is stored in adipose tissue.
- Chylomicrons deliver lipids to the liver
- Increased saturated fats delivered to the liver may increase hepatosteatosis and increase very low density lipoprotein (VLDL) secretion
- Monounsaturated fats delivered to the liver may not increase hepatosteatosis, and may not effect hepatic VLDL secretion
- Omega-3 fatty acids may decrease hepatosteatosis and may substantially decrease hepatic VLDL secretion
- During fasting and when fatty acids are needed for energy (i.e. fatty acids undergo beta oxidation to form acetyl CoA which enters the citric acid cycle in muscle, liver, etc.), stored free fatty acids in adipocytes are released into the circulation through increased activity of hormone sensitive lipase.
- Trans fatty acids reduce high density lipoprotein cholesterol, promote endothelial dysfunction, promote a proinflammatory response, and increase atherosclerotic vascular risk. Limited evidence suggests trans fatty acids promote insulin resistance.
- Trans fats may limit fat deposition in adipocytes, alter adipocyte plasma membrane composition, and increase insulin resistance in adipocytes.

Fig. 2 The contribution of fat intake to adiposopathy.

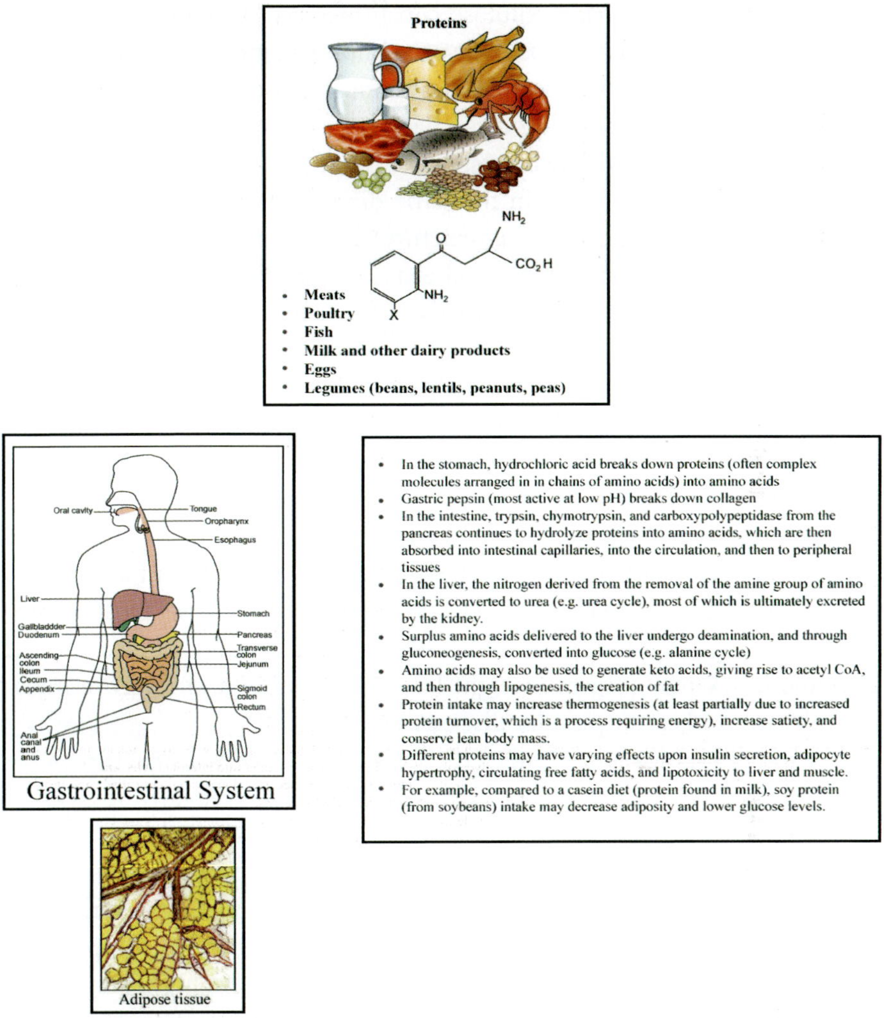

Fig. 3 The contribution of protein intake to adiposopathy.

effects and prevent, and in some cases improve, metabolic disease — significantly due to their effects upon adiposopathy.

Carbohydrates are organic compounds composed of one carbon per two hydrogens per one oxygen ($C_nH_{2n}O_n$). Simple sugars are often utilized for short term cellular energy, polysaccharide glycogen for intermediate energy storage, and polysaccharide cellulose constitutes the structural support for plant cell walls. Glucose is an example of a simple sugar monosaccharide with a formula of $C_6H_{12}0_6$. Other "simple sugars" include monosaccharides such as fructose and galactose, as well as disaccharides (e.g. sucrose, maltose, lactose). "Complex carbohydrates" include starches (long polymers of glucose molecules with bond attachments in the same direction), glycogen (polymer of glucose molecules with branching structure), and cellulose (long polymers of glucose molecules with bond alternating in opposite directions). Carbohydrate consumption, particularly of foods high in the glycemic index, increases insulin levels, which simulates adipogenesis and adipocyte lipogenesis. If insulin-induced adipocyte lipogenesis results in excessive adipocyte hypertrophy, particularly in individuals with impaired proliferation of new fat cells, then existing adipocytes may become further enlarged, dysfunctional, and pathogenic. The lack of further energy storage in adipocytes/adipose tissue promotes free fatty acid and fat storage in other adipose tissue depots and other body organs. This helps explain why intake of foods higher in the glycemic index may increase visceral adiposity, increase hepatosteatosis, and worsen metabolic disease,[8,9] and why intake of foods lower in the glycemic index may improve weight loss maintenance and improve metabolic disease and cardiovascular disease risk factors.[10,11]

Fats are organic compounds with the basic triglyceride structure being three fatty acids attached to a glycerol backbone. The free fatty acids associated with triglycerides may be saturated (no double bonds) or unsaturated (one or more double bonds). While a reduction in total caloric intake is the most consistent nutrient determinant of weight loss through nutritional intervention, a comparison of diets that varied in composition suggests that compared to high carbohydrate intake, high fat intake may cause greater long term weight loss, greater reduction in percent body fat, greater reduction in triglycerides, and greater reduction in blood pressure.[12] This is not to say, however, that all fats are equivalent in their effects upon adiposopathy and metabolic disease. High saturated fat intake can worsen dyslipidemia and insulin resistance, relative to monounsaturated and polyunsaturated fat intake. Perhaps the best example of the physiologic differences in fats might be found with polyunsaturated fats composed of fish oils rich in omega-3 fatty acids, which, in contrast to saturated fats, may reduce hepatosteatosis (probably through increased hepatic fat beta oxidation and inhibition of triglyceride synthesis) and hypertriglyceridemia (through reduced very low density lipoprotein secretion and increased lipoprotein lipase activity).[13]

Proteins are organic compounds composed of nitrogen-containing amino acids, arranged in a linear chain and folded into a tertiary or quarternary structure. Higher intake of dietary proteins can reduce adiposity, and decrease adipocyte size.[14] Specifically, if substituted for carbohydrates and saturated fats, protein intake may: (1) avoid the adverse effects of carbohydrates and saturated fats upon adipocyte and adipose tissue function and metabolic disease; (2) increase satiety and promote thermogenesis, and thus be potentially useful for maintaining or losing total and visceral body fat; and (3) preserve muscle mass, particularly in older individuals,[15] which in turn may increase resting energy expenditure and decrease total and visceral body fat. In these ways, substituting protein intake for some carbohydrate and saturated fat intake may favorably affect adiposopathy.

FUNCTIONAL FOODS AND ADIPOSOPATHY

A "functional food" can be defined as a consumed nutrient that has medicinal effects (Table 1). While many functional foods or their components have sound scientific evidence for their potential benefits, much more data is needed regarding: (1) the definitive benefits or risks of such foods, particularly at amounts (or "dose") that might reasonably be consumed, and (2) whether consuming the isolated components of a food (as often occurs in mechanistic and physiologic studies) provides the same benefits as when the component is incorporated in the food itself. As with drugs, the biologic effects of functional foods upon adiposopathy may be beneficial or adverse.

Retinoids and Carotenoids

Dietary vitamin A is required for retinal function, tissue growth, and immunity. Preformed, active retinoids (retinol, retinal, and retinoic acid) are absorbed in the intestine as retinol, and then mainly stored as fat-soluble retinyl esters in the liver, with some retinol stored in adipocytes. The provitamin carotenoids must be converted to retinol, with beta carotene being stored in adipose tissue (and the liver).

Table 1 Examples of Foods with Reported Biologic (Functional) Properties

Chemical Identity	Description	Examples of Food Sources
Carotenoids are 40-carbon tetraterpenoids, which are composed of 4 monoterpenoids (each monoterpenoid has 2 isoprene units of 5 carbons each). They contain vitamin A and serve as antioxidants.	Carotenoids are xanthophylls (containing oxygen) or carotenes (containing no oxygen) made by plants, algae, and fungi for photosynthesis and to protect chlorophyll from photodamage.	Alpha-carotene from carrots; Beta-carotene from fruits and vegetables; Lutein from green vegetables; Lycopene from tomatoes and tomato products; Zeaxanthin from eggs, citrus, and corn.
Dietary fibers are complex carbohydrates contained in roughage or indigestible content of plant foods. Insoluble fibers do not dissolve in water, and include	Insoluble fiber is metabolically inert, and may aid in easing defecation by absorbing water through the gastrointestinal track.	Insoluble fiber from vegetables, fruit and root vegetable skins, wheat bran, seeds, and nuts.
cellulose, hemicellulose, and lignin. Soluble fibers include gums, mucilages, and pectin.	Soluble fiber is soluble in water, and fermented in the (large) intestine by bacterial flora to form short chain fatty acids which may improve intestinal wall cellular function and improve nutrient metabolism.	Soluble fiber gums are found in oats, legumes, guar, and barley; mucilages are found in seeds, psyllium, marshmallows, flaxes, and agar; and pectin is used in jams/jellies and found in citrus fruits and carrots.
Flavonoids are ubiquitous plant metabolites having a ringed, polyphenolic chemical backbone structure similar to flavones.	Flavonoids may have antiallergic, anti-inflammatory, antioxidant, antimicrobial, antiplatelet, and antineoplastic activities.	Anthocyanidins from fruits such as berries; Chalcones from fruits such as apples, pears, berries, and tomatoes; Flavonols from fruits and vegetables (e.g. onion, kale); Flavanols include catechin found in green tea, black tea, raisin, red wine, and dark chocolate;

(*Continued*)

Table 1 (Continued)

Chemical identity	Description	Examples of Food Sources
		Flavones, found in cereals and herbs (e.g. celery, parsley); Flavanones from citrus (grapefruit, lemons, limes, etc.); Isoflavones from beans (e.g. soybeans, tofu).
Glucosinolates are water-soluble organic thioethers containing sulfur and nitrogen, which break down into indoles and isothiocyanates.	While glucosinolates may be thyroid-goitrogenic, they may also have antioxidant and antineoplastic properties.	Glucosinolates are found in cruciferous (mustard family) vegetables such as mustard greens, cabbages, broccoli, cauliflower, and brussel sprouts.
Mono- and polyunsaturated fatty acids are fatty acids with one or more than one double bond, respectively. Omega-3 fatty acids are polyunsaturated fatty acids that have a double bond at the n-3 or "omega" position, which is the third bond from the methyl end.	Polyunsaturated fatty acids may lower low density lipoprotein cholesterol, and have other potential vascular benefits if substituted for high saturated fat intake. Monounsaturated fatty acids may lower low density lipoprotein cholesterol, and have other potential vascular benefits if substituted for high saturated fat intake, and raise high density lipoprotein cholesterol levels, especially if substituted for high carbohydrate intake. Omega-3 fatty acids can substantially lower triglyceride levels, and have a number of other health benefits, such as reducing cardiovascular events.[203,204]	Monounsaturated fatty acids from olive oil, canola oil, and avocados, nuts, and seeds; Polyunsaturated fatty acids from vegetable oils, nuts, and seeds; Omega-3 fatty acids from fish oils, especially from cold-water marine fish.

(Continued)

Table 1 (*Continued*)

Chemical Identity	Description	Examples of Food Sources
Phenol (carbolic acid) is a single-benzene-ringed structure with a hydroxyl group.	Phenolic foods are found in plants. An example of a common phenol would be salicylate. Some phenolic foods may serve as antioxidants and antimicrobials.	Caffeic acid from coffee, green and black tea; Capsaicin from chilli peppers; Propolis (sticky resin from trees), which is used by bees as a component of hives; Curcumin is an extract of turmeric (curry), which is an Indian spice member of the ginger family used for coloring and flavoring, and thought to have antioxidant, anti-inflammatory, and antineoplastic properties; Catechin (previously described under "flavonoids") is polyphenolic, as are grape seeds. Other phenolic compounds are found in white tea, berries, spices, cannabinoids, and herbs.
Phytoestrogens are plant-derived substances with structural similarities to estrogen, and include isoflavones, flavones, and coumestans, often found in soy and soy products (which also provide essential amino acids such as arginine and glutamine), as well as ligans as found in flaxseeds.	Phytoestrogens act as female sex hormones. Men are not adversely affected by phytoestrogens, with regard to testosterone and sperm count/motility. Soy proteins may provide amino acids that may not be sufficiently produced during stress (e.g. intensive exercise).	Phytoestrogens are found in nuts, flaxseeds, soybeans and soy-based products/isoflavones, tofu, cereals (e.g. wheat germ, rice bran), breads, oats, barley, legumes, lentils, beans (e.g. dried beans), meat products, vegetables (e.g. carrot), fruits (e.g. apples, pomegranates), ginseng, hops, beer, and bourbon.

(*Continued*)

Table 1 (Continued)

Chemical Identity	Description	Examples of Food Sources
	Soy protein may reduce cholesterol and triglyceride levels, improve glucose metabolism, and may have antineoplastic activity.	
Plant stanols and sterols are plant-derived esters with structural similarities to cholesterol.	Plant stanols and sterols (phytostanols and phytosterols) may block intestinal absorption of cholesterol. Sitostanol obtained from "tall oils" previously discarded as a waste by-product of wood pulp processing (derived from the Swedish "*talloja*," meaning "pine oil") is commercially available as a margarine (Benecol). Sitostanol is then esterified with canola-sourced fatty acids to render it fat-soluble. Other marketed dietary margarines may contain naturally occurring unsaturated sterols (mainly sitosterol) from soybean oil (Take Control or Promise Active).	Plant sterols (e.g. sitosterol, stigmasterol, campesterol) are consumed largely through vegetable oils, cereals, fruits, vegetables, seeds, and nuts. Plant stanols (e.g. beta sitostanol, compestanol) are less consumed and typically found in corn, wheat, rye, and rice. Plant stanol esters may also be derived from wood pulp of pine trees.

(*Continued*)

Table 1 (*Continued*)

Chemical Identity	Description	Examples of Food Sources
Prebiotic (oligosaccharides) and probiotics (lactobacilli) are agents that influence the bacterial flora of the gastrointestinal tract, with bacteria constituting about 50% of feces content.	Prebiotics are indigestible oligosaccharides that may stimulate intestinal bacterial growth, mainly colonic lactobacilli and bifidobacteria. Probiotics such as lactobacilli may provide protection against micro-organisms such as yeast and pathogenic bacteria.	Fructo-oligosaccharides (short chain polymers of fructose units not hydrolyzed in the small intestine, but degraded by colon flora) are found in asparagus, Jerusalem artichoke, leeks, onions, and soybeans; Lactobacilli are found in yogurt, other dairy products, and sauerkraut.
Saponins are glucoside-based, multiringed, multipolycylic aglycones attached to one or more sugar side chains which contain both fat-soluble and water-soluble components and create foam when shaken.	Saponins are natural detergents found in plants, especially desert plants (e.g. yucca and soapbark trees), which may have antioxidant properties and lower cholesterol by binding cholesterol and bile acids.	Saponins are found in soybeans, soy foods, soy-protein-containing foods, yam, asparagus, red onions, tomatoes, spinach, oats, potatoes, herbs, and ginseng.
Thiols/sulfides: thiols (mercaptans) are sulfur analogs of alchohol; sulfides are thioester sulfur analogs.	Thiols and sulfides may reduce cholesterol and may have anticancer properties.	Thiols are found in onions and garlic oil. Sulfides are found in garlic, leeks, onions, and scallions.
Tannins are polyphenolic compounds that link, bind, and precipitate proteins/ amino acids.	Nonnutritional uses for tannins "tanning" raw animal hides, which helps preserve the skin in the creation of leather products. Otherwise, tannins are reported to potentially reduce feeding and digestion, as well as antioxidant effects, with varied reports regarding anti- or procancer effects.	Tannins are found in grape seeds, beer, fruit juices, pomegranates, persimmons, berries, nuts, smoked foods (e.g. mesquite),herbs, legumes, and chocolate. (Polyphenolic catechins found in tea do not have tanning properties, and are pseudotannins.)

Beta carotene helps account for the yellowish appearance of adipose tissue. The potential effects of vitamin A upon adiposopathy are complex. Some reports suggest that vitamin A is adipogenic, and increases the number of smaller (presumably more functional) adipocytes.[16] Retinol saturase is a member of the carotenoid/retinoid oxidoreductase family, which facilitates saturation of retinol intermediaries. Retinol saturase promotes adipogenesis, but is downregulated with obesity.[17] Retinoic acid may activate PPAR delta activity, which might improve adiposopathy through an increase in fatty acid oxidation, lipolysis, fat weight loss, and possibly adipogenesis.[18] However, other reports suggest that retinoic acid may suppress adipogenesis, possibly through inhibiting the activity of retinoid X receptor and peroxisome proliferator-activated receptor (PPAR) gamma activity.[19] Some retinoids may also increase visceral adiposity, worsen glucose sensitivity, and cause dyslipidemia.[20] This helps explain why, particularly among predisposed individuals, the acne drug isotretinoin can often cause dyslipidemia and hyperinsulinemia.[21]

Dietary Fiber

Dietary fiber consumption is variably reported to improve body weight and metabolic disease. Likely due to increased satiety and decreased caloric intake, an increase in fiber intake may improve adiposopathy. Consumption of whole grain cereals versus low fiber foods may have favorable effects upon waist circumference and metabolic cardiovascular risk factors.[22] Some dietary fibers may inhibit or delay carbohydrate digestion and absorption, with improved glucose control in patients with diabetes mellitus.[23] If a delay in glucose absorption decreases net insulin release, then this may theoretically reduce adipocyte hypertrophy, which may favorably affect adiposopathy. Beta glucan (such as found in barley) is an example of a soluble that reportedly reduces visceral fat, ameliorates a fatty liver, improves lipid levels, and increases insulin sensitivity.[24,25]

Flavonoids

As with other functional foods, the effects of flavonoids upon adipose tissue function are complex. Flavonoids may stimulate lipolysis, which is associated with decreased adipocyte size, decreased visceral fat accumulation, increased adipocyte apoptosis, and apoptosis, with a theoretical decrease in adipogenesis.[26] However, in physiologic amounts consumed by humans, flavonoids have limited effects upon adipogenesis.[27] Some evidence exists that flavonoids may enhance adiponectin secretion, independent of PPAR gamma mechanisms, which may help account for flavonoids' favorable effects upon inflammation and insulin sensitivity.[28] Catechin is a commonly consumed flavanol found in green tea, black tea, raisins, red wine, and dark chocolate, which may reduce fat accumulation in early adipocyte development, increase adipocyte differentiation, and increase insulin sensitivity,[29] possibly through increased PPAR activation.[30]

Glucosinolates

Not all functional foods have favorable effects upon adiposopathy. While potentially antineoplastic at lower doses, higher doses of glucosinolates are goitrogens. In extreme cases, glucosinolates may promote hypothyroidism.[31] This has relevance to adiposopathy, because among the many nonadipocyte/nonadipose-tissue factors that increase adipogenesis are catecholamines, lipoproteins, lipids, proteins, and hormones.[32] Thyroid hormone is a prime example of a nonadipocyte/nonadipose tissue factor important for adipogenesis.[33] Hypothyroidism is associated with reduced adipocyte number, and increased adipocyte size; the opposite is true of hyperthyroidism.[34] Given that reduced adipocyte number and increased adipocyte size are anatomic manifestations of adiposopathy, it is not surprising that hypothyroidism (as might be caused by excessive glucosinolate intake) is associated with visceral adiposity, fatty liver, insulin resistance, and (sometimes profound) dyslipidemia.[35]

Mono- and Polyunsaturated Fatty Acids

A rise in the consumption of saturated fats increases saturated fat storage in adipocytes, which may result in fat cell hypertrophy and adiposopathic pathophysiologic processes (such as a decrease in adiponectin) leading to metabolic disease.[36] A rise in trans fat intake may increase visceral adiposity, and worsen metabolic disease.[37] Conversely, particularly when substituted for saturated or trans fats, monosaturated fatty acids may increase lypolysis,[38] decrease proinflammatory factors,[39] increase anti-inflammatory factors such as adiponectin,[40] and improve metabolic disease. Non-omega-3 polyunsaturated fatty acids (e.g. vegetable oils) have variable effects upon adiponectin levels. While polyunsaturated omega-3 fatty acids in the form of dietary fish oils may not cause clinical weight loss,[41] they may reduce visceral adipose tissue depots,[42] improve triglyceride storage (through decreased lipolysis and possibly an increase in adipogenesis as evidenced by an increase in PPAR gamma activity), improve adipocyte secretory functions (such as reduced leptin), and reduce adipocyte and adipose tissue inflammatory responses including raised adiponectin levels.[43]

Phenols

Curcumin, an extract of a turmeric member of the ginger family (and the primary ingredient in curry) is an example of a phenol. It may increase fatty acid oxidation, reduce lipogenesis, improve adipocyte endocrine function, improve adipocyte and adipose tissue immune responses, reduce body weight, inhibit adipogenesis, and improve metabolic disease.[44,45] This illustrates that when one is assessing its pathogenic potential, the anticipated clinical consequences of reduced adipogenesis must take into account fat weight changes. If fat weight gain is accompanied by impaired adipogenesis, then an excessive increase in the size of existing adipocytes, and an increase in energy overflow to other adipose tissue depots and other body organs are often pathologic. Conversely, if adipogenesis is diminished as the result of weight loss through negative caloric balance via weight loss therapies and/or bariatric surgery, then this is more likely a natural, physiologic compensatory response.[46] This would be supported by reports of the effects of other phenols, wherein increased intra-adipocyte lipolysis and decreased intracellular adipocyte lipid droplets are associated with decreased adipogenesis, and anti-inflammatory effects.[47,48] The reduced or attenuated pathogenic inflammatory responses from adipose tissue associated macrophages with phenols (polyphenols) may improve adiposopathy and metabolic disease.[49]

Phytoestrogens

Perhaps the most well-described phytoestrogen is soy. The effects of soy phytoestrogen (e.g. genistein) upon adipogenesis may be dose-dependent, and may vary depending upon age and gender.[50] Because of estrogenic effects, it is perhaps not surprising that soy may reduce subcutaneous abdominal and total abdominal fat,[51] although men may not experience feminizing effects.[52] While more definitive data is needed, evidence suggests that soy may facilitate fat weight loss, improve adiposopathy, and improve metabolic disease.[53]

Plant Stanols/Sterols

Little evidence supports that plant stanols/sterols affect body weight, adipogenesis, adipocyte size, visceral adiposity, or improve adiposopathy. However, diets enriched in plant sterols/stanols do improve dyslipidemia,[54,55] with a reduction in cholesterol due to competitive inhibition of intestinal cholesterol micelle incorporation and absorption. This illustrates that many functional foods improve metabolic disease through mechanisms largely independent of adiposopathy.

Prebiotics and Probiotics

Agents that affect intestinal bacterial flora may affect adipogenesis through pathways including intestinal–central nervous system signaling,[56] energy harvest from the diet, alterations in tissue fatty acid composition, and intestinal endocrine/immune signaling.[57] Bacterial colonization of the gastrointestinal tract of germ-free mice increases fat mass and adipocyte size, and worsens metabolic disease. Prebiotics that stimulate the growth of more "favorable" bacteria (e.g. *Lactobacillus*, *Bifidobacterium*) instead of the potentially more adiposopathic bacteria (e.g. *Bacteroidetes*, often found to be disproportionately high in overweight individuals) may alter gastrointestinal endocrine and immune responses, leading to anorexigenic activities that could conceivably improve adiposopathy.[58]

Saponins

Saponins (e.g. ginseng) may increase adipogenesis via interaction with, or activation of PPAR gamma and glucocorticoid receptors.[59,60] Some saponins may also decrease adipocyte size, reduce adipose tissue inflammation,[61] reduce visceral adiposity,[62] reduce a fatty liver, and result in improvement in parameters such as lowered cholesterol levels and more favorable glucose metabolism.[63] Additionally, saponins may ameliorate the macrophage-mediated pathogenic inflammatory effects of adiposopathy.[64]

Thiols/Sulfides

Among the more-studied thiols/sulfides is garlic. The effect of garlic and/or its derivatives upon adipogenesis is unclear.[65,66] Garlic may more favorably affect metabolic processes (e.g. lipid and glucose metabolism) when compared to phenols such as ginger, or turmeric (curry).[67] Another thiol/sulfide food is onion. In a meta-analysis, the reduction in body weight and glucose lowering in diabetes mellitus rats were significant for onion extracts, but not for garlic.[68]

Tannins

Procyanidins are an example of a tannin from grape seeds which induce lipolysis of fat cells.[69] Procyanidins and other tannins inhibit adipogenesis, yet improve glucose metabolism.[70,71] This highlights the potential complexities of functional foods. For reasons previously mentioned, many (if not most) functional foods require more data to conclusively prove that they provide health benefits; the same applies to their potential effects upon adiposopathy. In the case of tannins, if the lipolysis of adipocytes results in smaller fat cells, then this might explain the reduction in adipogenesis and the improvement in adipose tissue function as demonstrated by improved glucose metabolism. This latter explanation is supported by data from a pine-bark-derived procyanidin (flavangelnol), which is associated with an increase in autonomic neurotransmission, an increase in brown adipose tissue temperature, and a reduction in body weight,[72] which would be expected to improve adipose tissue function.[73]

DIETARY SUPPLEMENTS AND ADIPOSOPATHY

Dietary supplements are nutrient components derived from food products that are concentrated in a liquid or pill (capsule) formulation. As with studies of the components of functional foods, it is unclear that extracting components of food products and administering them as supplements provides the same health benefits as if the components were consumed in their natural food. Nonetheless, many dietary supplements are generally thought to have health effects similar to their food origin, as described before regarding garlic (thiols/sulfides), fish oils (omega-3 polyunsaturated fatty acids), ginseng (saponins), and green tea extract (flavonoids).

Bitter Orange

Citrus aurantium is a citrus tree whose extract of its bitter orange peel is reported to facilitate weight loss due to the presence of synephrine alkaloids, which may suppress appetite (and thus decrease caloric intake) and increase the metabolic rate.[74] Molecular components of *Citrus aurantium* may improve adipocyte function through increased PPAR gamma activity and increased adiponectin secretion.[75] While these effects might be expected to favorably affect adiposopathy, *Citrus aurantium* has potential adverse effects related to the presence of contents with structural and functional similarities to epinephrine. So while bitter orange is a dietary supplement sometimes used as a substitute for the banned ephedra (a plant-derived supplement containing the stimulant ephedrine), it has similar potential for adverse effects (e.g. increase in blood pressure), and thus potential for an increased risk of cardiovascular events.[76]

Carnitine

Carinitine is a lysine and methionine amino acid compound required for transport of long chain fatty acids from the cytosol to the mitochondria for beta oxidation, which is an aerobic process wherein fatty acids are metabolized to form acetyl-CoA, which may then enter the citric acid cycle. Carnitine may increase adipocyte lipolysis, decrease adipoctye triglyceride content (and thus, presumably, decrease adipocyte size), and decrease adipogenic gene expression.[77] It may also improve strenuous exercise training performance by attenuating deleterious effects of hypoxic training and cellular damage,[78] which may allow greater energy expenditure, and facilitate weight reduction. However, neither animal nor human data consistently supports carnitine supplementation as promoting weight loss,[79,80] leading some reviewers to suggest that carnitine has unproven clinical benefit.[81]

Chitosan

Chitosan is a dietary supplement derived from chitin (a component of crustacean/arthropod exoskeletons) and is a starch structurally similar to cellulose, being composed of carbohydrate oligosaccharides. It is sometimes reported to impair the gastrointestinal absorption of fat, decrease body weight, improve adipokine gene expression, and reduce glucose levels.[82] The decrease in lipid accumulation in adipocytes with chitosan is associated with theoretically, a compensatory decrease in adipogenesis, such as through a decrease in PPAR gamma activity.[83] Chitosan may also reduce visceral and hepatic fat, with lowered triglyceride levels.[84] As with many if not most dietary supplements, more data is necessary in order to determine chitosan's safety and efficacy.[85] Some reviews suggest that chitosan's safety and efficacy are unproven.[86]

Collagen Hydrolysate

Collagen hydrolysate represents the hydrolysis of collagen (bone and cartilage) resulting in an end-product consisting of a variety of digestible peptide proteins, derived from a variety of animals. It is provided in protein dietary supplements. It is also employed to form gelatin (as used in medicinal capsules). Ingested proteins increase thermogenesis, as occurs to a lesser degree with carbohydrates, with little-to-no thermic food effects with ingested fats.[87] Collagen peptides may also affect markers of nuclear receptors, with a reduction in free fatty acids and inflammatory markers.[88] To the extent that collagen hydrolysate (if substituted for other nutrient intakes) increases net thermogenesis, and improves osteoarthritis and other joint disorders (thus potentially increasing physical activity, which may further increase energy expenditure),[89] then, as with other instances of protein intake, this may reduce or blunt adiposity, which in turn may diminish or limit adipocyte hypertrophy and visceral fat accumulation. Such effects may improve the metabolic abnormalities associated with adiposopathy, and may help explain rare reports suggesting that collagen peptides may improve metabolic disease.[90]

Conjugated Linoleic Acid

Linoleic acid is an omega-6, polyunsaturated fatty acid found in many vegetable oils, and used in the biosynthesis of arachidonic acid. When linoleic acid isomers contain at least one pair of double bonds separated by a single bond, the molecular form of linoleic acid is described as "conjugated." Although considered both a cis and a trans fatty acid, conjugated linoleic acid is thought to offer health benefits, and is provided as a dietary supplement. Conjugated linoleic acid may decrease adipocyte size, produce favorable effects upon adipokines, increase insulin sensitivity,[91] and decrease body weight; but these effects vary among studies, as do its effects upon inflammatory responses and glucose metabolism.[92] As with many, if not most dietary supplements, more data is necessary in order to determine conjugated linoleic acid's safety and efficacy.[85]

Garcinia Cambogia

Garcinia cambogia is a small fruit indigenous to India; its hydroxycytric acid extract potentially suppresses appetite and/or increases satiety, and potentially increases the metabolic rate. A meta-analysis suggests that *Garcinia cambogia* extract may cause short term fat weight loss[93] and inhibit lipogenesis in adipocytes (due to inhibition of the citric acid cycle). But while hydroxycytric acid may attenuate body weight gain and visceral fat accumulation, it may not have lasting beneficial effects upon hypertriglyceridemia and hyperinsulinemia. Rather, hydroxycytric acid supplementation may lead to liver pathology, possibly due to inhibition of lipid storage in adipocytes resulting in a fatty liver,[94] or potentially due to a yet-undefined hepatotoxicty, although limited evidence supports liver toxicity.[95] As with many, if not most dietary supplements, more data is necessary in order to determine *Garcinia cambogia*'s safety and efficacy.[85]

Glucomannan and Guar Gum

Glucomannan (from konjac roots) is a hemicellulose, polysaccharide food additive, emulsifier, and thickener. Guar gum is also a polysaccharide (composed of galactose and mannose) derived from guar beans, which are ground into a powder and provided as a supplement. When mixed with water, these agents become viscous,[96] and may block intestinal absorption of fats and increase a sense of fullness (satiety). Some glucomannan-containing supplements are reported to reduce body weight, waist circumference, and percent body fat, as well as improve lipid levels.[97] Guar gum is reported to cause weight loss, and reduce cholesterol and glucose levels, as well as cardiovascular risk.[98] Other reviews suggest that these supplements have uncertain clinical benefit.[99]

Guggulipid

Guggulipid is derived from the gummy resin of the mukul myrrh tree, which may increase the activity of both PPAR alpha and gamma receptors, resulting in increased adipocyte differentiation and improvement in both lipid and glucose metabolism in animals.[100] However, some controlled clinical trials in humans have not supported guggulipid supplementation as improving lipid levels.[101]

Hoodia

Hoodia supplements are derived from the South African plant *Hoodia gordonii*, traditionally used as a hunger and thirst suppressant for long hunting trips.[102] An often-described Hoodia molecule is the pregnane glycoside P57, which is thought to interact with appetite centers in the hypothalamus.[103] Insufficient information is available for determining Hoodia's weight loss efficacy and safety,[104] much less on adipocyte function, adiposopathy, and metabolic disease.

Irvingia Gabonensis

Extracts of the seeds of *Irvingia gabonensis* (African mango) are provided as a weight loss supplement. Animal studies suggest that *Irvingia gabonensis* seed extracts may promote weight loss with, theoretically, a compensatory decrease in adipogenesis through downregulation of PPAR gamma, decreased leptin, and increased adiponectin.[105] One randomized prospective human trial supported *Irvingia gabonensis* seed extract supplementation as reducing body weight, decreasing waist circumference, improving lipid levels, lowering glucose levels, reducing inflammatory markers (such as c-reactive protein), decreasing leptin, and increasing adiponectin.[106] More data is necessary in order to determine the safety and efficacy of *Irvingia gabonensis* supplementation.[85]

Pyruvate

Pyruvate is the salt form of pyruvic acid, which is a three-carbon molecule derived from glucose metabolism. Theoretically, increased pyruvate increases cellular metabolism and improves physical exercise endurance — which may facilitate weight loss. However, little is known about the effects of pyruvate supplementation on adipogenesis and adiposopathy. Also, some clinical trials suggest that calcium pyruvate supplementation does not alter body composition or physical exercise performance and may negatively affect some lipid levels,[107] and generally support the conclusion that pyruvate supplements have limited, if any, clinical benefit.[108]

MINERALS AND ADIPOSOPATHY

Dietary "minerals" are chemical elements required by living organisms that traditionally exclude the four basic elements of carbon, hydrogen, nitrogen, and oxygen found in organic molecules. Among the most common minerals described to facilitate weight loss are calcium, chromium, manganese, and zinc. However, no definitive clinical trial evidence exists to suggest that supplementation of these minerals promotes clinically meaningful weight loss. Where dietary minerals may have clinically significant relevance is in post-bariatric-surgery patients, who (depending upon the procedure) may develop macro- and micronutrient deficiencies that could impair multiple metabolic functions of adipocytes, adipose tissue, as well as other body organs.[109]

Calcium

Calcium is a biomineral cofactor for enzymes, a component of bone, and is important for cellular signaling and for maintaining potential differences across cellular membranes. While increased calcium intake may reduce systolic blood pressure, evidence supporting calcium supplementation in reducing body weight and improving glucose or lipid metabolism is inconsistent.[110] Some reports suggest that calcium supplementation may augment weight loss with caloric restriction,[111] including visceral weight loss.[112] Animal studies suggest that increased dietary calcium may increase thermogenesis, increase adipocyte lipolysis, decrease body weight, and reduce adipogenesis.[113] Studies vary widely regarding calcium supplementation (without vitamin D) and human health, such as the risk for cardiovascular disease.[114,115] Perhaps a reasonable conclusion is that calcium-supplementation alone will not consistently result in weight loss. However, if weight loss is achieved through negative caloric balance, calcium-enriched foods substituted for less healthy foods may augment fat weight loss, improve fat function, facilitate improvement in adiposopathy, and enhance improvement in metabolic diseases.

Chromium

Chromium is an essential trace element important for carbohydrate and lipid metabolism, as well as other metabolic processes. It may improve glucose metabolism via potentiation of insulin-stimulated adipogensis and glucose uptake in target organs, such as adipose tissue (and muscle).[116] Chromium supplementation (chromium picolinate) in patients with type 2 diabetes mellitus may have favorable effects upon body weight and visceral adiposity, with improvements in insulin sensitivity and glucose control.[117]

Manganese

Manganese is an essential trace element cofactor for enzymes, associated with mitochondria, and is important for many biologic metabolic processes. Manganese superoxide dismutase is a principal antioxidant enzyme that neutralizes toxic effects of reactive oxygen species. While clinical trial evidence is generally lacking with regard to the effects of manganese supplementation on weight loss and adipose tissue function, disruption in adipocyte manganese metabolism could theoretically worsen fat cell mitochondrial function, and promote cellular damage due to increased reactive (and potentially toxic) intracellular molecules containing oxygen. Other mechanisms by which manganese deficiency may worsen adiposopathy include possible impairment of adipocyte glucose transporter function, resulting in insulin resistance.[118]

Zinc

Zinc is an essential element found as a cofactor for many enzymes related to metabolism. Zinc finger transcription factors (such as Kruppel-like factors) may facilitate adipogenesis through enhancing the action of PPAR gamma activity.[119] Mice studies suggest that zinc-alpha(2)-glycoprotein may increase adipocyte lipolysis, and improve adipocyte function.[120] Human studies suggest that zinc ions influence adipose tissue function through regulating leptin secretion, free fatty acid release, and glucose uptake, with differences in zinc-transporting proteins found in visceral and subcutaneous adipocytes among lean versus obese individuals.[121] However, rat studies suggest that zinc supplementation does not alter short term food intake, even with zinc deficiency.[122] Little clinical trial evidence supports zinc supplementation as improving adiposopathy and metabolic disease.

VITAMINS AND ADIPOSOPATHY

Vitamins

Perhaps the most well-known dietary supplements are vitamins, which are essential organic nutrients of varying structures found in plant and animal food required in small amounts for biologic functions. Choline and inositol are water-soluble nutrients often grouped with B vitamins. Because they can be produced in small amounts by animals, some might not consider choline and inositol true vitamins. Choline is a quaternary saturated amine found in phospholipids (phosphatidylcholine or lecithin), and serves to provide structural integrity to membranes, and aid in cell signaling, nerve impulse transmission, as well as lipid transport and metabolism. Inositol is a carbohydrate whose functions include cellular signaling and secondary messaging. While choline-deficient diets may increase hepatosteatosis and insulin resistance, little evidence supports choline or inositol supplementation beyond correction of deficiencies as promoting weight loss or improving adiposopathy.

Otherwise, 13 different vitamins with different functions are recognized, which include their vitamers (compounds with similar biologic activity), and identified by letters, e.g. A (retinol/beta carotene), B1 (thiamine), B2 (riboflavin), B3 (niacin), B5 (pantothenic acid), B6 (pyridoxine), B9 (folic acid), B12 (cyanocobalamin), C, D, E, H (biotin), and K. Although vitamins may affect adipocyte function, as with minerals, no definitive clinical trial evidence exists to suggest that vitamin supplementation beyond correction of vitamin deficiency promotes clinically meaningful weight loss. Also as with minerals, where dietary vitamins may have clinically significant relevance is in post-bariatric-surgery patients, who (depending upon the procedure) may develop macro- and micronutrient deficiencies that could impair multiple metabolic functions of adipocytes, adipose tissue, as well as other body organs.[123]

Perhaps the vitamin most reported to have clinically relevant health effects upon adipose tissue is vitamin D, which is a fat-soluble vitamin important for calcium metabolism. Ultraviolet radiation from sunlight converts 7-dehydrocholesterol in the skin into D_3 (cholecalciferol), which is also obtained from consuming foods of animal origin. Vitamin D_2 (ergocalciferol) is a form of dietary vitamin D found in plants. Both vitamin D_3 and vitamin D_2 are reported as 25-hydroxyvitamin D, which is converted to the more active vitamin 1,25-dihydroxyvitamin D (calcitriol) by the kidneys via signaling by parathyroid hormone. While 25-hydroxyvitamin D is often a predictor of 1,25-dihydroxyvitamin D in overweight and obese patients, obesity-associated decreases in 25-hydroxyvitamin D levels may sometimes be associated with a compensatory relative increase in 1,25-dihydroxyvitamin D levels, although levels most often remain in the normal range.[124]

Increased body fat increases the total body storage of fat-soluble vitamins (such as vitamin D), and may decrease their blood levels. However, another important cause of lowered vitamin D blood levels is decreased sun exposure due to: (1) increased availability of dwellings and clothing that block sunlight skin exposure; (2) increased sheltered or enclosed work and leisure activity during the day; (3) medical recommendations to avoid sunlight (or use topical sunscreen protection) so as to prevent skin pathology (e.g. skin damage and cancer); and (4) less leisure physical activity in general — which might otherwise occur in sunlight. This may be especially true in overweight individuals. The relative contribution of increased fat depots that contain ("hide" or sequester) this fat-soluble vitamin[125] and/or lack of sunlight exposure to diminished circulating vitamin D levels is unclear. Both mechanisms may contribute.

Regarding adiposopathy, vitamin D regulates intracellular adipocyte calcium levels, which may affect adipose tissue function. Some studies suggest that increased 1,25-hydroxyvitamin D levels may stimulate lipogenesis and inhibit lipolysis. Thus, the relative increase in 1,25-hydroxyvitamin D levels often found in obese individuals may theoretically further promote obesity. However, while 1,25-hydroxyvitamin D may be relatively increased in obese versus nonobese individuals, absolute 1,25-hydroxyvitamin D levels may be decreased with obesity, such that a relative increase in 1,25-vitamin D is unlikely to be important in increasing fat mass.[126] However, the relative increase in 1,25-hydroxyvitamin D may adversely affect fat function, in that 1,25-vitamin D may decrease adipogenesis through decreased PPAR gamma activity,[127] which may worsen adiposopathy. The link between vitamin D and adiposopathy is further supported by the finding that low vitamin D levels are found in patients with metabolic syndrome,[128] arterial hypertension,[129] and increased cardiovascular disease risk and risk factors.[130] The link between vitamin D and adiposopathy is not supported by findings that, over time, asymptomatic primary hyperparathyroidism (which would be expected to increase 1,25-hydroxyvitamin D levels) does not seem to worsen metabolic disease, such as insulin insensitivity.[131]

While much needs to be learned about the interactions and relative contributions of vitamin D, calcium, and sunlight to adiposity and adiposopathy, data supports that the combined intake of calcium and vitamin D may increase fat oxidation rates, increase diet induced thermogenesis, and reduce spontaneous caloric intake.[132] When substituted for less healthy, obesigenic foods, dairy calcium (e.g. yoghurt) substitution may facilitate fat loss and reduce visceral adiposity, lowering 1,25-vitamin D levels, and reducing intra-adipocyte calcium, which may upregulate lipid utilization and decrease lipogenesis.[133,134] Increased nutritional intake of vitamin D may improve glucose metabolism in type 2 diabetes mellitus patients.[135] Thus, vitamin D levels raised through increased oral intake and ultraviolet-B irradiance may have important health benefits not only for adiposopathy, but also for a number of illnesses, such as musculoskeletal disease, many types of cancer, cardiovascular disease, diabetes mellitus, infectious disease, autoimmune disease, and brain disorders.[136]

LACK OF SUNLIGHT EXPOSURE AND ADIPOSOPATHY

From a historical perspective, concomitant with the increasing proportions and prevalence of persons who are overweight or obese over the past several decades, and in addition to increased caloric intake and decreased physical activity, one of the sentinel adaptive human lifestyle changes over the past few decades has been diminished sunlight exposure (see previous vitamin D discussion). Beyond considerations of vitamin D metabolism, diminished sunlight exposure may have other adverse metabolic consequences that promote adiposity and adiposopathy.[137]

In most cases, the number of skin melanocytes does not significantly differ among races. However, variations in skin color obviously exist, and are determined by the amount of carotene in the epidermis and subcutaneous (adipose) tissue, the amount of oxygen bound to hemoglobin in the dermal blood cells, and the amount, type, and grouping of melanin pigment in epidermal melanocytes. Greater grouping of melanosomes (organelles containing melanin) in melanocytes results in lighter skin, while more dispersed melanosomes result in darker skin. Exposure of epidermal melanocytes to the ultraviolet light of sunlight stimulates the transcription factor p53, which increases pro-opiomelanocortin (POMC) production. As with the brain, POMC in the skin is broken down into melanocortins,[138] which are a group of peptide hormones including adrenocorticotropic hormone (ACTH) and melanocyte-stimulating hormone (MSH), which activate melanocortin-1 receptors.[139,140] Increased melanocortin-1 receptor activity in melanocytes through melanocortins (MCs) such as alpha MSH stimulates melanin production (eumelanin is the most common biologic form of melanin), increases skin pigmentation ("tanning"), and protects against sun-induced deoxynucleic acid damage and photocarcinogenesis. Other physiologic effects of melanocortin agonism include catabolic/anorexigenic central nervous system processes which may favorably affect adiposopathy.[137,141]

An example of the physiologic relationship between melanocyte function and body fat involves Agouti mice, which have increased hypothalamic release of Agouti-related protein (AgRP). AgRP antagonizes central catabolic/anorexigenic MC receptors such as the MC4 (and MC3) receptors, found only in the brain. Just as impaired MC4 activity through MC4 receptor mutations may account for 0.5%–6% of severe cases of obesity in humans, blocking alpha MSH's activity on central catabolic/anorexigenic MC4 receptors causes Agouti mice to be overweight. Blocking peripheral alpha MSH's activity on skin melanin production in melanocytes also causes Agouti mice to have decreased black fur, with the emerging dominance of yellow fur pigment. Just importantly, the "yellow mouse obesity syndrome" described with Agouti competitive antagonism of melanocortins causes the adiposopathic consequences of hyperinsulinemia, insulin resistance, hyperglycemia, and hyperleptinemia.[142] Therefore, melanin metabolic pathways and body weight are interconnected, and this suggests that sunlight may play an important role in adipocyte and adipose tissue function.

Sunlight is mostly described as mainly harmful from a health standpoint. But it is unreasonable to believe that something as basic, as ubiquitous, and as timeless as sunlight would only promote detrimental health effects. This is made less plausible given that sunlight has important biologic effects such as enhanced vitamin D metabolism, important for musculoskeletal system health. Inadequate sunlight may have other adverse health effects, such as an increased prevalence of nonskin cancers, autoimmune disease, sleep disturbances, and affective disorders.[143] Conversely, sunlight and/or vitamin D may have favorable effects upon metabolic diseases associated with adiposopathy (e.g. glucose metabolism and blood pressure),[144] through mechanisms that are likely multifactorial.

Sunlight is composed of a spectrum of electromagnetic radiation, including gamma rays, X-rays, ultraviolet light (with shorter wavelength than visible and violet light), visible light, and infrared light (with longer wavelength than visible light). Most of the sun's gamma rays, X-rays, and ultraviolet waves are blocked by the earth's atmosphere. Excessive exposure of the integument to unblocked ultraviolet radiation which ultimately makes it to earth may unfavorably promote skin cancer and cataracts. But exposure of skin to the ultraviolet sun radiation spectrum may also increase melanocortin production. Those with darker skin may have raised basal melanocortin levels compared with those with lighter skin, while those with lighter skin may have greater increases in melanocortin levels after ultraviolet total body irradiation than more pigmented persons.[145]

From a cellular standpoint, adipocyte melanocortin receptor activity is a balance between the interactions of melanocortins and AgRP. Increased relative exposure to melanocortins decreases intra-adipocyte calcium, which favors lipolysis and suppresses lipogenesis, while increased relative exposure to the purified Agouti gene product increases intra-adipocyte calcium, which favors lipogenesis and suppresses lipolysis.[146,147] Similarly, as noted before, the relative increase in 1,25-hydroxyvitamin D found in obese individuals may also increase intra-adipocyte calcium, which again favors lipogenesis and suppresses lipolysis.[148] An increase in skin exposure to sunlight would be expected to raise the levels of 25-hydroxyvitamin D (which has vitamin D activity), decreasing the conversion to 1,25-vitamin D, and thus potentially decreasing intra-adipocyte calcium, decreasing lipogenesis, and increasing lipolysis. This sunlight–fat-cell physiologic connection has a logical justification. Reduced sunlight during the winter months may help facilitate increased fat storage for energy and increased thermal insulation, which may have historic advantages favoring survival during times before the year-round access to food and heated dwellings. Increased sunlight exposure during the summer months may reduce fat storage and decrease thermal insulation, which may also aid in survival, by reducing body weight for faster avoidance of predators and by avoidance of hyperthermia.

Another cellular mechanism accounting for melanocortins' decreased lipogenic and increased lipolytic effects is that melanocortins may increase thermogenesis in adipose tissue,[149] via increased sympathetic nervous system activity and increased uncoupling proteins,[150] with increased uncoupling proteins promoting "inefficient" heat release relative to molecular energy production. Finally, increased central melanocortin activity may increase thyroid hormone release, as well as have many other effects that promote catabolism.[137]

Sunlight spectra beyond ultraviolet light may also affect adiposopathy. Visible and infrared radiation may increase body temperature, indirectly by increasing the ambient earth temperature and directly by radiating the skin. It is true that cold ambient temperatures may induce compensatory physiologic responses (e.g. increased thyroid hormone and sympathetic nervous system activity) that increase resting energy expenditure.[151] However, these effects are minor and dependent upon persistent, prolonged cold exposure. Conversely, physical activity often increases during, warmer months compared to colder months (due to longer days, increased outside work, and increased leisure activity), resulting in greater energy expenditure during summer months.[152] Furthermore, while some absolute weight loss with strenuous physical exercise performed at higher temperatures may be attributable to salt loss and dehydration,[153] fat weight loss during sunny and warm summer months can also be attributable to the greater energy expenditure required to cool warm bodies during physical activities. This helps explain why fat oxidation, lipolysis, and free fatty acid mobilization are increased with hotter temperatures compared with physical activity performed at colder temperatures.[154] Thus, multiple spectra of sunlight have the potential to favorably affect adiposity and adiposopathy. The challenge to patients and clinicians is to find the correct balance of exposure that maximizes the potential beneficial effects of sunlight, while minimizing detrimental effects.[155]

DECREASED PHYSICAL ACTIVITY

Low level physical activity contributes to metabolic disease.[156] Minimal physical activity is sometimes involuntary and/or unavoidable, as might occur with physical limitations or disabilities. However, sedentary lifestyle is frequently a life choice, not unlike excessive caloric intake or cigarette smoking. The good news is that for able individuals increased physical activity can often improve adiposopathy, and thus improve metabolic health.

Total daily energy expenditure is determined by resting energy expenditure, nonresting energy expenditure (e.g. physical activity), and the thermal effect of food, which, as noted before, may be increased with protein intake (relative to fat intake). For those who do not engage in rigorous, routine exercise programs (beyond low level daily physical activity), resting energy expenditure is by far the greatest contributor to daily energy expenditure (Table 2). Unless it is due to an underlying disease (e.g. severe hypothyroidism, or a central nervous system disorder), having a low resting metabolic rate ("low metabolism") is not a major "cause" of obesity, despite a common belief otherwise.[157] Instead, resting energy expenditure is most dependent upon the body mass index, and the size of body organs. A higher body mass index is associated with higher (not lower) resting energy expenditure. On average, resting energy expenditure is greater in men than in women, because men are generally larger, have a higher body mass index, and an increase in the mass of major body organs such as skeletal muscle, compared to women. Table 2 describes the approximate daily metabolic rate for body organs and/or tissues.[158] Because adipose tissue contributes to resting energy expenditure, a reduction in body weight, including a reduction in adipose tissue, decreases resting metabolic expenditure. Such an effect may be mitigated by engaging in routine physical activity,[159] particularly if the physical activity is sufficient to avoid the loss of, or in fact, result in the gain of lean body mass, such as muscle mass.

Both dynamic and resistance exercise may reduce body fat, improve adiposopathy, and improve metabolic disease.[160] The amount of calories expended with each type of physical activity is variable, depending upon the duration, intensity, and body weight (with increased body weight increasing caloric expenditure for the same type and level of activity). Lower intensity physical activity results in less caloric expenditure than higher intensity physical activity. Sleeping and lying may burn ~70 kcal/h, sitting and standing ~100 kcal/h, and standing with activity (e.g. scrubbing or ironing) ~250 kcal/h. Regarding more dynamic physical activity, slow walking burns ~200 kcal/h, cycling and swimming ~ 500 kcal/h, and running ~700 kcal/h. Regarding resistance training (e.g. weightlifting), energy expenditure varies widely, generally ranging between 144 and 600 kcal/h, depending upon the intensity of the effort, and the rest time between the resistance activities.[161]

Table 2 Approximate Resting Energy Expenditure of Adult Body Organs[205,206]

Organ or Tissue	Daily Metabolic Rate (kcal/kg/day)	Daily Metabolic Rate (kcal/lb/day)	Body Organ Weight*	Total Energy Expenditure† (kcal/day)
Heart	440	200	0.3 kg/0.66 lb	132
Kidney	440	200	0.13 kg/0.29 lb (per kidney)	114 (per 2 kidneys)
Brain	240	109	1.3 kg/2.87 lb	312
Liver	200	91	1.5 kg/3.31 lb	300
Muscle	13	6	20 kg/44 lb women 30 kg/66 lb men	260 women 390 men
Adipose tissue‡	4.5	2	13 kg/28 lb women 16 kg/35 lb men	59 women 72 men
Residual mass‡	12	5	27.24 kg/60 lb women 30.6 kg men/67.3 lb	327 women 368 men
Total Resting energy expenditure§				~1500 women ~1700 men

*Body organ size and weight generally increase with the male gender, and an increased body mass index. Muscle and adipose tissue weight may vary substantially among individuals.

†Highly generalized figure, shown for illustrative purposes only. Individuals vary widely, depending upon differences in body organ weight — most applicable to organs with the highest total weight, such as muscle and adipose tissue.

‡Based upon 20% body fat and body weight 80 kg (176 lb), height 1.780 m (5 ft 10 in), and body mass index 25 kg/m^2 for men, and 63.6 kg (140 lb), 1.6 m (5 ft 3 in), and 25 kg/m^2 for women.

§Totals represent the addition of the rows above. For rough validation, using the Harris–Benedict formula for men [(13.75 × weight in kg) + (5 × height in cm) – (6.76 × age in years) + 66], the resting energy expenditure (resting metabolic rate) for a 40-year-old man with the metrics described above would be 1786 kcal/day. Using the Harris–Benedict formula for women [(9.56 × weight in kg) + (1.85 × height in cm) – 4.68 × age in years) + 655], the resting energy expenditure (resting metabolic rate) for a 40-year-old woman with the metrics described above would be 1372 kcal/day.

Dynamic physical exercise may contribute to negative caloric balance, reduce adipocyte and adipose tissue size, and improve adiposopathic responses. Resistance training increases total fat-free mass, increases muscle strength, increases resting energy expenditure, and mobilizes visceral and subcutaneous adipose tissue in the abdominal region.[162] The caveat here is the misconception that "every pound of muscle gained burns an additional 50 calories per day." This often-stated axiom implies that if 10 pounds of muscle are gained, the resting energy expenditure will increase by 500 kcal per day — roughly equivalent to negating the amount of kilocalories often found in one fast food hamburger sandwich per day. Unfortunately, little objective data supports this common quote. Instead, Table 2 describes how each pound of muscle actually "burns" 6 cal per day. Thus, if 10 pounds of fat are replaced by 10 pounds of muscle, then the resting energy expenditure will be expected to increase by only about 40 kcal per day (60 extra kcal per day from 10 pounds of added muscle; less the 20 kcal per day from 10 pounds of lost adipose tissue), which is roughly equivalent to two blocks of bubble gum, or a half-serving of a 170 g container of fat-free yoghurt per day. However, it should be kept in mind that, depending upon the intensity and duration, the amount of calories expended during resistance (anaerobic) physical exercise necessary to build 10 pounds of muscle may be as substantial as the amount of calories expended during dynamic (aerobic) physical exercise.

In addition to promoting weight loss through increased caloric expenditure, both dynamic and resistance exercise may help reduce weight regain after weight loss induced by improved nutrition, particularly with regard to regain of visceral fat.[163] This is important, because maintaining weight loss in overweight patients is often more challenging than initiating weight loss. While efficient in avoiding starvation through multiple redundant mechanisms, the human body it is not very efficient in self-regulating against excessive energy storage. Physiologic mechanisms to prevent starvation are advantageous with food scarcity. The lack of equivalent mechanisms to avoid weight gain is disadvantageous in the presence of food abundance and conveniences that contribute to limitations on physical activity (e.g. automobiles, elevators, escalators).

During active weight loss through negative caloric balance, resting metabolic expenditure may decrease below that anticipated for the body mass index, and may[164] or may not[165] return to what is appropriate for the body mass index upon stabilization of body weight. This contributes to fat regain. Additionally, nonresting energy expenditure may also decrease with weight loss as the result of improved skeletal muscle work efficiency.[166] Another mechanism contributing to fat regain after weight loss is reduced leptin levels, and it strongly favors positive caloric balance through increased appetite and perhaps other mechanisms.[167] Engaging in routine physical exercise during active weight loss may enhance central nervous system leptin (and insulin) sensitivity,[168] reduce so-called "leptin resistance,"[169] and better allow longer term weight loss maintenance and reduction in adiposopathy.[170]

Some reports suggest that weight loss through appropriate nutrition and physical exercise may preferentially decrease visceral adipose tissue relative to subcutaneous adipose tissue, with visceral fat loss sometimes changing even without changes in the body mass index.[171] However, this is variable, with the degree of relative fat depot loss most dependent upon the inherited, gender, and environmental nature of the individual, just as the relative degree of relative fat depot weight gain was dependent upon the inherited, gender, and environmental nature of the individual. Beyond the relative loss of fat depots is the interrelationship between absolute body weight and the relative loss or gain of weight with respect to fat and lean body mass. Engagement in an active resistance training program in adults may result in unchanged height, muscle gain, and no change in body weight. In this case, body fat and percent body fat would decrease, although the body mass index (measured in kg/height in meters squared) would not change. Furthermore, given that muscle weighs more than fat, and given that muscle gain is often most easily achieved at the start of a resistance exercise program, many who begin a resistance exercise program often experience an initial increase in total body weight. Because muscle is denser than fat, these same individuals may also become slimmer (as reflected by decreased waist circumference), unless their resistance exercise training substantially emphasizes core body muscle growth (through increase in abdominal and lower back/upper hip muscle mass). Overall, if exercise training is intended to substantially contribute to fat loss, then the routine assessment of percent body fat is of more relevance than body mass index measurements.

Regarding the anatomic manifestations of adiposopathy, whether it be a low calorie meal plan, bariatric surgery, or increased energy expenditure through increased physical activity, all such interventions generally produce fat loss in both the visceral and subcutaneous regions.[172–174] Exercise training reduces adipocyte size and improves adipocyte function,[175] which is consistent with the principle that reductions in adipocyte size and visceral adiposity improve adiposopathy and metabolic disease.[176,170] Furthermore, because subcutaneous adipose tissue is not always "protective," and may be potentially pathogenic, the loss of most any fat depot in overweight patients would be expected to have favorable health benefits.[177] The clinical point is that physical inactivity is an important potential contributor to adiposopathy, particularly if it facilitates adipocyte hypertrophy and visceral fat accumulation. Reversing physical inactivity by engaging in active daily activity, as well as through implementing a routine physical exercise program or other substantial increases in physical activity, can reduce body fat and improve many of the anatomic and functional abnormalities of adiposopathy, which in turn, improves metabolic disease.

A final consideration is that if reduced body fat is achieved, the metabolic effects upon an individual's health differ depending upon whether the low body fat is due to starvation or physical exercise training. Acute fasting for one week may actually raise cholesterol and triglyceride levels,[178] and fasting for 60 h increases circulating free fatty acid release from adipocytes, which worsens insulin resistance in muscle ("lipotoxicity"), even as glucose and insulin levels are low to low-normal as a result of fasting.[179] As opposed to the worsening insulin resistance seen with inadequate adipocyte and adipose tissue function found with starvation (and lipodystrophies), it is interesting that most anorexia nervosa patients appear to maintain sufficient adipocyte function, and thus tend to avoid the development of insulin resistance.[180]

Essential body fat is sometimes described as 10%–13% for women and 2%–5% for men. Athletes often have percent body fat 14%–20% for women and 6%–13% for men, with fit individuals having percent body fat 21%–24% for women and 14%–17% for men. "Acceptable" percent body fat is sometimes cited as 25%–31% for women and 18%–24% for men, with obesity often being greater than 32% body fat for women and greater than 25% body fat for men. However, these cutoff point percentages should be interpreted with caution, because: (1) they do not necessarily reflect the current or future metabolic health status of the individual; (2) percent body fat measures may differ widely, depending upon the methodology of how the percent body fat is assessed, and the expertise of those doing the measurements; (3) those with higher percent body fat but reduced waist circumference and reduced subcutaneous adipose tissue may not be at increased risk for adiposopathy and metabolic disease[181]; and (4) those with less percent body fat but a disproportional increase in visceral and abdominal subcutaneous fat may be at increased risk for adiposopathy and metabolic disease.[182,183]

National surveys suggest that average American women and men have body fat that ranges between 28% and 40%.[184] With bodybuilders, percent body fat may be as low as 6%–10% for women and as low as 2%–8% for men.[185] Also, while resistance training may be catabolic regarding fuel utilization (such as increased glucose and carbohydrate metabolism), resistance training is also anabolic with regard to hypertrophy of muscle, increasing insulin-sensitive muscle fibers (which contain greater oxidative and mitochondrial capacity), which in turn increases whole body insulin sensitivity.[186] Thus, just as with anorexia nervosa, the lower levels of body fat found in bodybuilders appear to be sufficient for avoiding insulin resistance, and the increase in muscle mass increases an important body tissue, contributing to enhanced insulin sensitivity.[187] When coupled with nutritional intake that may differ from that of nonbodybuilders,[188,189] the depletion of glycogen stores during training plus the high caloric expenditure per day (~3000–4000 kcal/day)[190] necessary for bodybuilding training helps explain why bodybuilders often have lower limits of normal glucose levels, despite low percent body fat.[191] Even among nonprofessional bodybuilders, individuals engaging in resistance training have increased insulin sensitivity, which may be advantageous to patients with type 2 diabetes mellitus.[192]

ADIPOSOPATHY TREATMENT

The intent of adiposopathy treatment is to reduce adipocyte hypertrophy and visceral (and periorgan) fat accumulation. Correction of adipocyte and adipose tissue endocrine and immune abnormalities improves clinical, metabolic disease, and can be accomplished with weight loss therapies such as improved nutrition, increased physical activity, pharmacotherapy, and bariatric surgery (Table 3). However, the first step of adiposopathy treatment is to identify those at greatest risk.

Particularly among those from the South and East Asian subcontinent, Asians have increased susceptibility to adiposopathy, as evidenced by the high prevalence of type 2 diabetes mellitus, the metabolic syndrome, and cardiovascular disease, which may be attributable to increased adipocyte size, decreased adipocytes, increased visceral adiposity, increased circulating free fatty acids, increased leptin levels, increased proinflammatory factors (e.g. increased C-reactive protein levels), and decreased anti-inflammatory factors (e.g. decreased adiponectin), all of which lead to increased insulin resistance and increased cardiovascular disease risk.[193] Asian individuals may develop adiposopathy and metabolic disease, even with modest weight gain. Similarly, for the same age and weight, men have higher rates of metabolic disease and cardiovascular disease than women. This is likely related to increased adipocyte hypertrophy and visceral fat accumulation.[194,195]

Other genetic disorders further emphasize the potential impact of adipose tissue function and dysfunction when one is attempting to medically justify treatment. Benign multiple symmetric lipomatosis increases fat accumulation in the peripheral subcutaneous adipose tissue regions of the arms, legs, shoulders, and neck. Yet, despite adiposity, such patients typically do not require treatment for metabolic abnormalities (e.g. glucose or lipid disorders). This is most likely because of the proliferation of small adipocytes, the peripheral subcutaneous adipose tissue location, and the increased secretion of anti-inflammatory adipokines, such as adiponectin.[196] Conversely, inherited lipodystrophy is manifested by a variable lack of body fat and impaired adipose tissue function (e.g. low adiponectin levels and inability to adequately store fat). The limited fat storage potential results in high circulating free fatty acids which contribute to lipotoxicity, and metabolic disorders (e.g. hyperglycemia and dyslipidemia),[197] which may require treatment. Thus, lipodystrophic patients with too little fat may require treatment for metabolic diseases most often associated with too much fat. Lipoatrophic mice have virtually no white adipose tissue. Yet they develop severe diabetes mellitus, which can be markedly improved through surgical implantation of adipose

Table 3 Examples of Treatments for Adiposopathy ("Sick Fat") and Their Effects Upon Illustrative and Selected Adipose Tissue Factors that may Contribute to Metabolic Disease

	May Affect Glucose Metabolism, BP and Lipid Metabolism				May Affect Glucose Metabolism	May Affect BP	May Affect Lipid Metabolism	
Intervention	Visceral adipose tissue*	Free fatty acids	Leptin	Adiponectin	TNF-α	Renin-anglotensin-aldosterone enzymes	Androgens	Estrogens
Weight loss interventions								
Appropriate nutrition and physical activity	↓	↓	↓	↑	↓	↓	↓(women), ↑(men)	↓/–(men)
Orilstat	↓	↓	↓	↑	↓	?	↓(women)	?
Sibutramine	↓	↓	↓	↑/–	?	?	↓(women)	?
Cannabinoid receptor antagonists	↓	↓	↓	↑	↓	?	?	?
Lorcaserin	↓	?	?	?	?	?	?	?
Phentermine	↓[105]	?	?	?	?	?	?	?
Topiramate	↓[119]	↓[172]	↓/–[119,130]	↑[119]	–[119]	?	?	
Phentermine/topiramate combination	↓	?	?	?	?	?	?	?
Laparoscopic adjustable gastric banding	↓	?	↓	↑	↓	↓	?	?
Gastric bypass	↓	↓†	↓	↑	↓	↓	↑(men)	↓(men)

(*Continued*)

Table 3 (Continued)

Intervention	May Affect Glucose Metabolism, BP and Lipid Metabolism				May Affect Glucose Metabolism	May Affect BP	May Affect Lipid Metabolism	
	Visceral adipose tissue*	Free fatty acids	Leptin	Adiponectin	TNF-α	Renin-anglotensin-aldosterone enzymes	Androgens	Estrogens
Diabetes mellitus therapies[‡]								
PPAR-γ agonists[§] (pioglitazone, rosiglitazone)	↓/–	↓	↓/–	↑	↓	–	↓	↓/–(men)
Metformin	↓	↓	↓	↑	–	↓	↓(women)	?
Sulfonylurea	↑	↓	↑/–	↑	↑/–/↓	?	?	?
Insulin	↑/–	↓	↑	↓	↓/–	↑	↓/–(women)	?
Glucagon-like peptide-1 (GLP-1)	↓	↓	↓/–	↑/–	–	?	↓	?

*Visceral adiposity may be approximated by measurements of waist circumference.

[†]Acutely (e.g., 1 month), free fatty acids may be increased during rapid weight loss.

[‡]Not all of the listed metabolic effects of anti-diabetes mellitus therapies are adipocyte or adipose tissue mediated.

[§]PPAR-γ agents may: (1) increase adipose tissue proliferation and differentiation, (2) favorably alter the visceral to subcutaneous adipose tissue deposition ratio. (3) reduce hepatic fat deposition, and (4) improve other aspects of adipose tissue function (Refs. 3, 23, 53). While some of the weight gain associated with PPAR-γ agents is due to fluid retention, much of the weight gain observed with these agents used to treat metabolic diseases traditionally associated with fat weight gain, are (paradoxically?) due to promoting increased amounts of functional adipose tissue.

[¶]Not all of the listed metabolic effects described with GLP-1 are yet proven to occur with GLP-1 agonists.

↑: Increased; ↓: Decreased; ?: Unreported; -:Neutral effect, PPAR-γ: Peroxisome proliferator-activated receptor-γ.

Date taken from Refs. 2, 3, 23, 171.

tissue. Adipose tissue implantation in these mice markedly reduces hyperglycemia and hyperinsulinemia, and improves muscle insulin sensitivity.[198]

Another example where fat loss is not always an effective treatment for adiposopathy is liposuction. Surgical removal of subcutaneous adipose tissue may not improve hyperglycemia, high blood pressure, or dyslipidemia.[199] Within the adipocentric paradigm, this is not unexpected given that the removal of subcutaneous adipose tissue may not only decrease the availability of functional adipocytes, but also cause a disproportional amount of visceral adipose tissue to remain, which would be pathogenic. Similarly, antiretroviral therapy treatment for the human immunodeficiency virus (HIV) may result in lipodystrophy wherein fat is lost but patients have worsening insulin resistance and dyslipidemia, which may be due to worsening of adiposopathy (i.e. greater loss of subcutaneous adipose tissue relative to visceral adipose tissue).[200]

Table 3 describes how many weight loss therapies improve adipocyte and adipose tissue function, which contributes to their ability to improve metabolic disease. It is the therapeutic focus on improving the functionality of fat rather than decreasing the amount of fat that helps explain why adding fat is a means of treating metabolic diseases.[170] Peroxisome-proliferator-activated receptor (PPAR) gamma agonists increase the recruitment, proliferation, and differentiation of functional fat cells in subcutaneous adipose tissue to a greater extent than visceral adipose tissue. Adding functional adipocytes and adipose tissue through increased adipogenesis helps account for how PPAR gamma agents increase body fat and yet improve metabolic disease (e.g. lower glucose levels, reduce hepatic steatosis, improve lipid parameters, and potentially reduce cardiovascular disease risk).[201,202]

CONCLUSION

Anatomically, adiposopathy is manifested by adipocyte hypertrophy, as well as visceral periorgan fat accumulation. Structurally, adiposopathy is manifested by extracellular matrix abnormalities and impaired angiogenesis, which may lead to adipocyte and adipose tissue hypoxia, mitochondrial and endoplasmic reticulum stress, and increased tissue oxidation. Functionally, adiposopathy is manifested by impaired adipogenesis, as well as pathogenic endocrine and immune responses. Adiposopathy is promoted by positive caloric balance, unhealthy nutritional intake, and sedentary lifestyle in genetically and environmentally susceptible individuals. These anatomic, structural, and functional abnormalities promote or worsen metabolic disease. The degree to which adiposopathy causes or worsens metabolic disease is dependent upon the interactions and crosstalk with other body tissues and other body organ systems. The treatment of adiposopathy includes reducing adipocyte hypertrophy, visceral adiposity, and periorgan adiposity. The ultimate goal of the treatment is not only to decrease the weight of the patients, but also to improve the health of the patients.

REFERENCES

1. World Health Organization, Global Strategy on Diet, Physical Activity and Health. http://www.who.int/dietphysicalactivity/publications/facts/obesity/en/ Accessed April 24, 2010.
2. Bays HE, Gonzalez-Campoy JM, Bray GA, *et al.* (2008) Pathogenic potential of adipose tissue and metabolic consequences of adipocyte hypertrophy and increased visceral adiposity. *Expert Rev Cardiovasc Ther* **6**: 343–368.
3. Bays HE, Fox KM, Grandy S. (2010) Anthropometric measurements and diabetes mellitus: clues to the "pathogenic" and "protective" potential of adipose tissue. *Metab Syndr Relat Disord* **8**: 307–315.
4. Bays HE, Fox KM, Grandy S. (2010) Anthropometric measurements and diabetes mellitus: clues to the "pathogenic" and "protective" potential of adipose tissue. *Metab Syndr Relat Disord* **8**: 307–315.
5. Raclot T, Holm C, Langin D. (2001) Fatty acid specificity of hormone-sensitive lipase. Implication in the selective hydrolysis of triacylglycerols. *J Lipid Res* **42**: 2049–2057.
6. Thampy KG. (1995) Hypercholesterolaemia of prolonged fasting and cholesterol lowering of re-feeding in lean human subjects. *Scand J Clin Lab Invest* **55**: 351–357.
7. Hoeks J, van Herpen NA, Mensink M, *et al.* (2010) Prolonged fasting identifies skeletal muscle mitochondrial dysfunction as consequence rather than cause of human insulin resistance. *Diabetes* **59**: 2117–2125.
8. Finley CE, Barlow CE, Halton TL, *et al.* (2010) Glycemic index, glycemic load, and prevalence of the metabolic syndrome in the Cooper Center Longitudinal Study. *J Am Diet Assoc* **110**: 1820–1829.

9. Leclercq IA, Horsmans Y. (2008) Nonalcoholic fatty liver disease: the potential role of nutritional management. *Curr Opin Clin Nutr Metab Care* **11**: 766–773.
10. Larsen TM, Dalskov SM, van BM, *et al.* (2010) Diets with high or low protein content and glycemic index for weight-loss maintenance. *N Engl J Med* **363**: 2102–2113.
11. Brand-Miller J, Millan-Price J, Steinbeck K, *et al.* (2009) Dietary glycemic index: health implications. *J Am Coll Nutr* **28(Suppl)**: 446S–449S.
12. Gardner CD, Kiazand A, Alhassan S, *et al.* (2007) Comparison of the Atkins, Zone, Ornish, and LEARN diets for change in weight and related risk factors among overweight premenopausal women: the A–Z Weight Loss Study: a randomized trial. *JAMA* **297**: 969–977.
13. Bays HE, Tighe AP, Sadovsky R, *et al.* (2008) Prescription omega-3 fatty acids and their lipid effects: physiologic mechanisms of action and clinical implications. *Expert Rev Cardiovasc Ther* **6**: 391–409.
14. Zhao S, Wang J, Song X, *et al.* (2010) Impact of dietary protein on lipid metabolism-related gene expression in porcine adipose tissue. *Nutr Metab (Lond)* **7**: 6.
15. Genaro PS, Martini LA. (2010) Effect of protein intake on bone and muscle mass in the elderly. *Nutr Rev* **68**: 616–623.
16. Arana A, Mendizabal JA, Alzon M, *et al.* (2008) The effect of vitamin A supplementation on postnatal adipose tissue development of lambs. *J Anim Sci* **86**: 3393–3400.
17. Schupp M, Lefterova MI, Janke J, *et al.* (2009) Retinol saturase promotes adipogenesis and is downregulated in obesity. *Proc Natl Acad Sci USA* **106**: 1105–1110.
18. Wolf G. (2010) Retinoic acid activation of peroxisome proliferation–activated receptor delta represses obesity and insulin resistance. *Nutr Rev* **68**: 67–70.
19. Lobo GP, Amengual J, Li HN, *et al.* (2010) Beta, beta-carotene decreases peroxisome proliferator receptor gamma activity and reduces lipid storage capacity of adipocytes in a beta,beta-carotene oxygenase 1–dependent manner. *J Biol Chem* **285**: 27891–27899.
20. Krupkova M, Janku M, Liska F, *et al.* (2009) Pharmacogenetic model of retinoic acid–induced dyslipidemia and insulin resistance. *Pharmacogenomics* **10**: 1915–1927.
21. Rodondi N, Darioli R, Ramelet AA, *et al.* (2002) High risk for hyperlipidemia and the metabolic syndrome after an episode of hypertriglyceridemia during 13-cis retinoic acid therapy for acne: a pharmacogenetic study. *Ann Intern Med* **136**: 582–589.

22. Maki KC, Beiseigel JM, Jonnalagadda SS, *et al.* (2010) Whole-grain ready-to-eat oat cereal, as part of a dietary program for weight loss, reduces low-density lipoprotein cholesterol in adults with overweight and obesity more than a dietary program including low-fiber control foods. *J Am Diet Assoc* **110**: 205–214.
23. Hannan JM, Ali L, Rokeya B, *et al.* (2007) Soluble dietary fibre fraction of *Trigonella foenum-graecum* (fenugreek) seed improves glucose homeostasis in animal models of type 1 and type 2 diabetes by delaying carbohydrate digestion and absorption, and enhancing insulin action. *Br J Nutr* **97**: 514–521.
24. Shimizu C, Kihara M, Aoe S, *et al.* (2008) Effect of high beta-glucan barley on serum cholesterol concentrations and visceral fat area in Japanese men — a randomized, double-blinded, placebo-controlled trial. *Plant Foods Hum Nutr* **63**: 21–25.
25. Choi JS, Kim H, Jung MH, *et al.* (2010) Consumption of barley beta-glucan ameliorates fatty liver and insulin resistance in mice fed a high-fat diet. *Mol Nutr Food Res* **54**: 1004–1013.
26. Zarrouki B, Pillon NJ, Kalbacher E, *et al.* (2010) Cirsimarin, a potent antilipogenic flavonoid, decreases fat deposition in mice intra-abdominal adipose tissue. *Int J Obes (Lond)* **34**: 1566–1575.
27. Andersen C, Rayalam S, la-Fera MA, *et al.* (2010) Phytochemicals and adipogenesis. *Biofactors* **36**: 415–422.
28. Wein S, Behm N, Petersen RK, *et al.* (2010) Quercetin enhances adiponectin secretion by a PPAR-gamma independent mechanism. *Eur J Pharm Sci* **41**: 16–22.
29. Sakurai N, Mochizuki K, Kameji H, *et al.* (2009) (-)-Epigallocatechin gallate enhances the expression of genes related to insulin sensitivity and adipocyte differentiation in 3T3-L1 adipocytes at an early stage of differentiation. *Nutrition* **25**: 1047–1056.
30. Shin DW, Kim SN, Lee SM, *et al.* (2009) (-)-Catechin promotes adipocyte differentiation in human bone marrow mesenchymal stem cells through PPAR gamma transactivation. *Biochem Pharmacol* **77**: 125–133.
31. Chu M, Seltzer TF. (2010) Myxedema coma induced by ingestion of raw bok choy. *N Engl J Med* **362**: 1945–1946.
32. Bays HE, Gonzalez-Campoy JM, Bray GA, *et al.* (2008) Pathogenic potential of adipose tissue and metabolic consequences of adipocyte hypertrophy and increased visceral adiposity. *Expert Rev Cardiovasc Ther* **6**: 343–368.
33. Obregon MJ. (2008) Thyroid hormone and adipocyte differentiation. *Thyroid* **18**: 185–195.

34. Levacher C, Sztalryd C, Kinebanyan MF, *et al.* (1984) Effects of thyroid hormones on adipose tissue development in Sherman and Zucker rats. *Am J Physiol* **246**: C50–C56.
35. Dimitriadis G, Mitrou P, Lambadiari V, *et al.* (2006) Insulin action in adipose tissue and muscle in hypothyroidism. *J Clin Endocrinol Metab* **91**: 4930–4937.
36. Funaki M. (2009) Saturated fatty acids and insulin resistance. *J Med Invest* **56**: 88–92.
37. Kavanagh K, Jones KL, Sawyer J, *et al.* (2007) Trans fat diet induces abdominal obesity and changes in insulin sensitivity in monkeys. *Obesity (Silver Spring)* **15**: 1675–1684.
38. Soriguer F, Moreno F, Rojo-Martinez G, *et al.* (2003) Monounsaturated n-9 fatty acids and adipocyte lipolysis in rats. *Br J Nutr* **90**: 1015–1022.
39. Rajaram S, Connell KM, Sabate J. (2010) Effect of almond-enriched high-monounsaturated fat diet on selected markers of inflammation: a randomised, controlled, crossover study. *Br J Nutr* **103**: 907–912.
40. Granados N, Amengual J, Ribot J, *et al.* (2011) Distinct effects of oleic acid and its trans-isomer elaidic acid on the expression of myokines and adipokines in cell models. *Br J Nutr* **105**(8): 1226–1234. Epub 2011 Jan 5.
41. Bays HE, Maki KC, Doyle RT, *et al.* (2009) The effect of prescription omega-3 fatty acids on body weight after 8 to 16 weeks of treatment for very high triglyceride levels. *Postgrad Med* **121**: 145–150.
42. Rokling-Andersen MH, Rustan AC, Wensaas AJ, *et al.* (2009) Marine n-3 fatty acids promote size reduction of visceral adipose depots, without altering body weight and composition, in male Wistar rats fed a high-fat diet. *Br J Nutr* **102**: 995–1006.
43. Puglisi MJ, Hasty AH, Saraswathi V. (2010) The role of adipose tissue in mediating the beneficial effects of dietary fish oil. *J Nutr Biochem.*
44. Ejaz A, Wu D, Kwan P, *et al.* (2009) Curcumin inhibits adipogenesis in 3T3-L1 adipocytes and angiogenesis and obesity in C57/BL mice. *J Nutr* **139**: 919–925.
45. Aggarwal BB. (2010) Targeting inflammation-induced obesity and metabolic diseases by curcumin and other nutraceuticals. *Annu Rev Nutr* **30**: 173–199.
46. Bays HE, Laferrere B, Dixon J, *et al.* (2009) Adiposopathy and bariatric surgery: is "sick fat" a surgical disease? *Int J Clin Pract* **63**: 1285–1300.
47. Freise C, Erben U, Neuman U, *et al.* (2010) An active extract of *Lindera obtusiloba* inhibits adipogenesis via sustained Wnt signaling and exerts anti-inflammatory effects in the 3T3-L1 preadipocytes. *J Nutr Biochem* **21**: 1170–1177.

48. Sohle J, Knott A, Holtzmann U, *et al.* (2009) White tea extract induces lipolytic activity and inhibits adipogenesis in human subcutaneous (pre)-adipocytes. *Nutr Metab (Lond)* **6**: 20.
49. Sakurai T, Kitadate K, Nishioka H, *et al.* (2010) Oligomerized grape seed polyphenols attenuate inflammatory changes due to antioxidative properties in coculture of adipocytes and macrophages. *J Nutr Biochem* **21**: 47–54.
50. Dang ZC. (2009) Dose-dependent effects of soy phyto-oestrogen genistein on adipocytes: mechanisms of action. *Obes Rev* **10**: 342–349.
51. Sites CK, Cooper BC, Toth MJ, *et al.* (2007) Effect of a daily supplement of soy protein on body composition and insulin secretion in postmenopausal women. *Fertil Steril* **88**: 1609–1617.
52. Messina M. (2010) Soybean isoflavone exposure does not have feminizing effects on men: a critical examination of the clinical evidence. *Fertil Steril* **93**: 2095–2104.
53. Cederroth CR, Nef S. (2009) Soy, phytoestrogens and metabolism: a review. *Mol Cell Endocrinol* **304**: 30–42.
54. Bays H, Stein EA. (2003) Pharmacotherapy for dyslipidaemia — current therapies and future agents. *Expert Opin Pharmacother* **4**: 1901–1938.
55. Theuwissen E, Plat J, van der Kallen CJ, *et al.* (2009) Plant stanol supplementation decreases serum triacylglycerols in subjects with overt hypertriglyceridemia. *Lipids* **44**: 1131–1140.
56. Muccioli GG, Naslain D, Backhed F, *et al.* (2010) The endocannabinoid system links gut microbiota to adipogenesis. *Mol Syst Biol* **6**: 392.
57. Musso G, Gambino R, Cassader M. (2010) Obesity, diabetes, and gut microbiota: the hygiene hypothesis expanded? *Diabetes Care* **33**: 2277–2284.
58. Morales P, Brignardello J, Gotteland M. (2010) [The association of intestinal microbiota with obesity.] *Rev Med Chil* **138**: 1020–1027.
59. Niu CS, Yeh CH, Yeh MF, *et al.* (2009) Increase of adipogenesis by ginsenoside (Rh2) in 3T3-L1 cell via an activation of glucocorticoid receptor. *Horm Metab Res* **41**: 271–276.
60. Shang W, Yang Y, Jiang B, *et al.* (2007) Ginsenoside Rb1 promotes adipogenesis in 3T3-L1 cells by enhancing PPARgamma2 and C/EBPalpha gene expression. *Life Sci* **80**: 618–625.
61. Uemura T, Hirai S, Mizoguchi N, *et al.* (2010) Diosgenin present in fenugreek improves glucose metabolism by promoting adipocyte differentiation and inhibiting inflammation in adipose tissues. *Mol Nutr Food Res* **54**: 1596–1608.
62. Lee HJ, Lee YH, Park SK, *et al.* (2009) Korean red ginseng (*panax ginseng*) improves insulin sensitivity and attenuates the development of diabetes in Otsuka Long-Evans Tokushima fatty rats. *Metabolism* **58**: 1170–1177.

63. Shi J, Arunasalam K, Yeung D, *et al.* (2004) Saponins from edible legumes: chemistry, processing, and health benefits. *J Med Food* **7**: 67–78.
64. Hirai S, Uemura T, Mizoguchi N, *et al.* (2010) Diosgenin attenuates inflammatory changes in the interaction between adipocytes and macrophages. *Mol Nutr Food Res* **54**: 797–804.
65. Lee JH, Kim KA, Kwon KB, *et al.* (2007) Diallyl disulfide accelerates adipogenesis in 3T3-L1 cells. *Int J Mol Med* **20**: 59–64.
66. Ambati S, Yang JY, Rayalam S, *et al.* (2009) Ajoene exerts potent effects in 3T3-L1 adipocytes by inhibiting adipogenesis and inducing apoptosis. *Phytother Res* **23**: 513–518.
67. Madkor HR, Mansour SW, Ramadan G. (2011) Modulatory effects of garlic, ginger, turmeric and their mixture on hyperglycaemia, dyslipidaemia and oxidative stress in streptozotocin-nicotinamide diabetic rats. *Br J Nutr* **105**(8): 1210–1217 Dec 10.
68. Kook S, Kim GH, Choi K. (2009) The antidiabetic effect of onion and garlic in experimental diabetic rats: meta-analysis. *J Med Food* **12**: 552–560.
69. Pinent M, Blade MC, Salvado MJ, *et al.* (2005) Intracellular mediators of procyanidin-induced lipolysis in 3T3-L1 adipocytes. *J Agric Food Chem* **53**: 262–266.
70. Pinent M, Blade MC, Salvado MJ, *et al.* (2005) Grape-seed derived procyanidins interfere with adipogenesis of 3T3-L1 cells at the onset of differentiation. *Int J Obes (Lond)* **29**: 934–941.
71. Muthusamy VS, Anand S, Sangeetha KN, *et al.* (2008) Tannins present in *Cichorium intybus* enhance glucose uptake and inhibit adipogenesis in 3T3-L1 adipocytes through PTP1B inhibition. *Chem Biol Interact* **174**: 69–78.
72. Tanida M, Tsuruoka N, Shen J, *et al.* (2009) Effects of flavangenol on autonomic nerve activities and dietary body weight gain in rats. *Biosci Biotechnol Biochem* **73**: 2374–2378.
73. Bays HE, Gonzalez-Campoy JM, Bray GA, *et al.* (2008) Pathogenic potential of adipose tissue and metabolic consequences of adipocyte hypertrophy and increased visceral adiposity. *Expert Rev Cardiovasc Ther* **6**: 343–368.
74. Haaz S, Fontaine KR, Cutter G, *et al.* (2006) *Citrus aurantium* and synephrine alkaloids in the treatment of overweight and obesity: an update. *Obes Rev* **7**: 79–88.
75. Liu L, Shan S, Zhang K, *et al.* (2008) Naringenin and hesperetin, two flavonoids derived from *Citrus aurantium*, up-regulate transcription of adiponectin. *Phytother Res* **22**: 1400–1403.
76. Fugh-Berman A, Myers A. (2004) *Citrus aurantium*, an ingredient of dietary supplements marketed for weight loss: current status of clinical and basic research. *Exp Biol Med (Maywood)* **229**: 698–704.

77. Lee MS, Lee HJ, Lee HS, *et al.* (2006) L-carnitine stimulates lipolysis via induction of the lipolytic gene expression and suppression of the adipogenic gene expression in 3T3-L1 adipocytes. *J Med Food* **9**: 468–473.
78. Karlic H, Lohninger A. (2004) Supplementation of L-carnitine in athletes: does it make sense? *Nutrition* **20**: 709–715.
79. Villani RG, Gannon J, Self M, *et al.* (2000) L-carnitine supplementation combined with aerobic training does not promote weight loss in moderately obese women. *Int J Sport Nutr Exerc Metab* **10**: 199–207.
80. Siegner R, Heuser S, Holtzmann U, *et al.* (2010) Lotus leaf extract and L-carnitine influence different processes during the adipocyte life cycle. *Nutr Metab (Lond)* **7**: 66–
81. Saper RB, Eisenberg DM, Phillips RS. (2004) Common dietary supplements for weight loss. *Am Fam Physician* **70**: 1731–1738.
82. Kumar SG, Rahman MA, Lee SH, *et al.* (2009) Plasma proteome analysis for anti-obesity and anti-diabetic potentials of chitosan oligosaccharides in ob/ob mice. *Proteomics* **9**: 2149–2162.
83. Cho EJ, Rahman MA, Kim SW, *et al.* (2008) Chitosan oligosaccharides inhibit adipogenesis in 3T3-L1 adipocytes. *J Microbiol Biotechnol* **18**: 80–87.
84. Shigematsu N, Asano R, Shimosaka M, *et al.* (2001) Effect of long term administration with *Gymnema sylvestre* R. BR on plasma and liver lipid in rats. *Biol Pharm Bull* **24**: 643–649.
85. Egras AM, Hamilton WR, Lenz TL, *et al.* (2011) An evidence-based review of fat modifying supplemental weight loss products. *J Obes* **2011**:
86. Saper RB, Eisenberg DM, Phillips RS. (2004) Common dietary supplements for weight loss. *Am Fam Physician* **70**: 1731–1738.
87. Karst H, Steiniger J, Noack R, *et al.* (1984) Diet-induced thermogenesis in man: thermic effects of single proteins, carbohydrates and fats depending on their energy amount. *Ann Nutr Metab* **28**: 245–252.
88. Zhu CF, Li GZ, Peng HB, *et al.* (2010) Effect of marine collagen peptides on markers of metabolic nuclear receptors in type 2 diabetic patients with/without hypertension. *Biomed Environ Sci* **23**: 113–120.
89. Bello AE, Oesser S. (2006) Collagen hydrolysate for the treatment of osteoarthritis and other joint disorders: a review of the literature. *Curr Med Res Opin* **22**: 2221–2232.
90. Zhu CF, Li GZ, Peng HB, *et al.* (2010) Treatment with marine collagen peptides modulates glucose and lipid metabolism in Chinese patients with type 2 diabetes mellitus. *Appl Physiol Nutr Metab* **35**: 797–804.
91. Noto A, Zahradka P, Yurkova N, *et al.* (2007) Dietary conjugated linoleic acid decreases adipocyte size and favorably modifies adipokine status and insulin sensitivity in obese, insulin-resistant rats. *Metabolism* **56**: 1601–1611.

92. Whigham LD, Watras AC, Schoeller DA. (2007) Efficacy of conjugated linoleic acid for reducing fat mass: a meta-analysis in humans. *Am J Clin Nutr* **85**: 1203–1211.
93. Onakpoya I, Hung SK, Perry R, *et al.* (2011) The use of garcinia extract (hydroxycitric acid) as a weight loss supplement: a systematic review and meta-analysis of randomised clinical trials. *J Obes* **2011**: Epub 2010 Dec 14.
94. Brandt K, Langhans W, Geary N, *et al.* (2006) Beneficial and deleterious effects of hydroxycitrate in rats fed a high-fructose diet. *Nutrition* **22**: 905–912.
95. Stohs SJ, Preuss HG, Ohia SE, *et al.* (2009) No evidence demonstrating hepatotoxicity associated with hydroxycitric acid. *World J Gastroenterol* **15**: 4087–4089.
96. Kristensen M, Jensen MG. (2011) Dietary fibres in the regulation of appetite and food intake. Importance of viscosity. *Appetite* **56**(1): 65–70. Epub 2010 Nov 27.
97. Lyon MR, Reichert RG. (2010) The effect of a novel viscous polysaccharide along with lifestyle changes on short-term weight loss and associated risk factors in overweight and obese adults: an observational retrospective clinical program analysis. *Altern Med Rev* **15**: 68–75.
98. Butt MS, Shahzadi N, Sharif MK, *et al.* (2007) Guar gum: a miracle therapy for hypercholesterolemia, hyperglycemia and obesity. *Crit Rev Food Sci Nutr* **47**: 389–396.
99. Saper RB, Eisenberg DM, Phillips RS. (2004) Common dietary supplements for weight loss. *Am Fam Physician* **70**: 1731–1738.
100. Cornick CL, Strongitharm BH, Sassano G, *et al.* (2009) Identification of a novel agonist of peroxisome proliferator–activated receptors alpha and gamma that may contribute to the anti-diabetic activity of guggulipid in Lep(ob)/Lep(ob) mice. *J Nutr Biochem* **20**: 806–815.
101. Szapary PO, Wolfe ML, Bloedon LT, *et al.* (2003) Guggulipid for the treatment of hypercholesterolemia: a randomized controlled trial. *JAMA* **290**: 765–772.
102. Vermaak I, Hamman JH, Viljoen AM. (2011) Hoodia gordonii: An up-to-date review of a commercially important anti-obesity plant. *Planta Med* **77**(11): 1149–1160. Epub 2011 Jan 21.
103. MacLean DB, Luo LG. (2004) Increased ATP content/production in the hypothalamus may be a signal for energy-sensing of satiety: studies of the anorectic mechanism of a plant steroidal glycoside. *Brain Res* **1020**: 1–11.
104. Whelan AM, Jurgens TM, Szeto V. (2010) Case report. Efficacy of Hoodia for weight loss: is there evidence to support the efficacy claims? *J Clin Pharm Ther* **35**: 609–612.

105. Oben JE, Ngondi JL, Blum K. (2008) Inhibition of *Irvingia gabonensis* seed extract (OB131) on adipogenesis as mediated via down-regulation of the PPARgamma and leptin genes and up-regulation of the adiponectin gene. *Lipids Health Dis* **7**: 44.
106. Ngondi JL, Etoundi BC, Nyangono CB, *et al.* (2009) IGOB131, a novel seed extract of the West African plant *Irvingia gabonensis*, significantly reduces body weight and improves metabolic parameters in overweight humans in a randomized double-blind placebo controlled investigation. *Lipids Health Dis* **8**: 7.
107. Koh-Banerjee PK, Ferreira MP, Greenwood M, *et al.* (2005) Effects of calcium pyruvate supplementation during training on body composition, exercise capacity, and metabolic responses to exercise. *Nutrition* **21**: 312–319.
108. Saper RB, Eisenberg DM, Phillips RS. (2004) Common dietary supplements for weight loss. *Am Fam Physician* **70**: 1731–1738.
109. Bays HE, Laferrere B, Dixon J, *et al.* (2009) Adiposopathy and bariatric surgery: is "sick fat" a surgical disease? *Int J Clin Pract* **63**: 1285–1300.
110. Chung M, Balk EM, Brendel M, *et al.* (2009) Vitamin D and calcium: a systematic review of health outcomes. *Evid Rep Technol Assess (Full Rep)* **183**: 1–420.
111. Zemel MB, Thompson W, Milstead A, *et al.* (2004) Calcium and dairy acceleration of weight and fat loss during energy restriction in obese adults. *Obes Res* **12**: 582–590.
112. Zemel MB, Richards J, Mathis S, *et al.* (2005) Dairy augmentation of total and central fat loss in obese subjects. *Int J Obes (Lond)* **29**: 391–397.
113. Shi H, Dirienzo D, Zemel MB. (2001) Effects of dietary calcium on adipocyte lipid metabolism and body weight regulation in energy-restricted aP2-agouti transgenic mice. *FASEB J* **15**: 291–293.
114. Bolland MJ, Avenell A, Baron JA, *et al.* (2010) Effect of calcium supplements on risk of myocardial infarction and cardiovascular events: meta-analysis. *BMJ* **341**: c3691.
115. Jackson G. (2010) Never mind the content, measure the impact: spin bowling for journal impact factors vs. the importance of patient impact. *Int J Clin Pract* **64**: 1461–1462.
116. Shindea UA, Sharma G, Xu YJ, *et al.* (2004) Insulin sensitising action of chromium picolinate in various experimental models of diabetes mellitus. *J Trace Elem Med Biol* **18**: 23–32.
117. Martin J, Wang ZQ, Zhang XH *et al.* (2006) Chromium picolinate supplementation attenuates body weight gain and increases insulin sensitivity in subjects with type 2 diabetes. *Diabetes Care* **29**: 1826–1832.

118. Baly DL, Schneiderman JS, Garcia-Welsh AL. (1990) Effect of manganese deficiency on insulin binding, glucose transport and metabolism in rat adipocytes. *J Nutr* **120**: 1075–1079.
119. Pei H, Yao Y, Yang Y, *et al.* (2011) Kruppel-like factor KLF9 regulates PPARgamma transactivation at the middle stage of adipogenesis. *Cell Death Differ* **18**: 315–327.
120. Russell ST, Tisdale MJ. (2011) Studies on the anti-obesity activity of zinc-alpha(2)-glycoprotein in the rat. *Int J Obes (Lond)* **35**(5): 658–665. Epub 2010 Sep 21.
121. Smidt K, Pedersen SB, Brock B, *et al.* (2007) Zinc-transporter genes in human visceral and subcutaneous adipocytes: lean versus obese. *Mol Cell Endocrinol* **264**: 68–73.
122. Jing MY, Sun JY, Wang JF. (2008) The effect of peripheral administration of zinc on food intake in rats fed Zn-adequate or Zn-deficient diets. *Biol Trace Elem Res* **124**: 144–156.
123. Bays HE, Laferrere B, Dixon J, *et al.* (2009) Adiposopathy and bariatric surgery: is "sick fat" a surgical disease? *Int J Clin Pract* **63**: 1285–1300.
124. Parikh SJ, Edelman M, Uwaifo GI, *et al.* (2004) The relationship between obesity and serum 1,25-dihydroxy vitamin D concentrations in healthy adults. *J Clin Endocrinol Metab* **89**: 1196–1199.
125. Wortsman J, Matsuoka LY, Chen TC, *et al.* (2000) Decreased bioavailability of vitamin D in obesity. *Am J Clin Nutr* **72**: 690–693.
126. Parikh SJ, Edelman M, Uwaifo GI, *et al.* (2004) The relationship between obesity and serum 1,25-dihydroxy vitamin D concentrations in healthy adults. *J Clin Endocrinol Metab* **89**: 1196–1199.
127. Kong J, Li YC. (2006) Molecular mechanism of 1,25-dihydroxyvitamin D3 inhibition of adipogenesis in 3T3-L1 cells. *Am J Physiol Endocrinol Metab* **290**: E916–E924.
128. Devaraj S, Jialal G, Cook T, *et al.* (2011) Low vitamin D levels in Northern American adults with the metabolic syndrome. *Horm Metab Res* **43**: 72–74.
129. Pilz S, Tomaschitz A. (2010) Role of vitamin D in arterial hypertension. *Expert Rev Cardiovasc Ther* **8**: 1599–1608.
130. Reddy VS, Good M, Howard PA, *et al.* (2010) Role of vitamin D in cardiovascular health. *Am J Cardiol* **106**: 798–805.
131. Ayturk S, Gursoy A, Bascil TN, *et al.* (2006) Changes in insulin sensitivity and glucose and bone metabolism over time in patients with asymptomatic primary hyperparathyroidism. *J Clin Endocrinol Metab* **91**: 4260–4263.
132. Chan SP-D, Soares M. (2011) Diet-induced thermogenesis, fat oxidation and food intake following sequential meals: influence of calcium and vitamin D. *Clin Nutr* **30**(3): 376–383.

133. Zemel MB. (2005) The role of dairy foods in weight management. *J Am Coll Nutr* **24**: 537S–546S.
134. Zemel MB, Richards J, Mathis S, *et al.* (2005) Dairy augmentation of total and central fat loss in obese subjects. *Int J Obes (Lond)* **29**: 391–397.
135. Nikooyeh B, Neyestani TR, Farvid M, *et al.* (2011) Daily consumption of vitamin D– or vitamin D+ calcium-fortified yogurt drink improved glycemic control in patients with type 2 diabetes: a randomized clinical trial. *Am J Clin Nutr* **93**(4): 764–771. Epub 2011 Feb 2.
136. Grant WB, Juzeniene A, Moan JE. (2011) Review Article: Health benefit of increased serum 25(OH)D levels from oral intake and ultraviolet-B irradiance in the Nordic countries. *Scand J Public Health* **39**: 70–78.
137. Bays H. (2006) The melanocortin system as a therapeutic treatment target for adiposity and adiposopathy. *Drugs R D* **7**: 289–302.
138. Bays HE. (2004) Current and investigational antiobesity agents and obesity therapeutic treatment targets. *Obes Res* **12**: 1197–1211.
139. Suzuki I, Kato T, Motokawa T, *et al.* (2002) Increase of pro-opiomelanocortin mRNA prior to tyrosinase, tyrosinase-related protein 1, dopachrome tautomerase, Pmel-17/gp100, and P-protein mRNA in human skin after ultraviolet B irradiation. *J Invest Dermatol* **118**: 73–78.
140. Schiller M, Brzoska T, Bohm M, *et al.* (2004) Solar-simulated ultraviolet radiation–induced upregulation of the melanocortin-1 receptor, proopiomelanocortin, and alpha-melanocyte-stimulating hormone in human epidermis *in vivo*. *J Invest Dermatol* **122**: 468–476.
141. Schulz C, Paulus K, Lobmann R, *et al.* (2010) Endogenous ACTH, not only alpha-melanocyte-stimulating hormone, reduces food intake mediated by hypothalamic mechanisms. *Am J Physiol Endocrinol Metab* **298**: E237–E244.
142. Moussa NM, Claycombe KJ. (1999) The yellow mouse obesity syndrome and mechanisms of agouti-induced obesity. *Obes Res* **7**: 506–514.
143. Mead MN. (2008) Benefits of sunlight: a bright spot for human health. *Environ Health Perspect* **116**: A160–A167.
144. Mead MN. (2008) Benefits of sunlight: a bright spot for human health. *Environ Health Perspect* **116**: A160–A167.
145. Bartelt RN, Altmeyer P, Stohr L, *et al.* (1985) [Endocrinological reactions following UV A whole body irradiation.] *Derm Beruf Umwelt* **33**: 50–55.
146. Kim JH, Kiefer LL, Woychik RP, *et al.* (1997) Agouti regulation of intracellular calcium: role of melanocortin receptors. *Am J Physiol* **272**: E379–E384.
147. Kim S, Moustaid-Moussa N. (2000) Secretory, endocrine and autocrine/paracrine function of the adipocyte. *J Nutr* **130**: 3110S–3115S.
148. Zemel MB. (2003) Mechanisms of dairy modulation of adiposity. *J Nutr* **133**: 252S–256S.

149. Richard D, Carpentier AC, Dore G, *et al.* (2010) Determinants of brown adipocyte development and thermogenesis. *Int J Obes (Lond)* **34**(Suppl 2): S59–S66.
150. Brito MN, Brito NA, Baro DJ, *et al.* (2007) Differential activation of the sympathetic innervation of adipose tissues by melanocortin receptor stimulation. *Endocrinology* **148**: 5339–5347.
151. Silva JE. (2011) Physiological importance and control of non-shivering facultative thermogenesis. *Front Biosci (Schol Ed)* **3**: 352–371.
152. Plasqui G, Westerterp KR. (2004) Seasonal variation in total energy expenditure and physical activity in Dutch young adults. *Obes Res* **12**: 688–694.
153. Tokizawa K, Yasuhara S, Nakamura M, *et al.* (2010) Mild hypohydration induced by exercise in the heat attenuates autonomic thermoregulatory responses to the heat, but not thermal pleasantness in humans. *Physiol Behav* **100**: 340–345.
154. Layden JD, Patterson MJ, Nimmo MA. (2002) Effects of reduced ambient temperature on fat utilization during submaximal exercise. *Med Sci Sports Exerc* **34**: 774–779.
155. Mead MN. (2008) Benefits of sunlight: a bright spot for human health. *Environ Health Perspect* **116**: A160–A167.
156. Lakka TA, Laaksonen DE, Lakka HM, *et al.* (2003) Sedentary lifestyle, poor cardiorespiratory fitness, and the metabolic syndrome. *Med Sci Sports Exerc* **35**: 1279–1286.
157. Flatt JP. (2007) Differences in basal energy expenditure and obesity. *Obesity (Silver Spring)* **15**: 2546–2548.
158. Wang Z, Ying Z, Bosy-Westphal A, *et al.* (2010) Evaluation of specific metabolic rates of major organs and tissues: comparison between men and women. *Am J Hum Biol* **23**(3): 333–338. Epub 2010 Dec 22.
159. Stiegler P, Cunliffe A. (2006) The role of diet and exercise for the maintenance of fat-free mass and resting metabolic rate during weight loss. *Sports Med* **36**: 239–262.
160. Fett CA, Fett WC, Marchini JS. (2009) Circuit weight training vs. jogging in metabolic risk factors of overweight/obese women. *Arq Bras Cardiol* **93**: 519–525.
161. Morgan B, Woodruff SJ, Tiidus PM. (2003) Aerobic energy expenditure during recreational weight training in females and males. *J Sports Sci Med* **2**: 117–122.
162. Tresierras MA, Balady GJ. (2009) Resistance training in the treatment of diabetes and obesity: mechanisms and outcomes. *J Cardiopulm Rehabil Prev* **29**: 67–75.

163. Hunter GR, Brock DW, Byrne NM, *et al.* (2010) Exercise training prevents regain of visceral fat for 1 year following weight loss. *Obesity (Silver Spring)* **18**: 690–695.
164. Weinsier RL, Nagy TR, Hunter GR, *et al.* (2000) Do adaptive changes in metabolic rate favor weight regain in weight-reduced individuals? An examination of the set-point theory. *Am J Clin Nutr* **72**: 1088–1094.
165. Rosenbaum M, Hirsch J, Gallagher DA, *et al.* (2008) Long-term persistence of adaptive thermogenesis in subjects who have maintained a reduced body weight. *Am J Clin Nutr* **88**: 906–912.
166. Goldsmith R, Joanisse DR, Gallagher D, *et al.* (2010) Effects of experimental weight perturbation on skeletal muscle work efficiency, fuel utilization, and biochemistry in human subjects. *Am J Physiol Regul Integr Comp Physiol* **298**: R79–R88.
167. Rosenbaum M, Goldsmith R, Bloomfield D, *et al.* (2005) Low-dose leptin reverses skeletal muscle, autonomic, and neuroendocrine adaptations to maintenance of reduced weight. *J Clin Invest* **115**: 3579–3586.
168. Flores MB, Fernandes MF, Ropelle ER, *et al.* (2006) Exercise improves insulin and leptin sensitivity in hypothalamus of Wistar rats. *Diabetes* **55**: 2554–2561.
169. Bjorbaek C. (2009) Central leptin receptor action and resistance in obesity. *J Investig Med* **57**: 789–794.
170. Bays H, Blonde L, Rosenson R. (2006) Adiposopathy: How do diet, exercise, weight loss and drug therapies improve metabolic disease? *Expert Rev Cardiovasc Ther* **4**: 871–895.
171. Kay SJ, Fiatarone Singh MA. (2006) The influence of physical activity on abdominal fat: a systematic review of the literature. *Obes Rev* **7**: 183–200.
172. Kopelman PG. (1997) The effects of weight loss treatments on upper and lower body fat. *Int J Obes Relat Metab Disord* **21**: 619–625.
173. O'Leary VB, Marchetti CM, Krishnan RK, *et al.* (2006) Exercise-induced reversal of insulin resistance in obese elderly is associated with reduced visceral fat. *J Appl Physiol* **100**: 1584–1589.
174. Ross R, Janssen I, Dawson J. *et al.* (2004) Exercise-induced reduction in obesity and insulin resistance in women: a randomized controlled trial. *Obes Res* **12**: 789–798.
175. Miyazaki S, Izawa T, Ogasawara JE, *et al.* (2010) Effect of exercise training on adipocyte-size-dependent expression of leptin and adiponectin. *Life Sci* **86**: 691–698.
176. Bays HE, Gonzalez-Campoy JM, Bray GA, *et al.* (2008) Pathogenic potential of adipose tissue and metabolic consequences of adipocyte hypertrophy and increased visceral adiposity. *Expert Rev Cardiovasc Ther* **6**: 343–368.

177. Bays HE, Fox KM, Grandy S. (2010) Anthropometric measurements and diabetes mellitus: clues to the "pathogenic" and "protective" potential of adipose tissue. *Metab Syndr Relat Disord* **8**: 307–315.
178. Thampy KG. (1995) Hypercholesterolaemia of prolonged fasting and cholesterol lowering of re-feeding in lean human subjects. *Scand J Clin Lab Invest* **55**: 351–357.
179. Hoeks J, van Herpen NA, Mensink M, *et al.* (2010) Prolonged fasting identifies skeletal muscle mitochondrial dysfunction as consequence rather than cause of human insulin resistance. *Diabetes* **59**: 2117–2125.
180. Karczewska-Kupczewska M, Straczkowski M, Adamska A, *et al.* (2010) Insulin sensitivity, metabolic flexibility, and serum adiponectin concentration in women with anorexia nervosa. *Metabolism* **59**: 473–477.
181. Karelis AD, St-Pierre DH, Conus F, *et al.* (2004) Metabolic and body composition factors in subgroups of obesity: what do we know? *J Clin Endocrinol Metab* **89**: 2569–2575.
182. Bays HE, Fox KM, Grandy S. (2010) Anthropometric measurements and diabetes mellitus: clues to the "pathogenic" and "protective" potential of adipose tissue. *Metab Syndr Relat Disord* **8**: 307–315.
183. Bays HE, Gonzalez-Campoy JM, Bray GA, *et al.* (2008) Pathogenic potential of adipose tissue and metabolic consequences of adipocyte hypertrophy and increased visceral adiposity. *Expert Rev Cardiovasc Ther* **6**: 343–368.
184. St-Onge MP. (2010) Are normal-weight Americans over-fat? *Obesity (Silver Spring)* **18**: 2067–2068.
185. Bazzarre TL, Kleiner SM, Litchford MD. (1990) Nutrient intake, body fat, and lipid profiles of competitive male and female bodybuilders. *J Am Coll Nutr* **9**: 136–142.
186. Brooks N, Layne JE, Gordon PL, *et al.* (2007) Strength training improves muscle quality and insulin sensitivity in Hispanic older adults with type 2 diabetes. *Int J Med Sci* **4**: 19–27.
187. Stone MH, Fleck SJ, Triplett NT, *et al.* (1991) Health- and performance-related potential of resistance training. *Sports Med* **11**: 210–231.
188. Lambert CP, Frank LL, Evans WJ. (2004) Macronutrient considerations for the sport of bodybuilding. *Sports Med* **34**: 317–327.
189. Bazzarre TL, Kleiner SM, Litchford MD. (1990) Nutrient intake, body fat, and lipid profiles of competitive male and female bodybuilders. *J Am Coll Nutr* **9**: 136–142.
190. Steen SN. (1991) Precontest strategies of a male bodybuilder. *Int J Sport Nutr* **1**: 69–78.

191. Kleiner SM, Bazzarre TL, Litchford MD. (1990) Metabolic profiles, diet, and health practices of championship male and female bodybuilders. *J Am Diet Assoc* **90**: 962–967.
192. Ishii T, Yamakita T, Sato T, *et al.* (1998) Resistance training improves insulin sensitivity in NIDDM subjects without altering maximal oxygen uptake. *Diabetes Care* **21**: 1353–1355.
193. Bays HE, Gonzalez-Campoy JM, Bray GA, *et al.* (2008) Pathogenic potential of adipose tissue and metabolic consequences of adipocyte hypertrophy and increased visceral adiposity. *Expert Rev Cardiovasc Ther* **6**: 343–368.
194. Tchoukalova YD, Koutsari C, Karpyak MV, *et al.* (2008) Subcutaneous adipocyte size and body fat distribution. *Am J Clin Nutr* **87**: 56–63.
195. McCarty MF. (2003) A paradox resolved: the postprandial model of insulin resistance explains why gynoid adiposity appears to be protective. *Med Hypotheses* **61**: 173–176.
196. Chen K, Xie Y, Hu P, *et al.* (2010) Multiple symmetric lipomatosis: substantial subcutaneous adipose tissue accumulation did not induce glucose and lipid metabolism dysfunction. *Ann Nutr Metab* **57**: 68–73.
197. Bays HE, Gonzalez-Campoy JM, Bray GA, *et al.* (2008) Pathogenic potential of adipose tissue and metabolic consequences of adipocyte hypertrophy and increased visceral adiposity. *Expert Rev Cardiovasc Ther* **6**: 343–368.
198. Gavrilova O, Marcus-Samuels B, Graham D, *et al.* (2000) Surgical implantation of adipose tissue reverses diabetes in lipoatrophic mice. *J Clin Invest* **105**: 271–278.
199. Mohammed BS, Cohen S, Reeds D, *et al.* (2008) Long-term effects of large-volume liposuction on metabolic risk factors for coronary heart disease. *Obesity (Silver Spring)* **16**(12): 2648–2651. Epub 2008 Sep 25.
200. Villarroya F, Domingo P, Giralt M. (2009) Drug-induced lipotoxicity: lipodystrophy associated with HIV-1 infection and antiretroviral treatment. *Biochim Biophys Acta* **1801**(3): 392–399. Epub 2009 Sep 30.
201. Lincoff AM, Wolski K, Nicholls SJ, *et al.* (2007) Pioglitazone and risk of cardiovascular events in patients with type 2 diabetes mellitus: a meta-analysis of randomized trials. *JAMA* **298**: 1180–1188.
202. Bays H, Rodbard HW, Schorr AB, *et al.* (2007) Adiposopathy: treating pathogenic adipose tissue to reduce cardiovascular disease risk. *Curr Treat Options Cardiovasc Med* **9**: 259–271.
203. Bays H. (2008) Rationale for prescription omega-3-acid ethyl ester therapy for hypertriglyceridemia: a primer for clinicians. *Drugs of Today* **44**: 205–246.

204. Bays H. (2010) Fish oils in the treatment of dyslipidemia and cardiovascular disease. In: Kwiterovich Jr, PO (ed.), *The Johns Hopkins Textbook of Dyslipidemia.* Lippincott Williams and Wolters Kluwer, Philadelphia.
205. Wang Z, Ying Z, Bosy-Westphal A, *et al.* (2010) Evaluation of specific metabolic rates of major organs and tissues: comparison between men and women. *Am J Hum Biol* **23**(3): 333–338. Epub 2010 Dec 22.
206. Janssen I, Heymsfield SB, Wang ZM, *et al.* (2000) Skeletal muscle mass and distribution in 468 men and women aged 18–88 yr. *J Appl Physiol* **89**: 81–88.

CHAPTER 5

Telomeropathies

Rodrigo T. Calado and Neal S. Young**

INTRODUCTION

Telomeres are the structural ends of linear chromosomes. They serve to protect chromosomes from being recognized by the DNA repair machinery as double-stranded breaks[1] (Fig. 1). When a DNA strand is disrupted, the DNA repair machinery identifies and repairs the damage, by either homologous recombination or nonhomologous end joining. The telomere makes the "natural ends" of linear DNA unrecognizable by the DNA repair machinery, thus preventing end-to-end fusions and translocation of chromosomes.[2]

In humans, telomeres are composed of hundreds to thousands of hexameric TTAGGG nucleotide repeats in the leading strand (CCCTAA in the lagging strand) that are covered by a complex of at least six proteins, collectively termed "shelterin."[3,4] Shelterin proteins bind to the telomeric repeat DNA sequence, serving as a protective shield for the DNA structure. The very extremity of telomeres ends as a 3′ overhang, which folds back and anneals the telomere sequence upstream, thus forming a lariat (T loop).

However, the ends of linear chromosomes pose problems. During cell division, the DNA polymerase is unable to fully duplicate the very end of

*Hematology Branch, National Heart, Lung, and Blood Institute, National Institutes of Health, Bethesda, Maryland, USA. E-mail: calado@nhlbi.nih.gov.

the single DNA molecule, so that a newly synthesized strand is shorter in comparison with the template DNA. With repeated cell divisions, telomeres progressively erode, eventually reaching a critically short length that signals cell proliferation arrest, cell senescence, or apoptosis via the p53 pathway.

Cells with high proliferative capacity, such as embryonic and adult stem cells, germ cells, or activated lymphocytes, maintain the lengths of their telomeres by adding telomeric repeats to the 3′ end of the leading strand. This DNA synthesis is catalyzed by the telomerase reverse transcriptase (TERT), which uses an RNA molecule (TERC) as a template for extension[1] (Fig. 1). TERT is a 1132-amino-acid-long protein that contains telomerase-specific and the reverse transcriptase domains.

Nevertheless, telomerase does not fully prevent telomere erosion with aging. In peripheral blood leukocytes, telomeres are on average

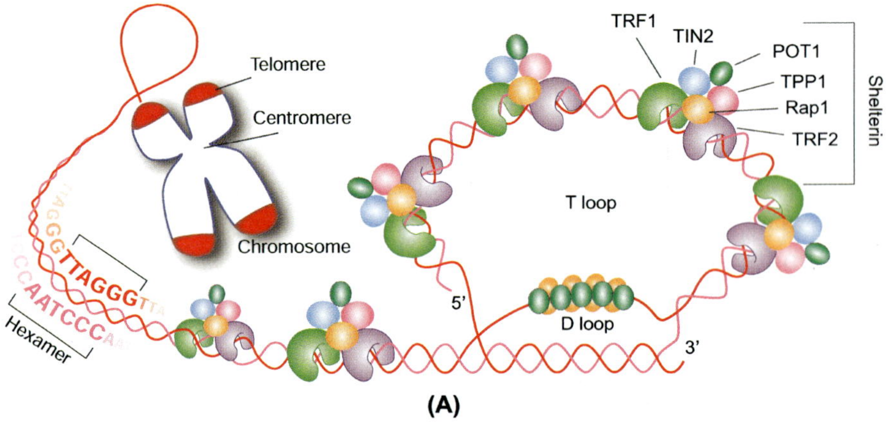

Fig. 1 (**A**) Telomere structure. Telomeres are located at the ends of linear chromosomes and are composed of tandem DNA repeat sequences (TTAGGG in the leading strand and CCCTAA in the lagging, strand). The nucleotide sequence is coated by a complex of protective proteins collectively called shelterin (TRF1, TRF2, TIN2, POT1, TPP1, and RAP1). The telomeric 3′ end terminates as a single strand that folds back, forming a lariat (T loop). (**B**) Telomerase complex. The enzyme telomerase reverse transcriptase (TERT), its RNA component (TERC) and the protein dyskerin (encoded by the *DKC1* gene), and other proteins (NHP2, NOP10, and GAR1) are components of the telomerase complex. TERT uses the CCCUAA template in the RNA component to catalytically add hexamers to the 3′ end of the leading strand.

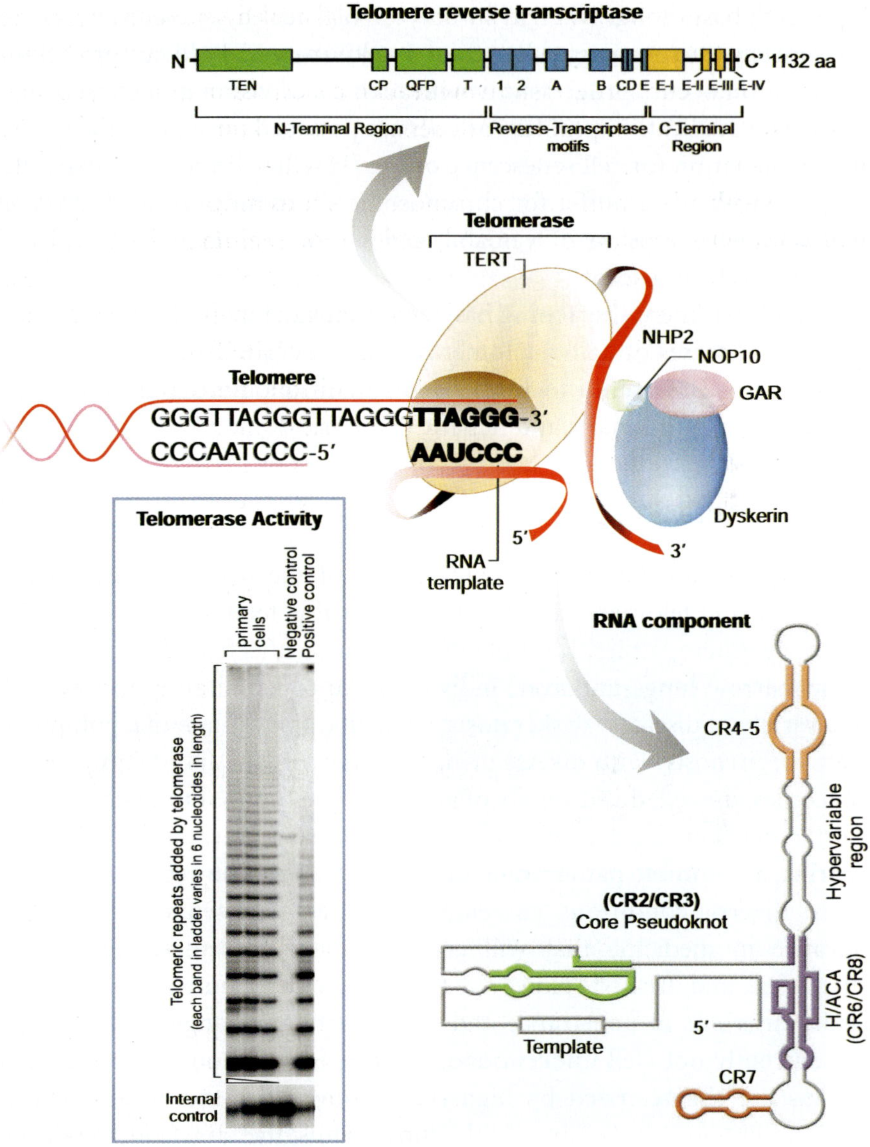

(B)

Fig. 1 (*Continued*)

11–12 kilobases long, whereas in 80-year-old healthy subjects, telomeres are shortened to an average length of 5–6 kilobases. In cell culture, telomere attrition occurs progressively with each cell division as a consequence of the end-replication problem as cells are passaged, providing the molecular explanation for cell senescence or the "Hayflick limit." Telomeres also may be viewed as a buffer for chromosome shortening with cell division, precluding the erosion of valuable genes more centromerically located within the chromosome.

While telomere shortening has been associated with physiologic aging and with diseases of aging, telomerase and shelterin gene mutations that disrupt the cell's ability to properly maintain telomeres are etiologic in hematologic, pulmonary, and hepatic diseases.

TELOMEROPATHIES

Defects in genes that encode components of the telomere maintenance apparatus and telomere capping and the ensuing telomere erosion underlie the telomeropathies. Multiple organs can be affected (the integument, bone marrow, lungs, and liver) individually or concomitantly and manifest as a variety of diseases (dyskeratosis congenita, aplastic anemia, pulmonary fibrosis, cirrhosis) with diverse principal target organs and distinct clinical outcomes. Indeed, this diversity of anatomic sites undoubtedly contributed to the fact that, until very recently, these diseases were not suspected of sharing a common pathophysiologic mechanism. The reclassification of these diverse conditions as telomeropathies represents a conceptual advance in medicine that will greatly impact how these diseases are diagnosed and treated. However, the factors determining the observed wide variations in penetrance and the severity and range of phenotypes are currently not well understoood. In particular, although some affected organs are characterized by high cell turnover (i.e. the bone marrow), other target organs are not (i.e. the lung). Thus, the target organ selectivity of these conditions remains unexplained, posing important questions for future investigations. Below we summarize each of the telomeropathies in turn, with an emphasis on their genetics and clinical manifestations.

Dyskeratosis Congenita

Dyskeratosis congenita is an inherited genodermatosis (i.e. a skin condition of genetic origin) characterized by the classical triad of ungual dystrophy, reticular skin pigmentation, and oral leukoplakia, accompanied by aplastic anemia (Fig. 2). The disease usually appears during infancy and

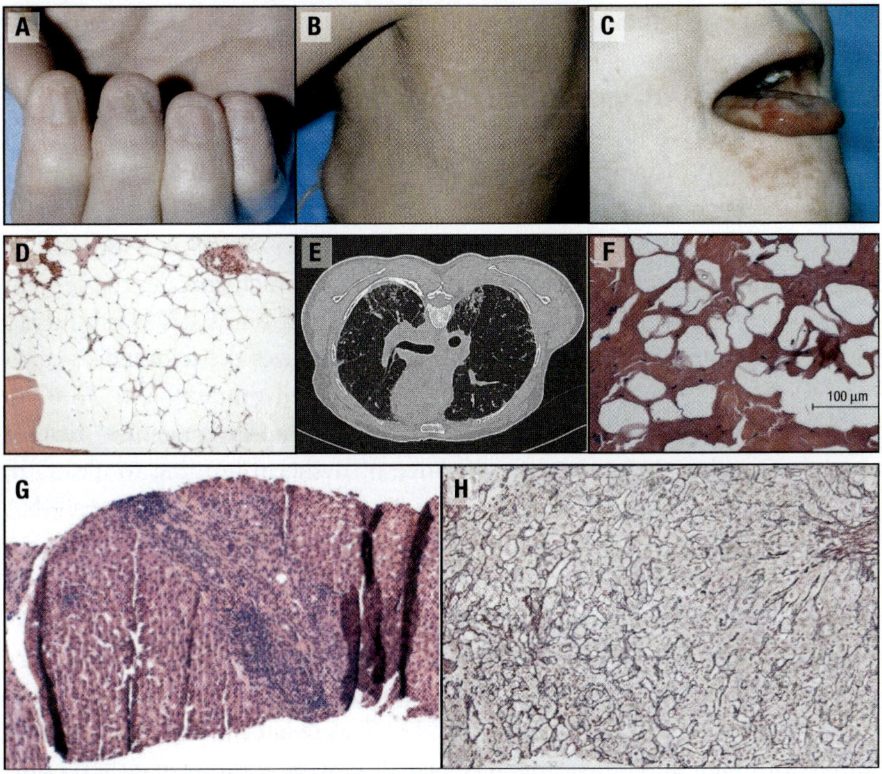

Fig. 2 Telomeropathies. Dyskeratosis congenita is a genodermatosis characterized by a classical triad of (**A**) ungual dystrophy, (**B**) reticular skin hyperpigmentation, and (**C**) oral leukoplakia. (**D**) Aplastic anemia, characterized by bone marrow in which hematopoietic tissue is replaced by fat and fibrotic stroma, is a hematologic manifestation of telomere diseases. (**E**) In the lungs, telomere erosion clinically manifests as idiopathic pulmonary fibrosis, with fibrotic zones alternating with less affected parenchyma. (**F**) Radiologically, it is determined by diffuse fibrosis predominantly in the subpleural regions. (**G**) In the liver, telomere dysfunction results in fibrosis accompanied by marked inflammation. (**H**) Alternatively, nodular regenerative hyperplasia may occur as a consequence of telomere shortening.

early childhood, and mucocutaneous abnormalities typically develop before bone marrow failure appears. Some rare patients present clinical findings of dyskeratosis congenita in their third of fourth decade of life. Along with bone marrow failure, the lungs are affected in approximately 20% of patients characteristically as idiopathic pulmonary fibrosis (see section below), and gastrointestinal disorders, especially liver disease, are observed in 7% of cases. The most common types of liver disease in dyskeratosis congenita are cirrhosis and nodular regenerative hyperplasia (the latter is the most common noncirrhotic cause of portal hypertension).[5] Patients with dyskeratosis congenita also may have esophageal or urethral stricture, short stature, or hypogonadism.

The pattern of inheritance may be either X-linked or autosomal, dominant or recessive. Linkage analysis of extensive pedigrees identified Xq28 as a genetic region associated with X-linked dyskeratosis congenita. Subsequent studies identified the relevant disease gene, *DKC1*, which encodes the protein dyskerin. Dyskerin is a small nucleolar protein that binds RNA; it binds to TERC, the telomerase RNA component, and stabilizes the telomerase enzymatic complex. Further genetic screening identified mutations in *TERC* in some families with autosomal dominant dyskeratosis congenita, further emphasizing the mechanistic significance of dyskerin binding to TERC in this disease. Dyskerin was initially thought to be mediate disease via RNA processing and pseudouridylation, but subsequent evidence involving other components of the telomerase complex and telomere-capping proteins in the etiology of the disease demonstrated that dyskerin deficiency leads to dyskeratosis congenita due to abnormal telomere maintenance.

As the link between the disease and lesions in components of the telomerase complex was clear, mutations in *TERT* were later indentified in autosomal dominant and autosomal recessive disease. Mutations in either *TERC* or *TERT* reduce telomerase enzymatic activity which catalyzes telomere elongation by haploinsufficiency, indicating that telomerase expression is tightly regulated and a 50% reduction in enzyme function is enough to cause significant loss and cause disease. In addition, mutations in *NOP10* and *NHP2* in autosomal recessive disease were identified. NOP10 and NHP2 are accessory proteins that give stability to the telomerae complex. In addition, mutations in *TINF2*, which encodes one of the shelterin proteins TIN2, also are etiologic in dyskeratosis congenita (Figs. 1A and 1B).

A peculiar characteristic of autosomal dominant dyskeratosis congenita with mutations in either *TERC* or *TERT* is disease anticipation: symptoms and signs appear at an earlier age and are more severe in successive generations in some pedigrees. Anticipation in dyskeratosis congenita results from the affected subjects' inheriting not only the genetic mutation but also short telomeres directly from the germ cells of an affected parent; telomeres show accelerated attrition in the following generation, because they are shorter in the embryo and telomere repair is impaired during development.

Dyskeratosis patients have a high predisposition to cancer. The risk of acute myeloid leukemia is 200 times that in the normal population. In addition, patients with telomerase mutations may develop myelodysplasia or *de novo* acute myeloid leukemia. Among leukemia cases, there is no particular subtype that correlates with telomerase mutations, but patients usually have abnormal karyotypes, including inv(16), trisomy 8, t(15;17), or complex karyotypes. Squamous cell carcinoma also is common in the disease, especially in the tongue. The rate of head and neck squamous cell carcinoma in patients with dyskeratosis congenita is over 1000 times higher compared to the general population.[6] The frequency of telomerase mutations in patients with squamous cell carcinoma of the oropharynx has not been determined.

There is no consensus as to the diagnostic criteria for dyskeratosis congenital.[7] Since the clinical outcome and treatment options are different for the diverse clinical manifestations, not all patients with short telomeres and telomerase mutations should be classified as having dyskeratosis congenita, which has a notoriously poor prognosis. The diagnosis of dyskeratosis congenita should be reserved for patients with genodermatosis [at least two of the following: (1) ungual dystrophy, (2) leukoplakia, and (3) skin reticular pigmentation] and bone marrow failure, with or without visceral organ involvement.

The measurement of the telomere length of peripheral blood leukocytes is critical in the diagnosis of dyskeratosis congenita. Quantitative polymerase chain reaction (qPCR) measures the average telomere length in the entire leukocyte population and is a useful screening method. Flow cytometry combined with fluorescence *in situ* hybridization (flow–FISH) can determine the average telomere length in individual cells and in cell subpopulations, such as granulocytes, lymphocytes, T cells, and B cells.

Granulocyte telomere length is very sensitive in the diagnosis of dyskeratosis congenita, but it lacks specificity since other bone marrow failure syndromes also show telomere shortening in granulocytes.[8] Lymphocyte telomere length is more specific to telomere diseases, but it does not distinguish between aplastic anemia due to telomerase mutations and dyskeratosis congenita. If telomeres are short, sequencing of telomerase genes (*DKC1*, *TERT*, *TERC*, *TINF2*, *TCAB1*) is warranted. However, no causal genetic lesion can be identified in a significant proportion of dyskeratosis congenita patients, and a mutation is not required for diagnosis. These patients may have mutations in unknown genes involved in telomere maintenance, or epigenetic changes modulating telomerase function. Mutation identification is helpful in genetic counseling and in donor selection for hematopoietic stem cell transplant. Some family members behave as "silent carriers" and, although they may have no clinical manifestation, they are not suitable as donors for hematopoietic stem cell transplant.[9]

The treatment of dyskeratosis congenita has not been systematically investigated. In general, patients are treated for life-threatening manifestations. The main cause of death is aplastic anemia, and the marrow failure can be cured by hematopoietic stem cell transplant if a suitable sibling donor is available. Transplanted patients appear to have an unusually high rate of lung and liver complications. If a donor is not available, androgens may be helpful in approximately 50% of patients. Androgens directly stimulate telomerase activity in hematopoietic stem cells, and they may ameliorate telomere attrition and improve marrow function by this mechanism.[10] In hematopoietic cells, androgens are converted into estrogens via aromatase, bind to estrogen receptor alpha and bind to known estrogen receptor elements in the *TERT* promoter region, thus stimulating transcription of the wild-type allele, protein expression, and telomerase function.

Dyskeratosis Congenita Variants

There are at least two variants of dyskeratosis congenita that are more severe in their manifestations and prognosis. The Hoyeraal–Hreidarsson syndrome, a form of dyskeratosis congenita of early infancy, has in addition to bone marrow failure characteristic neurological deficits, including ataxia due to cerebellar hypoplasia, microcephaly, and other disease manifestations, including immunodeficiency, intrauterine growth retardation, and developmental delay. These patients may not have mucocutaneous features. Hoyeraal–Hreidarsson syndrome is caused by mutations in either *DKC1*.[11] or *TINF2* genes.[12]

Revesz syndrome also appears in early infancy, with physical findings similar to those in Hoyeraal–Hreidarsson syndrome, but with hallmark exudative retinopathy. Patients with Revesz syndrome have thin sparse hair and reticular skin pigmentation. To date, only mutations in *TINF2* have been found in this disease.[12]

Bone Marrow

Aplastic anemia

Histologically, aplastic anemia is characterized by a reduction in the number of hematopoietic progenitors in the bone marrow and replacement by fat cells (Fig. 2). The hypoplastic bone marrow and diminished hematopoiesis results in peripheral blood pancytopenia. Most cases of acquired aplastic anemia are immune-mediated. However, 5%–10% of cases are due to mutations in *TERT* or *TERC*. Aplastic anemia may be the sole clinical manifestation of mutations in *TERT* or *TERC* (9;13), but with *TINF2* or *DKC1* mutations, aplastic anemia is accompanied by integument abnormalities and other visceral organ damage (lungs or liver).

The clinical manifestations of aplastic anemia resulting from telomerase mutations and telomere shortening, which represent approximately 10% of cases, are indistinguishable from those of immune-mediated acquired aplastic anemia (90% of cases). However, some features suggest a telomere disease: pancytopenia that is moderate and often chronic; a family history of aplastic anemia, leukemia, lung, or liver disease; unresponsiveness to immunosuppressive therapy; and relative preservation of the neutrophil counts to levels above 500 cells/μL. Patients and clinically healthy relatives who have the same mutation often have erythrocyte macrocytosis. Telomere length in peripheral blood leukocytes of patients with aplastic anemia and telomerase mutations is short, typically below the 10th percentile for age and sometimes below the 1st percentile (Fig. 3) — very strong indicators of

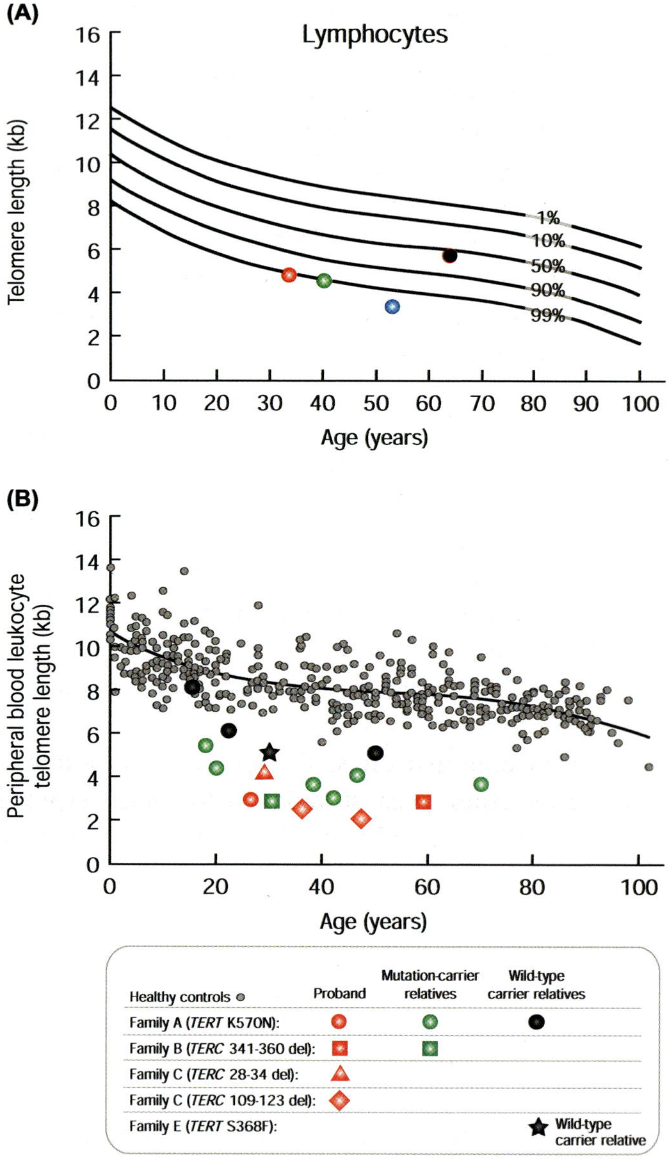

Fig. 3 Telomere length in patients with telomere disease. (**A**) Patients with aplastic anemia and telomerase mutations have short telomeres (reproduced from Ref. 13). (**B**) Patients with telomerase mutations and liver disease also have very short telomeres. Mutation carriers in the absence of clinically evident disease also have very short telomeres (reproduced from Ref. 20).

a telomere disease. As in dyskeratosis congenita, short telomeres in lymphocytes (in comparison with neutrophil telomere length) are more specific of a telomeropathy. In cases where aplastic anemia is the sole clinical manifestation, telomere length measurement by either qPCR or flow–FISH is a useful test for identifying a telomere disease and distinguishing it from acquired immune-mediated aplastic anemia.

Telomerase mutations cause deficient telomerase catalytic function and deficient telomere elongation, thus leading to accelerated telomere attrition and short telomeres.[14] Eroded telomeres activate the p53 and p21 pathways, which induce apoptosis and/or cell senescence, limiting the hematopoietic stem and progenitor cell capacity to maintain adequate hematopoiesis and blood cell production. Patients with telomerase mutations have a profound quantitative deficiency in the number of hematopoietic progenitor and stem cells. Decreased telomerase repair may also qualitiatively affect the ability of the marrow to respond to stress by regeneration. Prosenescent and defective hematopoietic progenitors with short telomeres clinically translate into marrow failure.

Patients with telomerase mutations may not respond to conventional immunosuppressive therapy administered for immune-mediated marrow failure. The major therapeutic options are related hematopoietic stem cell transplantation (providing that the sibling donor is not a mutation carrier) and androgens. Androgens stimulate telomerase expression and activity in hematopoietic cells and may alleviate telomere erosion.[10]

Lungs

Pulmonary fibrosis

The lungs may be the sole organ affected by excessive telomere attrition. The clinical feature of pulmonary telomere disease is idiopathic pulmonary fibrosis, which is the most common type of interstitial lung disease. Telomerase mutations are found in up to 15% of familial idiopathic pulmonary fibrosis and in 3% of sporadic cases.[15,16] Histologically, there is patchy fibrosis of the lungs and interstitial inflammation, normal lung alternating with fibrosis, inflammation, and collagen deposition with fibroblast foci (Fig. 2). High resolution computerized tomography shows interstitial pneumonitis with peripheral and basal fibrosis resulting in a characteristic pattern termed "honeycombing" (not seen in all patients). Enlarged mediastinal lymph nodes also are common. Emphysema is observed in some patients (Fig. 2). Pulmonary function testing shows reduced lung forced vital capacity, forced expiratory volume, and diffusion capacity.

A minority of patients with telomerase mutations may present with granulomatous pulmonary disease or pulmonary fibrosis that is not consistent with idiopathic pulmonary fibrosis.[15] Most patients with telomerase mutations also have an important environmental risk factor; in a large series, 96% of patients had exposure to smoking or other fibrogenic agents.[17] In some patients and pedigrees, idiopathic pulmonary fibrosis associates with liver cirrhosis and/or aplastic anemia.

As in aplastic anemia and dyskeratosis congenita, telomere length measurement in peripheral blood leukocytes contributes to the diagnosis of a telomere disease in idiopathic pulmonary fibrosis. Patients with idiopathic pulmonary fibrosis and a telomerase mutation have significantly shorter telomeres than do patients without a mutation.[18] Telomeres are usually below the 10th percentile and in younger patients below the 1st percentile for age, as in aplastic anemia. In addition, patients with idiopathic pulmonary fibrosis have shorter-for-age telomeres regardless of mutation status, suggesting that telomere attrition is a common molecular mechanism mediating disease.

Although the number of patients followed over time is still small, there is evidence that patients initiate lung involvement with alveolar inflammation, which can be detected in bronchoalveolar lavage many years before idiopathic pulmonary fibrosis is diagnosed.[19] Inflammation is characterized by an increased number of lymphocytes and high cytokine levels in the lavage fluid. How telomere shortening leads to idiopathic pulmonary fibrosis is unclear. It has been speculated that eroded telomeres limit the pneumocyte stem cell pool, which eventually may trigger a vicious circle of alveolar epithelium regeneration deficiency and alveolar disruption. Impairment of pneumocyte proliferation may be aggravated by tissue scarring caused by environmental factors, such as radiation or smoking, producing a profibrotic response.

Most patients who develop idiopathic pulmonary fibrosis are over the age of 50 years; males predominate and the prognosis is poor, with median overall survival of 3 years. Lung transplant is the only therapeutic option; prednisone is not helpful. Comprehensive clinical outcomes for lung transplant in patients with telomerase mutations are lacking.

In patients with dyskeratosis congenita, pulmonary fibrosis also can be severe and manifest in the first decades of life and is often a fatal complication after hematopoietic stem cell transplant.

Liver

Cirrhosis

Cirrhosis is not unusual in family members of patients with telomerase-mutant aplastic anemia or pulmonary fibrosis.[20] In some familial cases of cirrhosis, mutations in either *TERT* or *TERC* have been detected. In addition to a strong family history of aplastic anemia, patients present with cryptogenic liver cirrhosis. Histologically, liver disease is characterized by bridging fibrosis, moderate inflammation, and iron accumulation (Fig. 2).

In addition, in cirrhosis associated with chronic hepatitis virus C infection and alcohol abuse, telomerase mutations are present in 3%–7% of cases.[21,22] In such seemingly sporadic disease, telomerase mutations appear as a genetic risk factor for cirrhosis development in the context of an environmental insult. These patients have short telomeres (usually below the 10th percentile), but cirrhosis overall is associated with short telomeres, suggesting that telomere erosion is a key molecular event mediating disease. Telomeres are significantly short not only in peripheral blood leukocytes, but also in hepatocytes.[23] It is not clear how telomere erosion causes cirrhosis. Short telomeres may limit the regenerative capacity of hepatocyes, but also dysregulate stellate cells (the stem cell population that serves to regenerate hepatocytes lost to injury or normal cellular turnover) or the immune response. In murine models deficient for telomerase, liver injury secondary to CCl_4 results in increased regeneration defects and liver cirrhosis in comparison with animals with wild-type telomerase.[24] In addition, reconstitution of telomerase activity in the liver by exogenous delivery (gene transduction) abrogates liver cirrhosis and ameliorates liver function in these animals. The presence of inflammatory cells in the telomeropathic liver suggests a role of inflammation in cirrhosis development in the setting of dysfunctional telomeres. Abnormal telomeres in neutrophils, monocytes, or lymphocytes, key mediators of the inflammatory response, may elicit an abnormal and profibrotic response.

Nodular regenerative hyperplasia

Liver nodular regenerative hyperplasia also occurs in patients with telomerase mutations (Fig. 2).[20] Nodular hyperplasia is the most common noncirrhotic cause of portal hypertension, but its pathophysiology is unknown. The pathology is more common in the elderly, appearing as a necropsy finding in 5% of individuals over the age of 80 years.[25] Histologically, there are many small regenerative nodules clustered around the portal triads but minimal or no fibrosis.

TELOMERE SHORTENING AND CANCER

Cell culture and genetically engineered mouse experiments indicate that telomere erosion produces chromosomal instability via end-to-end fusions and break–fusion–bridge (BFB) cycles, thereby initiating a vicious circle of genomic instability that promotes cancer. Telomere dysfunction appears to be a form of genomic instability that is particularly relevant to the genesis of epithelial malignancies (i.e. carcinomas) but can also contribute to other cancers (lymphomas/leukemias). In BFB cycles, telomere erosion eventually leads to effectively naked chromosome ends that are sensed by the cell as DNA "breaks." Such breaks are repaired ("fused") through a specific DNA ligase IV-dependent repair pathway known as nonhomologous end joining (NHEJ). Such end-to-end fusions will lead to subsequent problems in the ensuing cell division if the ligated DNA ends come from different chromosomes. For example, a dicentric chromosome would lead to an anaphase "bridge" that will break at some point along its length, leading to more free ends, which would then need to be repaired by NHEJ, in turn resulting in more breaks, etc. Such a cascade of BFB cycles is believed to be an important trigger mechanism in the genesis of carcinomas, and to contribute to the exponential increase in (1) the carcinoma incidence with advancing age and (2) the rampant genomic instability that characterizes many carcinomas.[26] Such mechanisms also likely contribute to the increased cancer incidence observed in several of the telomeropathies, particularly dyskeratosis congenita.

Insights into telomere function and the cellular consequences have come from diverse model organisms. In budding yeast, deletion of telomerase results in progressive telomere shortening, increased genomic mutation, and chromosomal rearrangements, especially terminal deletions.[27] In mouse models deficient for telomerase and haploinsufficient for p53, eroded telomeres promote nonreciprocal translocations and an increased incidence of epithelial cancers.[28] Telomerase-deficient mice also have abnormal DNA repair and higher sensitivity to ionizing radiation.[29] In mouse models of uterine (endometrial) cancer, mice with genetically engineered telomere dysfunction exhibited distinct tumor histologies associated with genomic instability and poor prognosis, suggesting that telomere instability can influence disease progression and may be a signature of some types of carcinomas but not others.[30]

In humans, telomere attrition causes comparable chromosomal changes. Bone marrow cells of clinically healthy subjects with telomerase mutations and short telomeres feature increased numbers of cytogenetic abnormalities, including telomere associations (when chromosomes are associated via their telomeres), aneuploidy, translocations, and deletions that are likely ultimately due to end-to-end fusions and ensuing BFB cycles.[31]

Clinically, short telomeres promote cancer in humans. As described above, dyskeratosis congenita confers a very high risk of cancer.[6] In addition, patients with acquired aplastic anemia with shorter-for-age telomeres have an increased risk of evolution to myelodysplasia and acute myeloid leukemia, in comparison with aplastic anemia patients with longer telomeres.[32] The risk of cancer for individuals with shorter-for-age telomeres is observed in a wide variety of tumors and conditions. Patients with Barrett esophagitis and short leukocyte telomeres are at increased risk of developing esophagus adenocarcinoma compared to those with longer telomeres.[33] In colorectal cancer, the adjacent epithelium displays short telomeres, suggesting that eroded telomeres predispose epithelial cells to malignant transformation.[34] In a recent population-based prospective study, telomere length inversely correlated with cancer incidence and cancer mortality for many common tumors.[35] In this study, telomere length predicted cancer risk independent of age, sex, smoking status, or body mass index.

In addition to short telomeres, the telomerase gene loci also are risk factors for cancer development. Loss-of-function mutations in *TERT* are a risk factor for *de novo* and familial acute myeloid leukemia.[36,37] Genome-wide association studies identified the *TERT–CLPTM1L* locus on chromosome 5p15.33 as a major susceptibility region for a variety of cancers.[38,39] This association has been confirmed in multiple subsequent studies, including in Chinese patients, in whom short telomeres also were linked to cancer risk.[40] The *TERT–CLPTM1L* locus also has been implicated as a genetic risk factor for glioma[41]; and renal cell carcinoma,[42] but not melanoma[38] or breast cancer.[43] While in these large genome-wide association studies the specific risk is small (relative risk varying from 1.12 to 1.21), the risk is consistent across ethnic groups. Whether polymorphisms in *TERT* or *TERC* cause telomere shortening is controversial.[44,45]

In summary, there is both laboratory and clinical evidence that telomere attrition in humans produces telomere dysfunction and chromosomal instability, resulting in numerical and structural abnormalities in chromosomes, which may predispose to the development of malignant cells. In addition, short telomeres may limit proliferation of normal cells and select stem cells with defective DNA damage responses prone to genome instability. This conceptual framework rationalizes to a large extent the association between telomeropathies and cancer, as well as some of the principal disease manifestations of telomeropathies. Indeed, it appears that some types of cancers could be classified as telomeropathies. Recent findings that individuals with short telomeres are at an increased risk of developing cancer and cirrhosis are particularly intriguing, suggesting that even a state of relatively mild telomeropathy is a significant risk factor for cancer and other disease conditions that occur in patients with more overt telomeropathy. Further investigations into the biological basis of telomere length variation (genetic and environmental) in human populations and its relationship to diverse disease conditions are needed.

REFERENCES

1. Blackburn EH. (2001) Switching and signaling at the telomere. *Cell* **106**: 661–673.
2. de Lange T. (2005) Telomere-related genome instability in cancer. *Cold Spring Harb Symp Quant Biol* **70**: 197–204.

3. Moyzis RK, Buckingham JM, Cram LS, *et al.* (1988) A highly conserved repetitive DNA sequence, (TTAGGG)n, present at the telomeres of human chromosomes. *Proc Natl Acad Sci USA* **85**: 6622–6626.
4. de Lange T. (2005) Shelterin: the protein complex that shapes and safeguards human telomeres. *Genes Dev* **19**: 2100–2110.
5. Naber AH, Van HU, Yap SH. (1991) Nodular regenerative hyperplasia of the liver: an important cause of portal hypertension in non-cirrhotic patients. *J Hepatol* **12**: 94–99.
6. Alter BP, Giri N, Savage SA, Rosenberg PS. (2009) Cancer in dyskeratosis congenita. *Blood* **113**: 6549–6557.
7. Savage SA, Dokal I, Armanios M, *et al.* (2009) Dyskeratosis congenita: the first NIH clinical research workshop. *Pediatr Blood Cancer* **53**: 520–523.
8. Alter BP, Baerlocher GM, Savage SA, *et al.* (2007) Very short telomere length by flow fluorescence *in situ* hybridization identifies patients with dyskeratosis congenita. *Blood* **110**: 1439–1447.
9. Fogarty PF, Yamaguchi H, Wiestner A, *et al.* (2003) Late presentation of dyskeratosis congenita as apparently acquired aplastic anaemia due to mutations in telomerase RNA. *Lancet* **362**: 1628–1630.
10. Calado RT, Yewdell WT, Wilkerson KL, *et al.* (2009) Sex hormones, acting on the TERT gene, increase telomerase activity in human primary hematopoietic cells. *Blood* **114**: 2236–2243.
11. Knight SW, Heiss NS, Vulliamy TJ, *et al.* (1999) Unexplained aplastic anaemia, immunodeficiency, and cerebellar hypoplasia (Hoyeraal–Hreidarsson syndrome) due to mutations in the dyskeratosis congenita gene, *DKC1*. *Br J Haematol* **107**: 335–339.
12. Savage SA, Giri N, Baerlocher GM, *et al.* (2008) TINF2, a component of the shelterin telomere protection complex, is mutated in dyskeratosis congenita. *Am J Hum Genet* **82**: 501–509.
13. Yamaguchi H, Calado RT, Ly H, *et al.* (2005) Mutations in *TERT*, the gene for telomerase reverse transcriptase, in aplastic anemia. *N Eng J Med* **352**: 1413–1424.
14. Calado RT, Young NS. (2009) Telomere diseases. *New Engl J Med* **361**: 2353–2365.
15. Tsakiri KD, Cronkhite JT, Kuan PJ, *et al.* (2007) Adult-onset pulmonary fibrosis caused by mutations in telomerase. *Proc Natl Acad Sci USA* **104**: 7552–7557.
16. Armanios MY, Chen JJ, Cogan JD, *et al.* (2007) Telomerase mutations in families with idiopathic pulmonary fibrosis. *N Engl J Med* **356**: 1317–1326.
17. Diaz de Leon A, Cronkhite JT, Katzenstein AL, *et al.* (2010) Telomere lengths, pulmonary fibrosis and telomerase (TERT) mutations. *PLoS One* **5**: e10680.

18. Alder JK, Chen JJ, Lancaster L, *et al.* (2008) Short telomeres are a risk factor for idiopathic pulmonary fibrosis. *Proc Natl Acad Sci USA* **105**: 13051–13056.
19. El-Chemaly S, Ziegler SG, Calado RT, *et al.* (2011) Natural history of pulmonary fibrosis in two subjects with the same telomerase mutation. *Chest* **139**: 1203–1209.
20. Calado RT, Regal JA, Kleiner DE, *et al.* (2009) A spectrum of severe familial liver disorders associate with telomerase mutations. *PLoS One* **4**: e7926.
21. Hartmann D, Srivastava U, Thaler M, *et al.* (2011) Telomerase gene mutations are associated with cirrhosis formation. *Hepatology* **53**: 1608–1617.
22. Calado RT, Brudno J, Mehta P, *et al.* (2011) Constitutional telomerase mutations are genetic risk factors for cirrhosis. *Hepatology* **53**: 1600–1607.
23. Wiemann SU, Satyanarayana A, Tsahuridu M, *et al.* (2002) Hepatocyte telomere shortening and senescence are general markers of human liver cirrhosis. *FASEB J* **16**: 935–942.
24. Rudolph KL, Chang S, Millard M, *et al.* (2000) Inhibition of experimental liver cirrhosis in mice by telomerase gene delivery. *Science* **287**: 1253–1258.
25. Wanless IR. (1990) Micronodular transformation (nodular regenerative hyperplasia) of the liver: a report of 64 cases among 2,500 autopsies and a new classification of benign hepatocellular nodules. *Hepatology* **11**: 787–797.
26. DePinho RA, Wong KK. (2003) The age of cancer: telomeres, checkpoints, and longevity. *J Clin Invest* **111**: S9–S14.
27. Hackett JA, Feldser DM, Greider CW. (2001) Telomere dysfunction increases mutation rate and genomic instability. *Cell* **106**: 275–286.
28. Artandi SE, Chang S, Lee S-L, *et al.* (2000) Telomere dysfunction promotes non-reciprocal translocations and epithelial cancers in mice. *Nature* **406**: 641–645.
29. Wong KK, Chang S, Weiler SR, *et al.* (2000) Telomere dysfunction impairs DNA repair and enhances sensitivity to ionizing radiation. *Nat Genet* **26**: 85–88.
30. Akbay EA, Contreras CM, Perera SA, *et al.* (2008) Differential roles of telomere attrition in type I and II endometrial carcinogenesis. *Am J Pathol* **173**: 536–544.
31. Calado RT, Cooper JN, Padilla-Nash H, *et al.* (2011) Short telomeres result in chromosomal instability in hematopoietic cells and precede malignant evolution in human aplastic anemia. *Leukemia* (in press).
32. Scheinberg P, Cooper JN, Sloand EM, *et al.* (2010) Association of telomere length of peripheral blood leukocytes predicts hematopoietic relapse, malignant transformation, and survival in severe aplastic anemia. *JAMA* **304**: 1358–1364.

33. Risques RA, Baughan TL, li X, *et al.* (2007) Leukocyte telomere length predicts cancer risk in Barrett's esophagus. *Cancer Epidemiol Biomarkers Prev* **16**: 2649–2655.
34. O'Sullivan JN, Bronner MP, Brentnall TA, *et al.* (2002) Chromosomal instability in ulcerative colitis is related to telomere shortening. *Nat Genet* **32**: 280–284.
35. Willeit P, Willeit J, Mayr A, *et al.* (2010) Telomere length and risk of incident cancer and cancer mortality. *JAMA* **304**: 69–75.
36. Calado RT, Regal JA, Hills M, *et al.* (2009) Constitutional hypomorphic telomerase mutations in patients with acute myeloid leukemia. *Proc Natl Acad Sci USA* **106**: 1187–1192.
37. Kirwan M, Vulliamy T, Marrone A, *et al.* (2009) Defining the pathogenic role of telomerase mutations in myelodysplastic syndrome and acute myeloid leukemia. *Hum Mutat* **30**: 1567–1573.
38. Rafnar T, Sulem P, Stacey SN, *et al.* (2009) Sequence variants at the TERT–CLPTM1L locus associate with many cancer types. *Nat Genet* **41**: 221–227.
39. McKay JD, Hung RJ, Gaborieau V, *et al.* (2008) Lung cancer susceptibility locus at 5p15.33. *Nat Genet* **40**: 1404–1406.
40. Hosgood HD, III, Cawthon R, He X, *et al.* (2009) Genetic variation in telomere maintenance genes, telomere length, and lung cancer susceptibility. *Lung Cancer* **66**: 157–161.
41. Shete S, Hosking FJ, Robertson LB, *et al.* (2009) Genome-wide association study identifies five susceptibility loci for glioma. *Nat Genet* **41**: 899–904.
42. Chen M, Ye Y, Yang H, *et al.* (2009) Genome-wide profiling of chromosomal alterations in renal cell carcinoma using high-density single nucleotide polymorphism arrays. *Int J Cancer* **125**: 2342–2348.
43. Savage SA, Chanock SJ, Lissowska J, *et al.* (2007) Genetic variation in five genes important in telomere biology and risk for breast cancer. *Br J Cancer* **97**: 832–836.
44. Mirabello L, Yu K, Kraft P, *et al.* (2010) The association of telomere length and genetic variation in telomere biology genes. *Hum Mutat* **31**: 1050–1058.
45. Levy D, Neuhausen SL, Hunt SC, *et al.* (2010) Genome-wide association identifies OBFC1 as a locus involved in human leukocyte telomere biology. *Proc Natl Acad Sci USA* **107**: 9293–9298.

CHAPTER 6

FANC-BLM-Opathies: Recent Progress in the Understanding of Molecular Pathogenesis of Fanconi Anemia and Its Connection with Bloom Syndrome

*Toshiyasu Taniguchi**,†, *Kiranjit K. Dhillon**, *Maria Castella**, *Ronald S. Cheung*†,‡ *and Céline Jacquemont*†

INTRODUCTION

Fanconi anemia (FA) and Bloom syndrome (BS) are rare genetic disorders which have distinct clinical features but which share genomic instability and cancer susceptibility (Table 1). The molecular pathogenesis of these two disorders reveals a family of linked "FANC-BLM-opathies" whose central feature is defective DNA repair, and whose study is important for at least three reasons. First, it helps to establish strategies for diagnosing and treating

*Divisions of Human Biology and Public Health Sciences, Fred Hutchinson Cancer Research Center, Howard Hughes Medical Institute, 1100 Fairview Ave. N., C1-015, Seattle, WA 98109-1024, USA. Tel.: 206-667-7283; Fax: (206) 667-5815; E-mail: ttaniguc@fhcrc.org.

†Howard Hughes Medical Institute, Divisions of Human Biology and Public Health Sciences, Fred Hutchinson Cancer Research Center, Seattle, WA, USA.

‡Division of Pediatric Hematology/Oncology, Department of Pediatrics, University of Washington School of Medicine, Seattle, WA, USA.

Table 1 Comparison of Fanconi Anemia and Bloom Syndrome

	Fanconi anemia	Bloom syndrome
Clinical manifestations	aplastic anemia (bone marrow failure) developmental defects (short statue, abnormalities of the thumb or arm, skin, head, eyes, kidneys, and ears, developmental disability) male infertility cancer susceptibility (leukemia, liver tumors, head and neck squamous carcinomas, etc.)	short stature, high-pitched voice, distinct facial features facial rash, sun-sensitive skin lesions susceptibility to infections (moderate immune deficiency) diabetes, chronic lung disease male infertility cancer susceptibility (broad spectrum, at young age)
Cellular phenotype	sensitivity to interstrand crosslink (ICL)-inducing agents chromosome breakage after ICL treatment late S and G2 accumulation after ICL treatment	elevated frequency of sister chromatid exchange (SCE)
inheritance	autosomal recessive X-linked recessive (FA-B group only)	autosomal recessive
responsible genes	14 FA genes+ 1 FA-like disorder gene (See Table 2)	*BLM*

affected patients. Second, it improves our understanding of biologically important cellular processes, such as DNA repair. Third, it provides clues to understanding the pathogenesis of human cancer in the general population and to improve the treatment for patients with cancer. Therefore, although the numbers of patients with these genetic disorders are small, these diseases have fascinated clinical and basic science researchers alike.

Identification of the genes responsible for FA (14 FA genes, 1 FA-like disorder gene) and BS (*BLM*) revealed that the two disorders are connected at the molecular level. BLM and FA proteins comprise several multi-subunit protein machines that manage DNA replication stress and maintain chromosomal stability. In this chapter, we discuss recent progress in our understanding of the molecular pathogenesis of FA (Fig. 1). We also review current knowledge about the interaction between BLM and FA proteins.

FANCONI ANEMIA

FA is a rare autosomal or X-linked recessive disease characterized by developmental defects (such as radial aplasia, hyperpigmentation of the skin, and

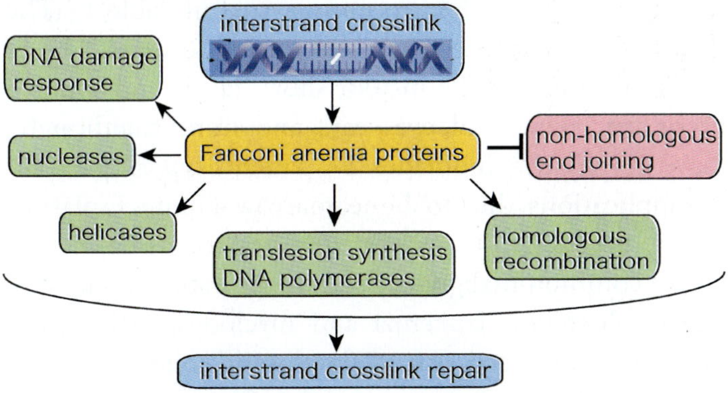

Fig. 1 The Fanconi anemia pathway coordinates interstrand crosslink repair. The main function of the FA pathway is to coordinate multiple DNA repair mechanisms and enzymes during interstrand crosslink (ICL) repair. The FA pathway activates DNA damage response and positively regulates homologous recombination, translesion DNA synthesis, endonucleases, and helicases, while it suppresses nonhomologous end joining during replication-dependent ICL repair.

growth retardation), childhood-onset aplastic anemia, cancer/leukemia susceptibility, and cellular sensitivity to interstrand DNA crosslink (ICL)-inducing agents (drugs that induce ICLs between the two complementary strands of the DNA helix), such as mitomycin C, diepoxybutane, melphalan, cisplatin, and cyclophosphamide (reviewed in Refs. 1–5). This disease was first described by the Swiss pediatrician Guido Fanconi, in 1927.

FA is a genetically heterogeneous disease, and can be divided into at least 14 subtypes or complementation groups (A, B, C, D1, D2, E, F, G, I, J, L, M, N, P). Since the first identification of an FA gene, *FANCC*, in 1992, a total of 14 FA genes have been identified (Table 2), although more FA genes may await identification.[3,4,6,7] In addition, one family with on FA-like disorder, whose responsible gene is *RAD51C/FANCO*, has been reported. This disorder may represent another complementation group, FA-O. The FA-B group shows X-linked recessive inheritance, while all of the other complementation groups show autosomal recessive inheritance. Overall, the incidence of FA is very low (1–5 per million), but its overall heterozygous carrier frequency is relatively high, estimated to be 1/300 in the United States and Europe. FA is prevalent in some ethnic groups: Ashkenazi Jews, Spanish gypsies, and South African Afrikaners present an incidence of FA up to 1 in 18,000.

The clinical features of FA are summarized in Table 1.[5] The median survival age of FA patients is now greater than 30 years. The congenital defects seen in patients with FA include short stature and abnormalities of the thumb or arm, skin, head, eyes, ears, and kidneys, although some FA patients have no physical anomalies. FA patients typically develop hematological complications due to bone marrow failure (aplastic anemia) within the first decade of life. At least 20% of patients develop malignancies. The most common malignancy seen in FA patients is leukemia, especially acute myelogenous leukemia and myelodysplastic syndrome. FA patients also have increased risk of developing solid tumors (head and neck squamous cell carcinoma, gynecological squamous cell carcinoma, esophageal carcinoma, liver tumors, brain tumors, skin tumors, renal tumors, and other tumors).[5] Patients of subtypes FA-D1 and FA-N are clinically distinct from other subtype patients, as they have increased risk of developing early-onset acute myelogenous leukemia, brain tumor (medulloblastoma), and renal tumor (Wilms' tumor).

Table 2 Complementation Groups and Responsible Genes for Fanconi Anemia

Subtype	Responsible gene	FA patients, estimated, %*	Chromosome location	Protein kDa	Requirement for FANCD2 monoubiquitination	Comment
A	*FANCA*	57	16q24.3	163	+	FA core complex
B	*FANCB (FAAP95)*	0.3	Xp22.31	95	+	FA core complex
C	*FANCC*	15	9q22.3	63	+	FA core complex
D1	*FANCD1/BRCA2*[§]	4	13q12-13	380	–	RAD51 recruitment
D2	*FANCD2*	3	3p25.3	155, 162	+	Monoubiquitinated protein, ID complex with FANCI, FAN1 recruitment
E	*FANCE*	1	6p21-22	60	+	FA core complex, direct binding to FANCD2
F	*FANCF*	2	11p15	42	+	FA core complex
G	*FANCG/XRCC9*	9	9p13	68	+	FA core complex
I	*FANCI*	rare	15q25-26	140, 147	+	Monoubiquitinated protein, ID complex with FANCD2
J	*FANCJ/BACH1/BRIP1*[¶]	1.6	17q22-q24	130	–	5′->3′ DNA helicase/ATPase, bind to BRCA1,TopBP1, and MLH1, regulate BLM protein stability

(Continued)

Table 2 (*Continued*)

Subtype	Responsible gene	FA patients, estimated, %*	Chromosome location	Protein kDa	Requirement for FANCD2 monoubiquitination	Comment
L	*FANCL/PHF9/POG (FAAP43)*	0.1	2p16.1	43	+	FA core complex, ubiquitin ligase
M[†]	*FANCM/Hef (FAAP250)*	rare	14q21.3	250	+	FA core complex, ATPase/ translocase, connect FA core complex and BLM complex, ATR activation
N	*FANCN/PALB2*[§]	rare	16p12.1	140	–	Bind to BRCA2 and BRCA1
O[‡]	*FANCO/RAD51C/ RAD51L2*[§]	rare	17q25.1	42	–	RAD51 recruitment, RAD51 paralog
P	*FANCP/SLX4/ BTBD12*	rare	16p13.3	200	–	Interact with SLX1, XPF-XRCC1 and MUS81-EME1

*Adapted from Taniguchi *et al.*, Blood 2006 with modifications.

[†] There is no pure FA-M patient. The only patient who was reported to have biallelic mutantions in FANCM, turned out to have biallelic mutations in FANCA as well (Singh *et al.*, Blood 2009).

[‡] There is only one family with patients with biallelic mutations in RAD51C.

Therefore, this complementation group (FA-O) is still provisional and called FA-like disorder rather than FA.

[§] Heterozygous mutation carriers have breast/ovarian cancer susceptibility.

[¶] Heterozygous mutation carriers have breast cancer susceptibility.

Importantly, inherited monoallelic mutations of some of the FA genes (*FANCD1*, *FANCJ*, *FANCN*, and *FANCO*) lead to breast and ovarian cancer susceptibility.[3] *BRCA1* and *BRCA2* are well-known breast/ovarian cancer susceptibility genes; heterozygous germline mutations in *BRCA1* and *BRCA2* predispose to breast/ovarian cancer and some other cancers. In fact, one of the FA genes, *FANCD1*, is identical to *BRCA2*; FA patients of the FA-D1 subtype have biallelic germline mutations in *BRCA2/FANCD1*. FA patients with biallelic germline mutations in *BRCA1* have not yet been reported. Inherited monoallelic mutations in *FANCJ* [also known as *BACH1* (BRCA1-Associated C-terminal Helicase 1), *BRIP1* (BRCA1 Interacting Protein C-terminal Helicase 1)] and *FANCN* [also known as *PALB2* (Partner And Localizer of BRCA2)] also confer breast cancer susceptibility (the relative risk of breast cancer associated with *BRIP1* and *PALB2* mutations was estimated at 2.0 and 2.3, respectively).[3] *FANCN/PALB2* is also a pancreatic cancer susceptibility gene. Furthermore, monoallelic pathogenic mutations in *FANCO/RAD51C* confer an increased risk for breast and ovarian cancer.[8] In contrast, mutations of other FA genes (*FANCA*, *FANCC*, *FANCD2*, *FANCE*, *FANCF*, and *FANCG*) in families with inherited breast cancer are rare.

The most prominent cellular phenotype of all forms of FA is hypersensitivity to ICL-inducing agents. Therefore, one of the important functions of the FA proteins is considered to be repair of ICL lesions. Cells derived from FA patients typically show increased chromosome breakage and marked late S phase (or G2 phase) accumulation after treatment with ICL-inducing agents. These cellular phenotypes are the current gold standard for the diagnosis of FA, and are now considered to reflect the role of FA proteins in the coordination of sensing and repairing ICL lesions during DNA replication in the S phase of the cell cycle.

BLOOM SYNDROME

BS is an autosomal recessive disease characterized by proportional dwarfism, immunodeficiency, hypersensitivity to sunlight, subfertility, and cancer predisposition.[9,10] This disease was first described by a New York dermatologist,

David Bloom, in 1954. The responsible gene for BS, *BLM*, was identified in 1995 and is located at chromosome 15q26.1. BS is a rare disease: only 265 cases worldwide have been registered in Bloom's Syndrome Registry as of the year 2009 (http://weill.cornell.edu/bsr/data_from_registry/). In the Ashkenazi Jewish population, the heterozygous carrier frequency of a particular *BLM* mutation (blmAsh) is relatively high (1 in 107) and the frequency of BS is estimated to be 1 in 48,000.[10] However, the frequency of the BS or mutant BLM alleles in the general population is unknown.

The clinical features of BS are summarized in Table 1.[10] Among the 265 registered cases, 122 have cancer. The malignancies found in BS patients are of diverse types, and include leukemia, lymphoma, various epithelial carcinomas (colorectal, skin, breast, esophageal, uterus, etc.), and rare tumors. The development of more than one primary cancer in one individual is frequent in BS patients. Carriers of heterozygous BLM mutations are reported to have a mildly increased risk for colorectal cancer, but a conflicting result has been published.[10]

The most prominent cellular phenotype of BS is marked increase in the rate of sister chromatid exchanges (SCEs).[9,10] This phenotype is considered to reflect the role of the BLM protein in suppressing crossover during homologous recombination (HR)-mediated DNA repair. This cellular phenotype is distinct from that of FA.

THE FANCONI ANEMIA PATHWAY: AN OVERVIEW

All of the 15 FA (and FA-like disorder) proteins (Table 2 and Fig. 2) and the breast/ovarian cancer susceptibility proteins, BRCA1 and BRCA2, cooperate in a pathway/network called the FA pathway, otherwise known as the FA-BRCA pathway, or the FA-BRCA network.[3,4]

The main function of the FA pathway is to coordinate multiple DNA repair mechanisms and enzymes during replication-dependent ICL repair (Fig. 1). ICLs, which covalently link the two strands of the DNA helix, are highly toxic lesions, because they block DNA strand separation, which is required for completion of DNA replication during the S phase as well as transcription throughout the cell cycle. Although the repair of ICLs in mammalian cells is not well understood, it is considered to involve a replication- and recombination-dependent

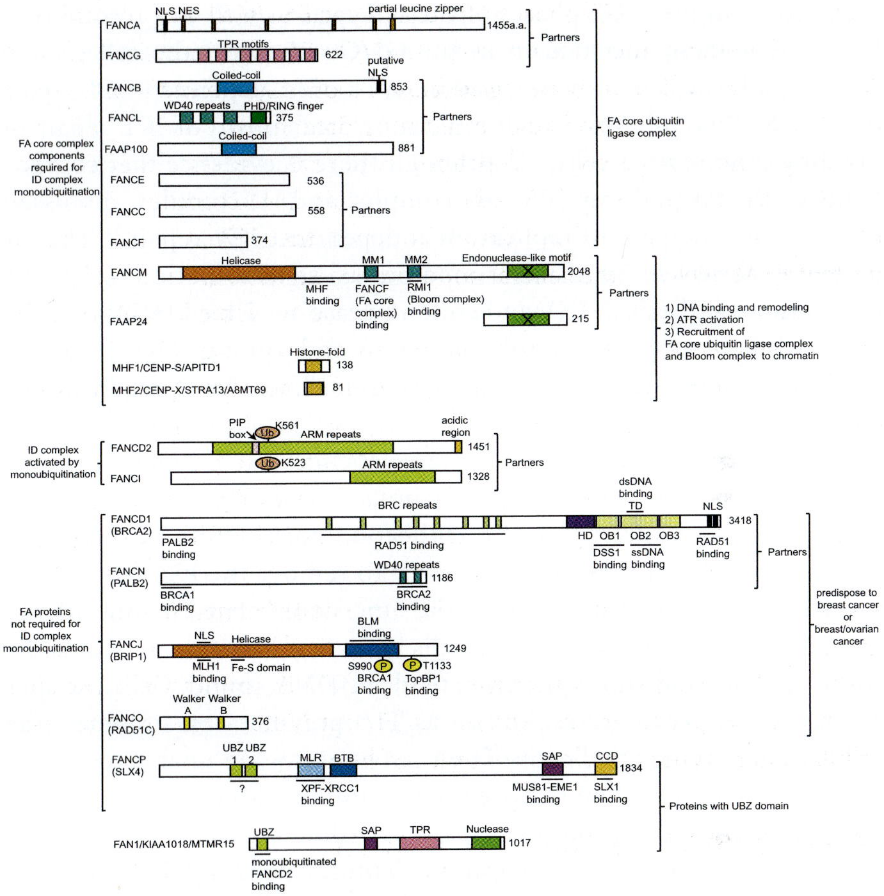

Fig. 2 Schematic representation of the 14 human Fanconi anemia proteins and related proteins (FAAP24, FAAP100, MHF1, MHF2, FANCO/RAD51C, FAN1). (Adapted with modifications from Fig. 2 of Ref. 2.) ARM, armadillo repeat; BRC, the internal repeat domains of BRCA2; CCD, conserved C-terminal domain; dsDNA, double-strand DNA; FAAP, FA-associated protein; HD, helical domain; MLR, MUS312/MEI9 interaction-like region; NES, nuclear export sequences; NLS, nuclear localization signals; OB, oligonucleotide/oligosaccharide-binding folds; PALB2, partner and localizer of BRCA2; PIP box, PCNA-binding motif; ssDNA, single-strand DNA; TD, tower domain; TPR, tetratricopeptide repeat motifs; Ub; ubiquitin; UBZ, ubiquitin-binding zinc finger.

mechanism in the S/G2 phases of the cell cycle as well as a recombination-independent mechanism in the G1/G0 phase of the cell cycle.[11] The FA pathway is mainly involved in replication-dependent ICL repair during the S/G2 phases, which is a major mechanism of ICL repair in dividing mammalian cells,[12,13] although there is evidence that components of the FA pathway (FA core complex and ID complex, discussed later) can participate in replication-independent ICL repair.[14] The FA pathway positively regulates homologous recombination (HR),[15,16] translesion synthesis (TLS), and endonuclease-mediated DNA processing,[3,4] while it suppresses nonhomologous end joining (NHEJ) during replication-dependent ICL repair[17,18] (Fig. 1). These DNA repair mechanisms will be discussed later.

One favored model of replication-dependent ICL repair is shown in Fig. 3.[12,13] Since an ICL covalently links the two strands of the DNA helix, the progression of DNA replication forks stalls near an ICL lesion during DNA replication (Fig. 3). To remove the ICL from DNA, specialized endonucleases first make "incisions" that cut one of the DNA strands near the ICL. Then, cells insert a nucleotide to bypass the adducted base and subsequently extend the DNA strand. Cells use specialized DNA polymerases, known as TLS polymerases, to bypass the lesion, since regular replicative DNA polymerases are unable to do this efficiently. The cells then remove the remaining adduct, presumably by using endonucleases and nucleotide excision repair, a DNA repair mechanism that removes damaged nucleotides. At this point, there still remains a DNA double strand break (DSB) that must be repaired. Mammalian cells have two major pathways to repair DSBs: HR and NHEJ. NHEJ mediates direct ligation of broken ends, while HR utilizes the undamaged homologous sequence as a template to repair DSBs. For the repair of DSBs arising from replication-dependent ICL repair, HR is preferred, and NHEJ should be suppressed. The FA pathway coordinates these complex and interrelated processes during ICL repair.

The FA pathway is activated by DNA damage. For the activation of the pathway, posttranslational modifications of proteins, such as phosphorylation and ubiquitination, play important roles. A current model of the FA

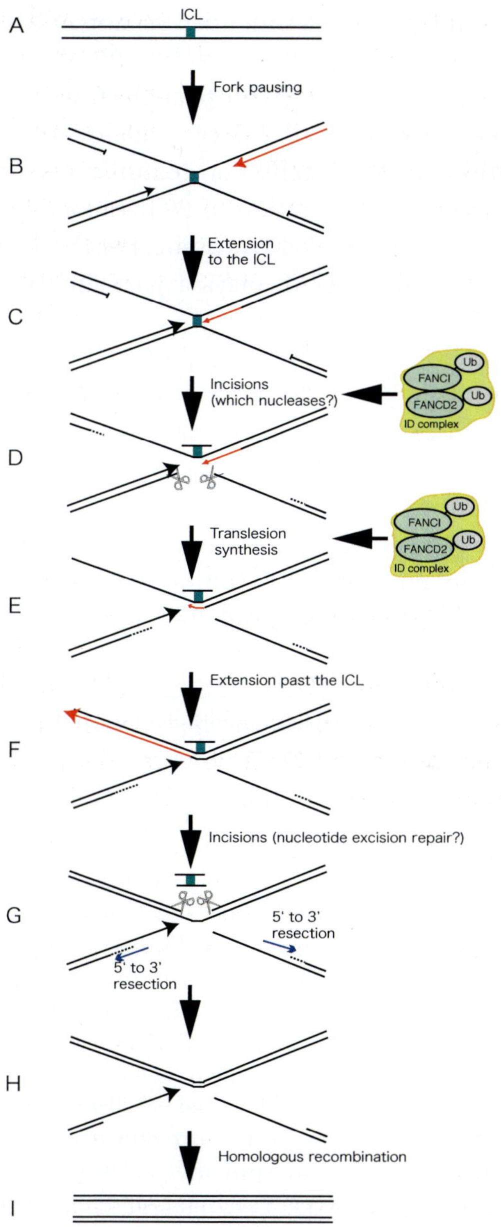

Fig. 3 Model of replication-dependent interstrand crosslink repair in *Xenopus* egg extracts; the double fork collision model. (Adapted with modifications from Fig. 7 of Ref. 12.) During DNA replication, if an interstrand crosslink (ICL) lesion exists (A), progression of two

pathway is shown in Fig. 4 and a simplified version of the model is shown in Fig. 5.

Two protein kinases, ATM (ataxia-telangiectasia-mutated) and ATR (ATM and RAD3-related), act as master regulators of the DNA damage response. Mutation in ATM causes a genomic instability syndrome, ataxia telangiectasia, while mutation in ATR causes a distinct genomic instability syndrome called Seckel syndrome. For the activation of the FA pathway, ATR, rather than ATM, plays a particularly important role[19] (Figs. 4 and 5).

Monoubiquitination of FANCD2 and FANCI is another critical step in the activation of the FA pathway.[20,21] Monoubiquitination is a post-translational modification in which one ubiquitin molecule is covalently attached to a target protein. Monoubiquitination of FANCD2 is required for recruitment of some nucleases to the site of DNA damage and repair, as will be discussed later. Monoubiquitination is sequentially mediated by three enzymes: E1 (ubiquitin-activating enzyme), E2 (ubiquitin-conjugating enzyme), and E3 (ubiquitin ligase). It can be

---

Fig. 3 (*Continued*) converging replication forks is blocked (B). The leading strands of the replication forks stall 20–40 nt from the ICL lesion. One of the leading strands is extended to within 1 nt of the ICL (C). Then, dual incisions on either side of the ICL occur (D). The incisions are dependent on FANCD2 monoubiquitination and FANCI.[13] Which nucleases are involved in this step is not clarified, but the FANCP/SLX4 nuclease complex (XPF-ERCC1, MUS81-EME1, SLX1) and FAN1 are possible candidates. Next, a translesion synthesis (TLS) polymerase (REV1 is a possible candidate) inserts a nucleotide to bypass the adducted base (E). This TLS step is also dependent on FANCD2 monoubiquitination and FANCI.[13] Then, the nascent strand is extended past the ICL by another TLS polymerase, pol ζ (REV3-REV7) (F). Subsequently, the ICL adduct is removed, possibly by nucleotide excision repair (G, H). Finally, the double strand break of the bottom duplex is repaired by homologous recombination (I). Prior to homologous recombination, 5′-to-3′ resection of double strand break ends will occur (G). Although this *Xenopus* egg extract–ICL plasmid system is currently the most sophisticated system for studying replication-dependent ICL repair, it should be noted that this is an *in vitro* system and may not necessarily reflect mechanisms happening *in vivo*. In particular, how frequently a double fork collision at the ICL lesion — which results in generation of a two-ended DSB — happens in mammalian cells is not clear, and a collision of a single replication fork with the ICL — lesion, which results in generation of a one-ended DSB — may occur more frequently.

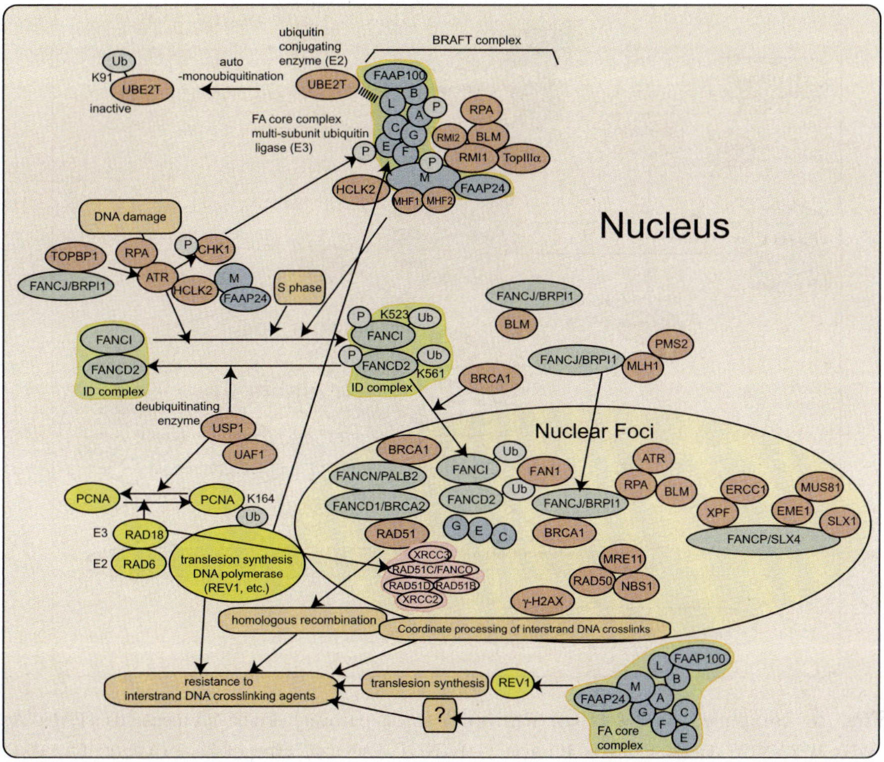

Fig. 4 Current model of the Fanconi anemia pathway. The FA pathway is a highly complicated network that coordinates multiple DNA repair mechanisms in ICL repair during DNA replication. (See text for details.)

reversed by a family of deubiquitinating enzymes, and therefore can serve as a reversible signal regulating various cellular processes, including DNA repair.

The 15 FA (and FA-like disorder) proteins can be classified into three groups, using the ability of monoubiquitination of FANCD2 and FANCI as a discriminator (Table 2 and Fig. 2): (1) FA core complex components required for FANCD2 and FANCI monoubiquitination (FANCA, FANCB, FANCC, FANCE, FANCF, FANCG, FANCL, FANCM), (2) "ID" complex components (FANCD2 and FANCI), and (3) FA proteins not required for monoubiquitination of FANCD2 or FANCI (FANCD1/BRCA2, FANCJ/

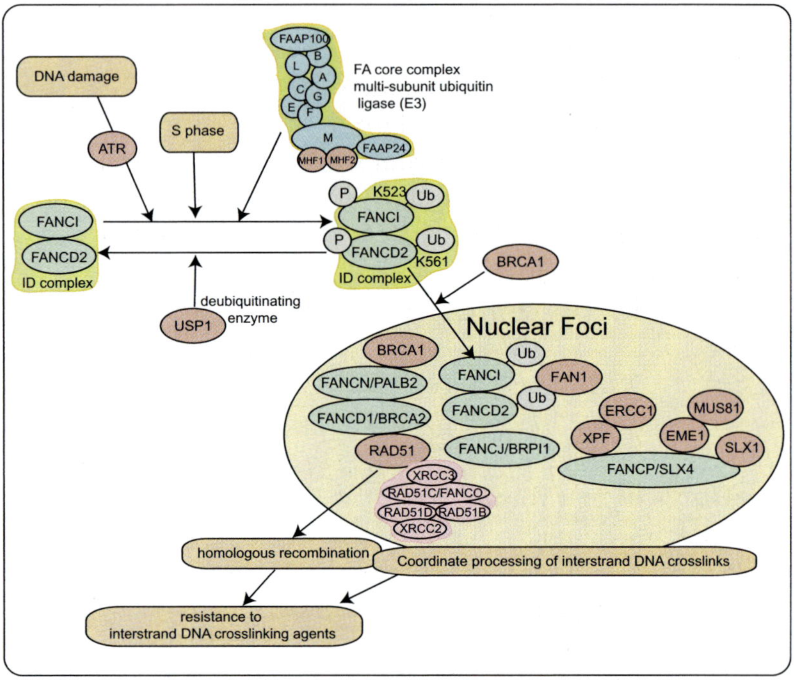

Fig. 5 Simplified model of the Fanconi anemia pathway. Eight FA proteins (FANCA, FANCB, FANCC, FANCE, FANCF, FANCG, FANCL, FANCM) along with FAAP100, FAAP24, MHF1, and MHF2 form a nuclear protein complex (FA core complex), which is considered to be a multi-subunit ubiquitin ligase (E3) for FANCD2 and FANCI. FANCI and FANCD2 form another protein complex, called the ID complex. FANCM has DNA translocase activity and may play a role in the recognition of DNA damage. In response to DNA damage or during the S phase, FANCD2 and FANCI are monoubiquitinated on specific lysine residues [lysine 561 (K561) for FANCD2; lysine 523 (K523) for FANCI] in an FA core complex-dependent manner. DNA-damage-induced monoubiquitination of FANCD2 also requires activation of ATR. ATR directly phosphorylates FANCI and possibly FANCD2, and these phosphorylations facilitate monoubiquitination of FANCD2. Monoubiquitinated FANCD2 and monoubiquitinated FANCI are translocated into nuclear foci at sites of DNA damage and colocalize with BRCA1, FANCD1/BRCA2, FANCN/PALB2, RAD51, FANCJ/BRIP1, FAN1, and other proteins. Monoubiquitinated FANCD2 recruits FAN1 nuclease at sites of DNA damage through interaction between the monoubiquitin on FANCD2 and the UBZ motif (a ubiquitin-binding domain) of FAN1. BRCA1, FANCN/PALB2, FANCD1/BRCA2, and RAD51 paralogs including FANCO/RAD51C are required for RAD51 loading on ssDNA, and facilitate homologous recombination. FANCP/SLX4 directly binds to three nucleases (SLX1, ERCC1-XPF, MUS81-EME1) and also has two UBZ motifs, which may interact with ubiquitinated proteins. USP1 deubiquitinates FANCD2 and FANCI. All of these factors are required for cellular resistance to interstrand DNA crosslinking agents.

BRIP1/BACH1, FANCN/PALB2, FANCO/RAD51C, FANCP/SLX4/BTBD12).[2]

Several steps define monoubiquitation of FANCD2 and FANCI in the FA pathway. First, eight FA proteins form a ubiquitin ligase complex (FA core complex) in the nucleus.[22] A component of the complex, FANCM, moves along DNA and binds to DNA structures that resemble stalled replication forks *in vitro*,[23] suggesting that FANCM may play a role in the recognition of lesions that leads to replication fork stalling, such as ICLs, and may facilitate the recruitment of the FA core complex. FANCI and FANCD2 form another protein complex, called the "ID complex."[21] In response to DNA damage, the ATR kinase becomes activated and phosphorylates FANCI[24] and possibly FANCD2.[19,25] Then, both FANCD2 and FANCI proteins become monoubiquitinated in an FA core complex-dependent manner.[20,21] The phosphorylation of FANCI and FANCD2 facilitates their monoubiquitination. Monoubiquitinated FANCD2 and FANCI accumulate in chromatin at sites of DNA damage and form nuclear foci (Fig. 6). A deubiquitinating enzyme, USP1, can remove ubiquitin from FANCD2 and FANCI[26] to turn off the pathway.

These protein machines can be observed in the appropriate subcellular localization, at the site of DNA damage. Many DNA repair proteins, such as BRCA1, BRCA2, PALB2, FAN1 (Fanconi-associated nuclease 1), RAD51, and FANCJ, colocalize with FANCD2-FANCI in DNA-damage-induced nuclear foci.[4] BRCA1 and FANCJ facilitate FANCD2 focus formation.[20,27] Monoubiquitinated FANCD2 recruits the FAN1 nuclease at sites of DNA damage through interaction between the monoubiquitin on FANCD2 and a ubiquitin-binding domain of FAN1.[4] FAN1 is critical for cellular resistance to ICL-inducing agents, but the mechanism has not been elucidated yet. BRCA1, PALB2, and BRCA2 form a protein complex.[28] The BRCA1-PALB2-BRCA2 complex and RAD51C regulate HR through loading of the RAD51 recombinase on single-stranded DNA (ssDNA).[8,29] PALB2 and RAD51C also play roles in later steps in HR.[8,30,31] FANCJ is a DNA-dependent ATPase and a 5′-to-3′ DNA helicase, which can bind to BRCA1.[32]

FANCP/SLX4 is a multidomain scaffold protein that interacts with three nucleases: SLX1, ERCC4/XPF-ERCC1, and MUS81-EME1.[33] Among them, MUS81-EME1 and ERCC4/XPF-ERCC1 play a critical role in ICL repair.[34]

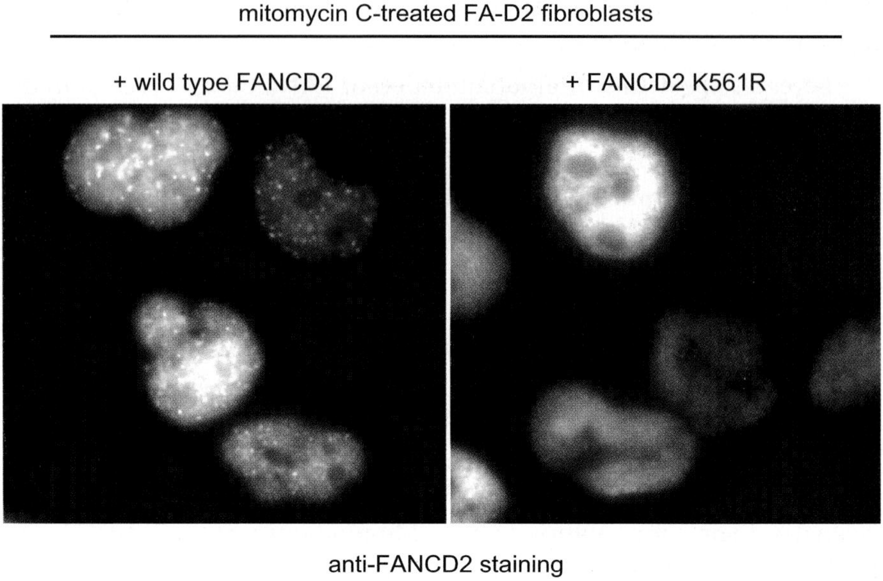

Fig. 6 Monoubiquitination of FANCD2 is required for nuclear focus formation of FANCD2. PD20 (FA-D2) fibroblasts were transduced with retroviral constructs encoding either FANCD2 (wild type) or FANCD2-K561R (nonubiquitinatable mutant). Anti-FANCD2 immunofluorescence of the indicated PD20F cells following mitomycin C treatment (160 ng/ml, 22 h) is shown. Wild-type FANCD2 forms mitomycin C-induced nuclear foci (dot-like structures in the nuclei), while the K561R mutant fails to form foci.

Thus, the FA pathway is a complex network of proteins with several types of enzymatic, structural, and regulatory activities (FA proteins, BRCA proteins, ubiquitin ligases, kinases, nucleases, helicases, DNA-binding proteins, a recombinase, translesion DNA polymerases, etc.) that collectively execute ICL repair. Next, we investigate the properties of the amazing FA core complex.

THE FA CORE COMPLEX

Eight FA proteins (FANCA, FANCB, FANCC, FANCE, FANCF, FANCG, FANCL, FANCM) along with FAAP100, FAAP24, MHF1 (FAAP16/CENP-S/APITD1), and MFH2 (FAAP10/CEMP-X/STRA13) form a nuclear protein complex (FA core complex)[22] (Figs. 4 and 5). The FA core complex has E3 ubiquitin ligase activity, which, as mentioned, is required for monoubiquitination of FANCD2 and FANCI.[22,35] One of the components of the FA core complex, FANCL, contains a variant of RING finger domain and ubiquitin ligase (E3) activity. FANCL directly associates with UBE2T, a ubiquitin conjugating enzyme (E2), which is also required for monoubiquitination of FANCD2 *in vivo*.[36] FANCL and UBE2T can monoubiquitinate FANCD2 *in vitro*, especially in the presence of FANCI.[37] Therefore, the FA core complex is a multi-subunit E3 ubiquitin ligase complex with FANCL as its catalytic subunit, and with FANCD2 and FANCI as substrates.

Although one of the important functions of the FA core complex is to monoubiquitinate FANCD2 and FANCI, the FA core complex has several other functions. First, it is required for the translocation of monoubiquitinated FANCD2 into chromatin.[38] Second, the recruitment of the error-prone TLS polymerase REV1 into nuclear foci after DNA-damaging UVC or cisplatin treatment depends on the FA core complex, but not on FANCD2 or FANCI.[39,40] Third, the FA core complex interacts with BLM, replication protein A (RPA) (RPA is a protein complex comprising of RPA1, RPA2, and RPA3, which can bind single-stranded DNA), RMI1 (BLAP75), RMI2 (BLAP18), and topoisomerase IIIα, to form a larger complex called BRAFT (BLM, RPA, FA, and topoisomerase IIIα),[22] and the formation of this BRAFT complex seems important for the recruitment of BLM protein to chromatin, as will be discussed later. In addition, it has been reported that FANCA, FANCB, FANCC, and FANCD1/BRCA2-deficient cells fail to arrest DNA synthesis in response to ICL whereas normal cells arrest DNA replication, suggesting that the FA core complex (and FANCD1/BRCA2) has a role in the activation of ICL-induced S phase checkpoint function (reviewed in Ref. 1).

FANCM SENSES STALLED REPLICATION FORKS, RECRUITS THE FA AND BLOOM COMPLEXES, AND ACTIVATES THE ATR-CHK1 PATHWAY

Among the FA proteins in the FA core complex, FANCM is unique. FANCM interacts not only with the FA core complex, but also with other proteins and complexes (Bloom complex, HCLK2, MHF1-MHF2, and FAAP24). It translocates along DNA and recruits interacting proteins to the sites of ICL damage (Fig. 7).[23,41,42] It also plays a critical role in the

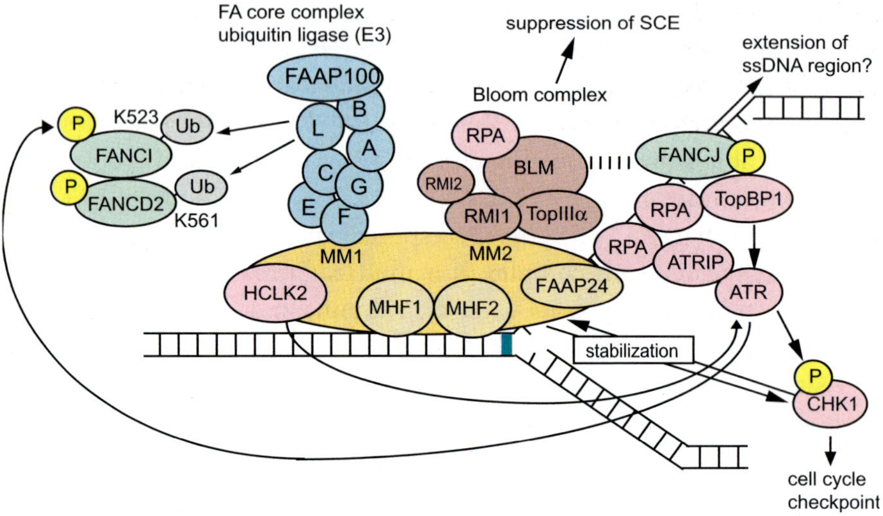

Fig. 7 Model of interaction of BLM and FA proteins at an ICL-blocked replication fork. (Adapted with modifications from Fig. 7 Ref. 54.) FANCM directly interacts with the FA core complex, Bloom complex, and HCLK2, and targets these factors to stalled replication forks. First, FANCM forms a protein complex with FAAP24, and with MHF1 and MHF2. This FANCM-FAAP24-MHF1-MHF2 complex binds to branched DNA structures with dsDNA and ssDNA, such as an ICL-blocked replication fork. FANCM recruits the FA core complex to chromatin though direct binding of FANCF and the FANCM-MM1 domain. The FA core complex serves as a ubiquitin ligase which monoubiquitinates FANCD2 and FANCI. FANCM recruits the Bloom complex (BLM, topoisomerase IIIα, RMI1, RMI2, RPA) to chromatin through direct binding of the FANCM-MM2 domain and RMI1. Both the Bloom complex and FANCM suppress sister chromatid exchanges (SCEs). FANCM-FAAP24 facilitates ATR activation by interacting with HCLK2,[49] promoting chromatin retention of TopBP1,[50] and promoting RPA recruitment to ICL-stalled replication forks.[51] ATR phosphorylates and activates CHK1. FANCM and activated CHK1 kinase stabilize each other after exposure to DNA-damaging agents.[47] Activated ATR phosphorylates FANCI and possibly FANCD2, and facilitates monoubiquitination of FANCD2-FANCI. FANCJ interacts with BLM, and BLM is degraded by a proteasome-mediated pathway when FANCJ is depleted.[92] TopBP1 directly interacts with FANCJ and this interaction is mediated by the C-terminal tandem BRCT domain of TopBP1 and phosphorylation at threonine 1133 of FANCJ.[91] Both FANCJ and TopBP1 facilitate RPA loading and ATR-dependent phosphorylation events in response to replication stress, suggesting that a specific interaction between TopBP1 and FANCJ may be required for the extension of ssDNA regions.[91]

activation of ATR kinase in response to DNA replication stress that occurs when cells are treated with ICL-inducing agents. In addition to ICL-inducing agents, FANCM-deficient cells are also sensitive to UVC (short wavelength UV) and the topoisomerase I inhibitor, camptothecin. In contrast, cells deficient in other FA proteins within the FA core complex are sensitive to ICL-inducing agents, but not to UVC or camptothecin. Interestingly, the only patient who was reported to have biallelic mutations in FANCM turned out to have biallelic mutations in FANCA as well.[43] Thus, there is no known pure FANCM mutant patient, and FANCM is not a bona fide FA gene, although it is called "FANC"-M.

FANCM contains DNA helicase motifs and a degenerate (i.e. nonfunctional) endonuclease motif (Fig. 2). It has been shown to have some enzymatic activities *in vitro*: ssDNA and double-stranded DNA (dsDNA)-stimulated ATPase activity, DNA translocase activity, an ATP-independent DNA-binding affinity to fork and four-way junction DNA, and an ATP-hydrolysis-dependent branch migration activity for four-way junction DNA in both directions.[23]

FANCM forms a protein complex with the proteins FAAP24, MHF1, and MHF2.[41,42,44] This FANCM-FAAP24-MHF1-MHF2 complex has DNA-binding activity and fork reversal activity (fork reversal is a reaction in which annealing of the nascent strands and pairing of the template strands form a four-way junction at blocked replication forks),[41,42,45] can remodel the branch point of stalled replication forks,[45,46] and can counteract the advancement of replication forks and prevent them from running into lesions and collapsing.[45,47] MHF1-MHF2 recruits FANCM to dsDNA,[41] while FANCM itself preferentially binds to branched DNA[46] and FAAP24 preferentially binds to ssDNA.[44] These binding specificities are considered to help the FANCM-FAAP24-MHF1-MHF2 complex bind to branched DNA structures with dsDNA and ssDNA, such as stalled replication forks.

The N-terminus of FANCM has a helicase domain with DNA translocase activity and ATPase activity. The ATPase-dependent activity of FANCM is not essential for the monoubiquitination of FANCD2.[43,48] Through FANCM's helicase domain, it directly interacts with HCLK2, which is a critical factor for ATR activation.[23] Remodeling of stalled replication forks by FANCM-FAAP24 is required to facilitate efficient activation of ATR signaling in response to replication stress. Furthermore, the ability of FANCM to translocate along DNA is essential for efficient activation of ATR signaling.[49] FANCM is reported to promote chromatin retention of the protein TopBP1, which plays a critical role in ATR-mediated checkpoint activation.[50] The FANCM-FAAP24 complex also promotes RPA recruitment to ICL-stalled replication forks, which is required for efficient ATR-mediated checkpoint activation.[51] FANCM also interacts with CHK1 kinase, which is directly phosphorylated and activated by ATR kinase: phosphorylated CHK1 can be degraded,[52] but FANCM[47] and HCLK2[53] stabilize CHK1 and, in turn, the CHK1 kinase stabilizes FANCM in the presence of replication inhibitors and DNA-damaging agents.[47]

FANCM has two highly conserved protein–protein interaction domains, named MM1 and MM2. The FANCM-MM1 domain directly binds to FANCF and, through this binding, FANCM recruits the FA core complex to chromatin.[54] The FANCM-MM2 domain directly binds to the RMI1 subunit of the Bloom complex (BLM, topoisomerase IIIα, RMI1, RMI2) and, through this binding, FANCM recruits the Bloom complex to chromatin. Consistently, both the Bloom complex and FANCM suppress sister chromatid exchanges (SCEs).

Thus, FANCM is a central player in sensing stalled replication forks, recruiting both the FA core complex and the Bloom complex, and activating the ATR-CHK1 pathway in response to DNA replication stress.

MONOUBIQUITINATION OF THE ID COMPLEX AND RECRUITMENT OF NUCLEASES

The human FANC<u>I</u> and FANC<u>D</u>2 proteins share modest amino acid identity (13%) and similarity (20%), and are considered to be paralogs[21] (Fig. 2). Together, they form the "ID" protein complex. In response to treatment with DNA-damaging agents, such as ionizing radiation (IR), ultraviolet light, ICL-inducing agents, and hydroxyurea, the FA core complex monoubiquitinates FANCD2 [on lysine 561 (K561)] and FANCI [on lysine 523 (K523)]. Monoubiquitination of FANCD2 and FANCI also occurs during the S phase of the cell cycle even in the absence of treatment with exogenous DNA-damaging agents. Monoubiquitination of FANCD2 and of FANCI are mutually dependent. Monoubiquitinated

FANCD2 and FANCI are targeted to nuclear foci (Fig. 6), presumably at the sites of DNA damage and repair, where they colocalize with multiple factors required for cellular resistance to ICL-inducing agents (BRCA1, PALB2, BRCA2, RAD51, FAN1, RPA, FANCJ, etc.).

A DNA-damage-activated signaling kinase, ATR, is required for DNA-damage-induced monoubiquitination and nuclear focus formation of FANCD2.[19,55] However, in the absence of DNA damage, monoubiquitination and focus formation of FANCD2 during the normal S phase do not require ATR.[19] Phosphorylation of FANCI by ATR is a critical event for both FANCD2 and FANCI monoubiquitination,[24] although FANCD2 may also be phosphorylated by ATR.[19,25] One of the substrates of the ATR kinase, the CHK1 kinase, also modulates FANCD2 monoubiquitination. CHK1 is required specifically for exogenous DNA-damage-inducible upregulation of monoubiquitination and focus formation of FANCD2, while basal monoubiquitination of FANCD2 is elevated in CHK1-depeleted cells.[53,56,57] Furthermore, CHK1-mediated phosphorylation on serine 331 of FANCD2 is reported to be required for DNA-damage-induced monoubiquitination of FANCD2.[58] CHK1 also directly phosphorylates FANCE on threonine 346 and serine 374 in response to DNA damage, especially UV-induced DNA damage,[57] although FANCE phosphorylation is not required for FANCD2 monoubiquitination or focus formation.

Another DNA-damage-activated signaling kinase, ATM, directly phosphorylates FANCD2 on serine 222, but this phosphorylation is not required for FANCD2 monoubiquitination or cellular resistance to ICL-inducing agents (reviewed in Ref 1). Instead, it is required for establishment of the IR-induced S phase checkpoint. This phosphorylation is dependent on NBS1, whose biallelic germline mutations cause a distinct genomic instability and cancer predisposition syndrome — Nijmegen breakage syndrome. NBS1 has a well-characterized role in ATM activation in response to IR, so it is not a surprise that its mutation also causes genomic instability. It forms a protein complex with MRE11 and RAD50 [MRE11-RAD50-NBS1 (MRN) complex], which plays a critical role in sensing DSB. Thus, FANCD2 has a role in ATM-dependent cell cycle checkpoint activation besides a role in the FA pathway. In addition, the MRN complex is reported to be crucial for stability of the FANCD2 protein.[59]

Monoubiquitination of FANCD2 is required for nuclear focus formation and chromatin localization of FANCD2, and is critical for cellular resistance to ICLs,[20,60] while monoubiquitination of FANCI is not required for nuclear focus formation and chromatin localization of FANCI and FANCD2, and plays a minor role in cellular resistance to ICLs.[24] The mechanism by which monoubiquitination regulates chromatin localization and focus formation of FANCD2 is still unknown: a receptor of monoubiquitinated FANCD2 has not been identified. However, recent identification of the FAN1 nuclease provides some clues to how monoubiquitinated FANCD2 regulates ICL repair.[61–64]

FAN1 has several domains, including a ubiquitin-binding (UBZ) domain and a C-terminus nuclease domain (Fig. 2). FAN1-deficient cells are hypersensitive to ICL-inducing agents. FAN1 is recruited to sites of DNA damage through direct binding of the UBZ domain to monoubiquitin on FANCD2. It has an endonuclease activity preferentially toward 5′ flap structures (a 5′ flap structure is a bifurcated DNA structure composed of double-stranded DNA and a displaced single strand with a 5′ end) and a 5′-3′ exonuclease activity.[61–64]

How FAN1 participates in ICL repair is not yet clear. In the replication-dependent ICL repair model shown in Fig. 3,[12,13] initial dual incisions to unhook the ICL lesion are dependent on FANCD2 monoubiquitination. The FANCP/SLX4 nuclease complex, which includes the proteins XPF-ERCC1 (a 3′ flap endonuclease), MUS81-EME1 (a 3′ flap endonuclease), and SLX1 [SLX1-SLX4; a Holliday junction (HJ) resolvase], may collaborate with FAN1 in this process. MUS81-EME1 (a 3′ flap endonuclease) has been implicated in the first incision, and FAN1 with its 5′ flap endonuclease activity may participate in the second incision to unhook the ICL.[61] XPF-ERCC1 has been implicated in this second incision, but there is a controversy.[65] It is unlikely that SLX1 will be important for this step, because SLX1-deficient human cells are not sensitive to ICL-inducing agents.[66] The fully unhooked ICL adduct must be removed, presumably by nucleotide excision repair (NER), but FAN1 may also be involved in this step.[61]

FAN1 is required for efficient HR in human cells,[63,64] and the 5′-3′ exonuclease activity of FAN1 may contribute to the processing of the DSB ends to generate a 3′ overhang for HR.[61] However, induction of cisplatin/mitomycin C-induced focus formation of RPA and RAD51 proteins is normal or even elevated in FAN1-depleted cells, and the disappearance of cisplatin/mitomycin C-induced RPA and RAD51 foci is substantially delayed in FAN1-depleted cells, suggesting that FAN1 is not required for resection, but is involved in a later stage of HR.[63,64] FAN1 also forms a protein complex with DNA mismatch repair proteins (MLH1, PMS1, PMS2, MLH3), but the functional significance of these interactions is unknown.[61–64] In a chicken B cell lymphoma cell line (DT40) that is popularly used for studies of DNA repair, FAN1 is not involved in regulation of HR, while FAN1 is still required for cellular resistance to ICL-inducing agents.[67] More surprisingly, DT40 cells deficient in both FAN1 and FANCC, or FAN1 and FANCJ, show increased cisplatin sensitivity compared to cells deficient in FAN1 alone, suggesting that FAN1 has some functions independent of the FA pathway.[67] Therefore, the precise mechanism by which FAN1 regulates ICL repair is still a mystery.

FANCP/SLX4 has tandem UBZ motifs (ubiquitin-binding domains)[33] (Fig. 2), and it is tempting to speculate that SLX4 may interact with some ubiquitinated proteins, such as FANCD2 and FANCI, and that the interaction leads to recruitment of the SLX4 complex to sites of DNA repair. Consistent with this speculation, at least in chicken DT40 cells, SLX4 is recruited to sites of DNA damage in an FANCD2-dependent manner.[68]

FANCD2 is itself reported to have a nuclease activity.[17] It is reported to have a 3′-5′ exonuclease activity, but does not have an obvious nuclease domain.[17] How this activity contributes to ICL repair has yet to be determined.

USP1 DEUBIQUITINATES THE ID COMPLEX AND PCNA: THE FA PATHWAY AND TRANSLESION SYNTHESIS

USP1, a deubiquitinating enzyme, together with its partner UAF1 (USP1-associated factor 1), removes ubiquitin from FANCD2 and FANCI, thus negatively regulating the FA pathway.[26] It also removes ubiquitin from monoubiquitinated PCNA (proliferating cell nuclear antigen), which is a processivity factor for DNA polymerases. Monoubiquitination of PCNA is mediated by RAD6 (an E2) and RAD18 (an E3), and is involved in a DNA polymerase switch from a replicative polymerase to translesion synthesis

(TLS) polymerases, such as Pol η, REV1, and Pol ζ (REV3-REV7), at the site of blocked replication forks.[69] TLS is a DNA damage tolerance mechanism that allows the bypass of damaged nucleotides during replication. In TLS, specialized error-prone DNA polymerases (TLS polymerases) are recruited to DNA lesions that would otherwise block the high-fidelity replicative polymerases (such as Pol δ and Pol ε). These TLS polymerases have ubiquitin-binding motifs (UBM and UBZ domains) that mediate their association with monoubiquitinated PCNA. The same factors involved in TLS are required for cellular resistance to ICL. Thus, these two mechanisms (the FA pathway and TLS) share a common shutoff mechanism mediated by USP1. Intriguingly, USP1-deficient cells show hypersensitivity to ICL-inducing agents, and persistent monoubiquitinated FANCD2 (but not PCNA) is responsible for this ICL sensitivity.[70] These results suggest that monoubiquitination of FANCD2 must be tightly regulated for cellular resistance to ICL-inducing agents.

Several connections between activation of the FA pathway and that of TLS have been suggested.[71] First, cells deficient in FA core complex genes are reported to be hypomutable (i.e. point mutations occur less frequently in these cells than in normal cells.) This phenotype is similar to the phenotype of TLS-polymerase-deficient cells. Second, cells deficient in either of the TLS polymerases REV1 and REV3 show crosslinker hypersensitivity that is epistatic to (i.e. is unaffected by) FANCC deficiency. Third, REV1 and FANCD2 colocalize in nuclear foci upon replication arrest. The recruitment of REV1 into nuclear foci depends on an intact FA core complex, but not on FANCD2 or FANCI.[39] Fourth, FANCD2 has a PIP box (a PCNA-binding domain), physically interacts with PCNA, and the PIP box is required for FANCD2 monoubiquitination.[72] FANCL also directly binds to PCNA.[73] Furthermore, in replication-dependent ICL repair modeled in *Xenopus* egg extracts (Fig. 3), the TLS step is dependent on FANCD2 monoubiquitination and FANCI.[13]

Interestingly, monoubiquitination of FANCD2 (and nuclear focus formation of FANCD2) may be facilitated by monoubiquitination of PCNA by Rad6 (HHR6A and HHR6B) and RAD18 (Fig. 4),[73–75] although Rad18 mutant murine fibroblasts are capable of supporting FANCD2 monoubiquitination in response to UVC irradiation,[39] and loss of RAD18 in DT40 cells does not significantly affect FANCD2 monoubiquitination.[76]

Monoubiquitinated PCNA facilitates FANCL-mediated monoubiquitination of FANCD2 in an *in vitro* ubiquitination assay.[73] Efficient monoubiquitination of FANCD2 is conferred by RAD18-mediated PCNA monoubiquitination and subsequent recruitment of a TLS polymerase (pol η), which leads to chromatin association of the FA core complex.[75] In turn, FANCD2 has been reported to be required for efficient UV-induced (but not cisplatin-induced) nuclear focus formation of pol η.[74] Thus, TLS activation and FA pathway activation appear to be connected, although the connection seems quite complex and there are some unresolved contradictions among reports.

FANCD1/BRCA2, FANCN/PALB2, AND FANCO/RAD51C MEDIATE RAD51 LOADING AND REGULATE HOMOLOGOUS RECOMBINATION

DNA double-stranded breaks (DSBs) can occur during ICL repair (Fig. 3). There are two major pathways for repairing DSBs: NHEJ and HR (Fig. 8). NHEJ mediates direct ligation of broken ends, while HR utilizes the undamaged homologous sequence (for example, a sister chromatid or homologous chromosome sequence) as a template to repair DSBs. In addition, there are two less characterized pathways for repairing DSBs: single-stranded annealing (SSA) and microhomology-mediated end joining (MMEJ) (see Fig. 8 legend).[77] How cells deploy these pathways during DSB repair is a highly active area of current research. Among these four pathways, HR plays a major role for the repair of DSBs arising during replication-dependent ICL repair.

The RAD51 recombinase is a critical factor for HR.[78] RAD51 polymerizes to form extended filaments on ssDNA and promotes the DNA strand exchange steps of HR. To initiate HR, DNA ends at the DSBs are digested to generate 3′ single-stranded tails, in a reaction called "resection." The resulting single-stranded DNA (ssDNA) immediately becomes coated and stabilized by a ssDNA-binding protein complex, RPA. RPA competes with the RAD51 recombinase for binding sites on the ssDNA. With the help of a number of accessory factors (mediators), RAD51 can displace RPA, and be loaded onto ssDNA. This presynaptic filament of the RAD51-ssDNA complex invades homologous duplex DNA to generate a structure called the D loop (Fig. 8). At the D loop,

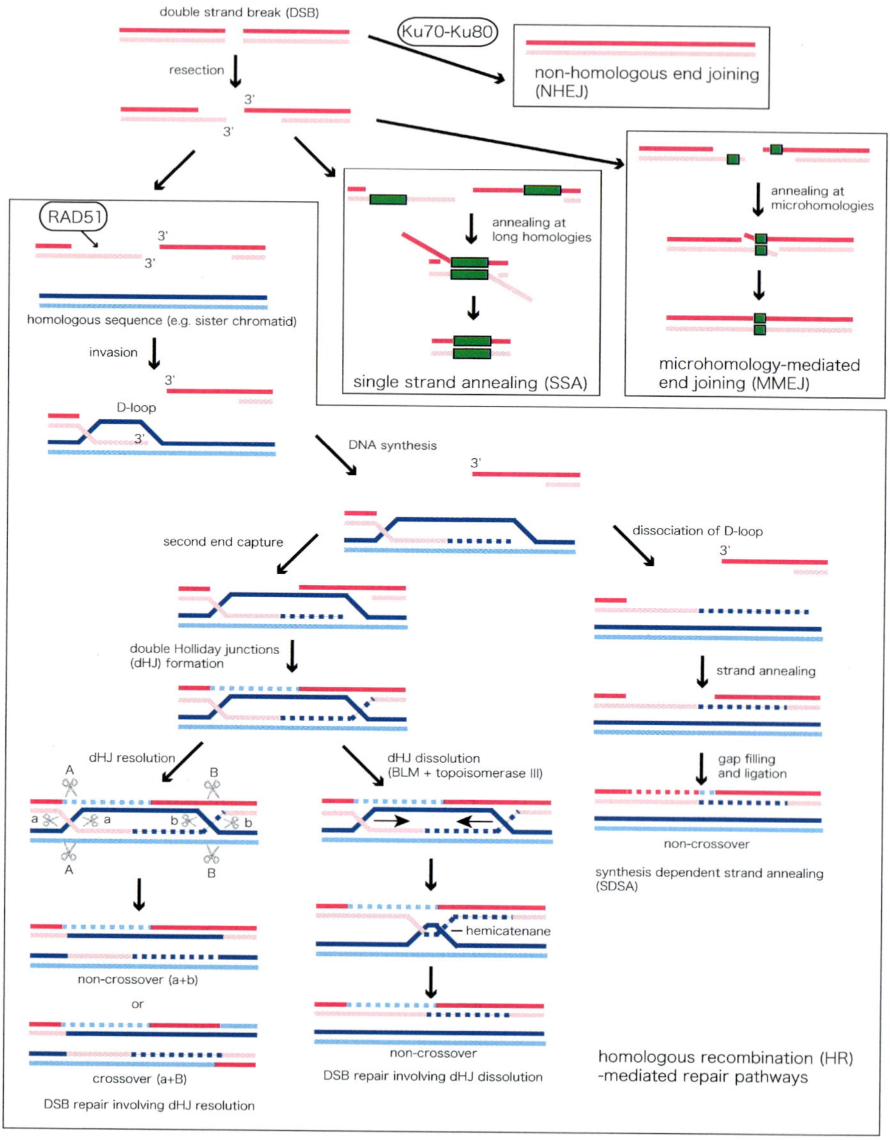

Fig. 8 Model of repair of double strand breaks in mammalian cells. (Adapted with modifications from Fig. 4 of Ref. 23, and from Fig. 1 of Ref. 77.) Mammalian cells have two major pathways for repairing DNA double-strand breaks (DSBs) [homologous recombination (HR) and nonhomologous end joining (NHEJ)] and two less characterized pathways [single-strand annealing (SSA) and microhomology-mediated end joining (MMEJ)].

DNA synthesis extends the invading strand. Then, the recombination process can be routed to one of the three pathways shown in Fig. 8: (1) synthesis-dependent strand annealing (SDSA), (2) DSB repair involving double Holliday junction (dHJ) resolution, or (3) DSB repair involving dHJ dissolution.

Three FA proteins (BRCA2, PALB2, RAD51C), which are not required for FANCD2-FANCI monoubiquitination, mediate RAD51 loading on ssDNA and facilitating HR.[28,29,79,80] Cells deficient in BRCA2, PALB2, or RAD51C are unable to efficiently form RAD51 foci, which is a marker of RAD51 loading on ssDNA. These cells are also deficient in HR, and consequently are sensitive to ICL-inducing agents.

BRCA2 directly binds to RAD51 and potentiates HR by promoting assembly of RAD51 onto ssDNA.[29] It is suggested to bind to the ssDNA-dsDNA junction of resected DNA ends of DSBs, thereby helping to load RAD51 onto ssDNA.

PALB2 directly binds to BRCA2 and regulates stability and localization of BRCA2.[28] It also directly binds to BRCA1, linking BRCA1 and BRCA2 to form a "BRCA complex,"[28] which plays a critical role in RAD51 loading.

Fig. 8 (*Continued*) During NHEJ repair, Ku70-Ku80 binds to DSB ends and prevents DNA end resection. HR, SSA, and MMEJ require DNA end resection to generate 3′ single-strand DNA (ssDNA) overhang. MMEJ is a Ku-independent end joining process mediated by base pairing between microhomologous sequences.[77] SSA is mediated by base pairing between longer homologous sequences.[77] Neither SSA nor MMEJ requires RAD51. In contrast, RAD51 is a critical factor for HR.[78] When DNA break ends are resected, the resulting ssDNA immediately becomes coated with RPA. With the help of a number of mediators, RAD51 can displace RPA, and be loaded onto ssDNA. This RAD51-ssDNA complex invades homologous duplex DNA (for example, sister chromatid) to generate a displacement loop (D loop). At the D loop, DNA synthesis occurs to extend the invading strand. Then, the recombination process will be routed to one of the three pathways: synthesis-dependent strand annealing (SDSA), DSB repair involving double Holliday junction (dHJ) resolution, and DSB repair involving dHJ dissolution. SDSA and DSB repair involving dHJ dissolution lead to generation of noncrossover products, while DSB repair involving dHJ resolution can lead to generation of either crossover products or noncrossover products, depending on how the dHJs are cut. BLM is involved in multiple steps in HR, including DSB resection and dHJ dissolution. In dHJ dissolution, BLM acts on the dHJ to promote convergent branch migration, resulting in creation of a hemicatenane. The hemicatenane is then unlinked by the single-strand DNA passing activity of topoisomerase IIIα.[105]

In addition, PALB2 preferentially binds D loop structures in DNA. It also physically interacts with RAD51 and enhances RAD51's ability to stimulate strand invasion and form the D loop.[30,31]

RAD51C is one of the five human "RAD51 paralogs," which include XRCC2, XRCC3, RAD51B/RAD51L1, RAD51C/RAD51L2, and RAD51D/RAD51L3.[8] These RAD51 paralogs form two distinct protein complexes: the RAD51B-RAD51C-RAD51D-XRCC2 complex and the RAD51C-XRCC3 complex. RAD51C has a role in both early and late stages in HR. It colocalizes with RAD51 in nuclear foci in response to treatment with IR or the replication inhibitor hydroxyurea (HU).[81] This RAD51C recruitment is dependent on the ATM kinase, NBS1, and RPA. RAD51C is also required for efficient RAD51 focus formation.[82] RAD51C and XRCC3 are part of a complex that has DNA branch migration and Holliday junction resolution activities.[83] Consistent with this role at a late stage in HR, RAD51C foci are retained after RAD51 focus disassembly.[81] Interestingly, RAD18 directly binds to RAD51C and recruits RAD51C at sites of DNA damage and regulates HR.[84] This function of RAD18 is distinct from its well-known function of ubiquitinating PCNA. The multiple connections between RAD18 and the FA pathway (Fig. 4) may explain an extreme cisplatin sensitivity of RAD18-deficient cells.

Judging from the critical role of HR in ICL repair, it is plausible that mutations of other RAD51 paralogs, as well as other genes involved in HR, may cause FA phenotypes.

FANCJ/BRIP1 HELICASE

FANCJ/BRIP1 was originally identified as the helicase BACH1 (BRCA1-associated C-terminal helicase 1). It possesses DNA-dependent ATPase and 5′-to-3′ DNA helicase activities, and belongs to the DEAH subfamily of superfamily2 (SF2) helicases [helicases with the aspartate–glutamate–alanine–histidine box motif (DEAH) domain] (summarized in Refs. 85 and 86). FANCJ helicase activity is required for cellular resistance to ICLs. *In vitro*, FANCJ can unwind simple duplex DNA substrates, but requires a minimal 5′ ssDNA tail of 15 nucleotides. FANCJ is able to release the third strand of the HR intermediate D loop structure irrespective of the DNA tail status. In contrast, it cannot unwind or branch-migrate synthetic Holliday junction structures. More detailed discussions about FANCJ helicase activity can be found elsewhere.[85,86]

In human cells, FANCJ is required for efficient HR, but this does not seem conserved in chicken DT40 cells or *C. elegans*. FANCJ directly binds to the BRCT domain of BRCA1 in human cells, when serine 990 of FANCJ is phosphorylated.[32] However, this BRCA1-FANCJ interaction is not required for cellular resistance to ICL-inducing agents, and serine 990 is not conserved in chickens.[85] In human cells, in the absence of the BRCA1-FANCJ interaction, the efficiency of HR is decreased, but pol η-dependent bypass is promoted and, overall, cells are resistant to ICL-inducing agents.[87]

FANCJ forms a protein complex with the DNA mismatch repair proteins MLH1-PMS2 (MutLα). It directly interacts with MLH1 through its helicase domain, and the FANCJ-MLH1 interaction is required for cellular resistance to ICLs.[88] Currently, how FANCJ-MLH1 interaction functions in the ICL response/repair remains unknown.

FANCJ also has a specialized role under replication stress. It directly binds to the RPA70 subunit of RPA, and RPA stimulates FANCJ to unwind duplex DNA substrates.[89] Guanine (G)-rich DNA can form G4 DNA (G-quadruplex DNA), and FANCJ has a well-defined role in unwinding G4 DNA substrates.[90] FANCJ can unwind G4 DNA substrates *in vitro*, and RPA stimulates this unwinding activity.[90] This activity is specific to FANCJ and is independent of the other FA proteins.

FANCJ also plays a role in the early steps of DNA replication checkpoint control.[91] (Fig. 7). This function is mediated by interaction between FANCJ and TopBP1, and involves FANCJ's DNA unwinding activity. TopBP1 directly interacts with FANCJ, and this interaction is mediated by C-terminal tandem BRCT domains of TopBP1 and phosphorylation at threonine 1133 of FANCJ. This FANCJ phosphorylation is catalyzed by CDK in the S phase. Both FANCJ and TopBP1 facilitate RPA loading and ATR-dependent phosphorylation in response to replication stress, suggesting that a specific interaction between TopBP1 and FANCJ may facilitate the extension of ssDNA regions.

Interestingly, FANCJ helicase physically and functionally interacts with the BLM helicase[92] (Fig. 7), and FANCJ and BLM colocalize in nuclear foci in response to replication stress. In the absence of FANCJ,

BLM is degraded by a proteasome-mediated process. Additionally, as with BLM-deficient cells, FANCJ-deficient cells show increased SCE and sensitivity to replication stress.

NONHOMOLOGOUS END JOINING AND THE FANCONI ANEMIA PATHWAY

Suppression of NHEJ during ICL repair is an important function of the FA pathway, although the precise mechanism is unknown and there are some discrepancies among reports, as described below.[17,18,93] During ICL repair, DSBs are generated (Fig. 3), and if NHEJ is not suppressed, processing of DSBs required for loading of HR factors can be inhibited and illegitimate repair may occur. Human FANCD2-deficient cells are highly sensitive to ICL-inducing agents, but abrogation of NHEJ by inhibition of DNA-PK or Ku80 restores resistance to ICL-inducing agents to wild-type levels.[18] Furthermore, in FANCD2-deficient cells, phosphorylated DNA-PKcs (DNA-PK catalytic subunit) is recruited to sites of ICL-induced (or hydroxyurea-induced) DNA damage, while in wild-type cells this does not occur.[18] These findings suggest that suppression of NHEJ during replication-dependent ICL repair is one of the most critical functions of FANCD2. If NHEJ is inhibited in human FANCA- or FANCC-deficient cells, the cells become only partially resistant to ICL-inducing agents, suggesting that the FA core complex not only participates in suppression of NHEJ, but also has other functions important for ICL resistance.[18] FANCD2 and FA core complex proteins may directly antagonize NHEJ by blocking or displacing NHEJ factors from DSB ends, or may coordinate the processing of ICLs to disfavor NHEJ and favor other pathways (HR, SSA, and MMEJ).[18] Consistent with this, decreased HR and SSA efficiencies are observed in FANCA-, FANCG-, and FANCD2-deficient cells.[1] Furthermore, FANCA-, FANCC-, and FANCD2-deficient FA cell lines may have decreased MMEJ (see caption of Fig. 8) efficiency.[77,94]

FANCC-deleted chicken DT40 cells are highly sensitive to cisplatin.[17] Interestingly, combined ablation of FANCC with an NHEJ factor, Ku70, but not with other NHEJ factors (DNA-PKcs, ligase IV), suppresses cisplatin sensitivity in the *FANCC*-deleted cells.[17] However, inconsistent with these data,[17,18] *Fancd2*-knockout mouse embryonic fibroblasts (MEFs) with DNA-PKcs deficiency are just as sensitive to an ICL-inducing agent as

Fancd2-knockout MEFs.[93] Thus, there may be tissue-specific requirements in cisplatin sensitivity in genetically deleted fibroblasts vs. other tissues.

THE FA PATHWAY IN SPORADIC HUMAN CANCER

There are numerous reports on alterations of the FA pathway in sporadic cancer, including promoter methylation of the *FANCF* gene in various types of cancers, *FANCC* or *FANCG* mutations in early-onset pancreatic cancer, and elevated expression of a splice variant of *FANCL* in various cancers[95] (reviewed in Ref. 1). These FA pathway alterations lead to inactivation of the FA pathway, which will cause genomic instability in premalignant or malignant cells and facilitate cancer progression.

Despite the potential roles of FA pathway deficiency during cancer progression, defects in the FA pathway may be exploited in treatment strategies. Because the FA pathway is critical for cellular resistance to many widely used anticancer ICL-inducing agents, such as cisplatin, carboplatin, melphalan, cyclophosphamide, and mitomycin C, inactivation of the pathway should confer chemosensitivity on these agents. Cancer cells with disrupted FA pathway integrity are also sensitive to poly(ADP-ribose) polymerase (PARP) inhibitors such as Olaparib (AZD-2281).[96] Furthermore, FA-pathway-deficient cells are sensitive to ATM inhibition,[97] suggesting that ATM inhibitors such as KU55933 may be useful in the treatment of FA-pathway-deficient cancers. Conversely, reactivation of the FA pathway in the FA-pathway-deficient cancer cells is a documented mechanism of acquired resistance to these cisplatin and PARP inhibitors.[98–104] Indeed, secondary mutations of *BRCA1/2* that restore the wild-type *BRCA1/2* reading frame were observed in platinum-resistant recurrent *BRCA1/2*-mutated ovarian carcinoma and are considered to be the mechanism of acquired cisplatin resistance.[99–104] Small molecule inhibitors of the FA pathway may therefore be useful as chemosensitizers in the treatment of FA-pathway-proficient, chemoresistant tumors.

FA-BLM INTERACTION

BLM is a 3′-5′ DNA helicase, and a member of the RecQ family of DNA helicases, which are defined as proteins sharing a homologous region with *Escherichia coli* RecQ.[105] In humans, there are five members of the RecQ helicase family: RECQ1, BLM, WRN (defective in Werner syndrome), RECQ4 (defective in Rothmund–Thompson syndrome, Baller–Gerold syndrome, and RAPADALINO), and RECQ5. All of the RecQ helicases exhibit a strand-annealing activity in addition to their ATP-dependent DNA helicase activities. BLM-deficient cells show elevated frequencies of chromosome breaks, interchanges between homologous chromosomes, and SCEs. BLM protein forms protein complex with topoisomerase IIIα, RMI1, RMI2, and RPA (known as the Bloom complex) (Fig. 7).[105]

The BLM helicase plays important roles in HR (Fig. 8).[105] First, BLM is involved in DNA end resection to generate ssDNA. Second, the BLM-topoisomerase IIIα-RMI1 complex can remove double Holliday junctions in a reaction called "dissolution," which results in only non-crossover products (see Fig. 8). This is the main reason why cells deficient in BLM show marked increase in the rate of SCE, which reflects increased generation of crossover products. Third, BLM stimulates DNA strand exchange activity of RAD51 if RAD51 is present in an active ATP-bound form, while it can disrupt RAD51-ssDNA filaments if RAD51 is present in an inactive ADP-bound form. This suggests that BLM disrupts RAD51 filaments, when cells are not fully prepared for the later steps of HR and RAD51 is present in an inactive ADP-bound form. Fourth, BLM may promote synthesis-dependent strand annealing (SDSA) (see caption of Fig. 8) by facilitating DNA synthesis of the invading strand in a D loop and by dissociating the invading strand from the D loop.[106] On the other hand, the dissociation of the invading strand from the D loop may prevent D loop structures from being converted into mature recombinants.[106] This may explain the hyper-recombination phenotype of BLM-deficient cells.

BLM also has several functions during DNA replication.[107] It can unwind G4 DNA structures which interfere with DNA replication. It is a substrate of the ATM and ATR kinases, and is required for the efficient restart of stalled replication forks after replication is blocked and for the suppression of dormant origin firing when the S phase checkpoint is activated. BLM may promote fork reversal (or fork regression) at stalled forks, since it catalyzes the regression of model fork structures *in vitro.*

BLM is proposed to play a role in the resolution of late-stage replication intermediates. In the late stage of replication, every fork inevitably collides with another fork that has started from a neighboring replication origin.[107] These converging replication forks generate late-stage replication intermediates that need to be resolved prior to completion of the S phase. By interacting with topoisomerase IIIα and RMI1, BLM appears to be involved in this process. Failure to resolve late-stage replication intermediates can cause abnormal sister chromatid disjunction during mitosis, which is reflected by increased anaphase bridges, lagging chromatin, and micronucleations observed in BLM-deficient cells.[108]

Anaphase bridges link sister chromatids during cell division, and are divided into three categories: (1) conventional DAPI (4′,6-diamino-2-phenylindole)-positive bridges (DAPI is the DNA-binding dye), (2) ultra-fine DAPI-negative DNA bridges (UFBs) (although they are not stained with DAPI, they contain DNA) originating from centromeric DNA, and (3) UFBs whose extremities localize with FANCD2/FANCI foci.[109,110] Many of these FANCD2/FANCI foci localize with common fragile sites. These UFBs are observed even in normal cells in mitosis. The BLM-topoisomerase IIIα-RMI1 complex localizes at these UFBs in early mitosis. FANCM is recruited to the UFBs, but at a later stage.[111] Therefore, both FA proteins and BLM have a role in regulating sister chromatid segregation during mitosis, preventing micronucleation (i.e. generation of micronuclei derived from chromosomal fragments and whole chromosomes lagging behind in anaphase) and chromosome abnormalities.[108,110,111]

FA proteins and BLM interact with each other in various ways. First, as discussed above, they cooperate at UFBs.[109–111] Second, the FA core and Bloom complexes are connected through the FANCM protein[54] (Fig. 7), and form a larger complex that has been called BRAFT.[35] Third, FANCJ physically and functionally interacts with BLM, and stabilizes BLM.[92] Fourth, BLM and FANCD2 colocalize and coimmunoprecipitate following treatment with ICL-inducing agents or agents inducing replication arrest. The FA core complex is required for BLM phosphorylation specifically in response to ICL-inducing agents, suggesting a role of activating DNA damage response to ICL. ICL-induced nuclear focus formation of BLM is dependent on FANCC, FANCG, and FANCD2. Interestingly, in chicken DT40 cells, increased SCE in the FANCC/BLM double mutant cells was similar to that in BLM single mutant cells, indicating a functional linkage between FANCC and BLM in suppressing SCEs (reviewed in Ref. 1). Therefore, the two chromosome instability syndromes, FA and BS, are connected at the molecular level.

CONCLUDING REMARKS

FA and BS are distinct genetic diseases, but identification of the responsible genes has enabled a dissection of several intertwined molecular interactions between FA proteins and BLM. There remain many critical unanswered questions related to the FA pathway. First, how FA proteins cooperate with other proteins to repair ICLs remains a central question in the field. Second, the intracellular signaling that regulates the FA pathway has not been completely elucidated. For example, it is well established that phosphorylation of several FA proteins by ATR kinase and CHK1 kinase is important for activation of the pathway, but which signals and phosphatases reverse these phosphorylations is not known. Third, while a defect in DNA repair can explain increased genomic instability and cancer risk in FA patients, why these patients develop aplastic anemia and have congenital dysmorphic defects is not understood. Fourth, effective measures to prevent tumors and aplastic anemia in FA patients have not been established. Finally, further elucidation of the role of the FA pathway in cancers is warranted. A better understanding of the molecular mechanisms by which the FA pathway and BLM operate should lead to better diagnosis and treatment for patients with these disorders, and a subset of sporadic cancer as well.

ACKNOWLEDGMENTS

We thank Drs. Koji Nakanishi and Maria Jasin for sharing data prior to publication. We also thank members of Taniguchi Lab for critical reading of the manuscript. We apologize that we cannot cite some of the important literature because of the limitation of space. This work is supported by the Howard Hughes Medical Institute, NIH/National Cancer Institute (R01CA125636 to T. T.), NIH/National Heart, Lung and Blood Institute (R21 HL092978 to T. T.), Chromosome Metabolism and Cancer Training

Grant (T32CA09657 to K. K. D.), and NIH/Ruth L. Kirschstein National Research Service Award (T32CA009351 to R. S. C.).

REFERENCES

1. Taniguchi T, D'Andrea AD. (2006) Molecular pathogenesis of Fanconi anemia: recent progress. *Blood* **107**: 4223–4233.
2. Wang W. (2007) Emergence of a DNA-damage response network consisting of Fanconi anaemia and BRCA proteins. *Nat Rev Genet* **8**: 735–748.
3. D'Andrea AD. (2010) Susceptibility pathways in Fanconi's anemia and breast cancer. *N Engl J Med* **362**: 1909–1919.
4. Kee Y, D'Andrea AD. (2010) Expanded roles of the Fanconi anemia pathway in preserving genomic stability. *Genes Dev* **24**: 1680–1694.
5. Alter BP. (2003) Inherited bone marrow failure syndromes. In: Nathan DG, Orkin SH, Ginsburg D, *et al.* (eds.), *Nathan and Oski's Hematology of Infancy and Childhood*, 6th edn. PA Saunders, Philadelphia: pp. 280–365.
6. Stoepker C, Hain K, Schuster B, *et al.* (2011) SLX4, a coordinator of structure-specific endonucleases, is mutated in a new Fanconi anemia subtype. *Nat Genet* **43**: 138–141.
7. Kim Y, Lach FP, Desetty R, *et al.* (2011) Mutations of the SLX4 gene in Fanconi anemia. *Nat Genet* **43**: 142–146.
8. Somyajit K, Subramanya S, Nagaraju G. (2010) RAD51C: a novel cancer susceptibility gene is linked to Fanconi anemia and breast cancer. *Carcinogenesis* **31**: 2031–2038.
9. Payne M, Hickson ID. (2009) Genomic instability and cancer: lessons from analysis of Bloom's syndrome. *Biochem Soc Trans* **37**: 553–559.
10. Amor-Guéret M. (2004) Bloom's syndrome. *Orphanet Encyclopedia*: http://www.orpha.net/data/patho/GB/uk-Bloomsyndrome.pdf
11. Muniandy PA, Liu J, Majumdar A, *et al.* (2010) DNA interstrand crosslink repair in mammalian cells: step by step. *Crit Rev Biochem Mol Biol* **45**: 23–49.
12. Raschle M, Knipscheer P, Enoiu M, *et al.* (2008) Mechanism of replication-coupled DNA interstrand crosslink repair. *Cell* **134**: 969–980.
13. Knipscheer P, Raschle M, Smogorzewska A, *et al.* (2009) The Fanconi anemia pathway promotes replication-dependent DNA interstrand cross-link repair. *Science* **326**: 1698–1701.
14. Shen X, Do H, Li Y, *et al.* (2009) Recruitment of Fanconi anemia and breast cancer proteins to DNA damage sites is differentially governed by replication. *Mol Cell* **35**: 716–723.

15. Nakanishi K, Cavallo F, Perrouault L, *et al.* (2011) Homology-directed Fanconi anemia pathway cross-link repair is dependent on DNA replication. *Nat Struct Mol Biol* **18**: 500–503.
16. Nakanishi K, Yang YG, Pierce AJ, *et al.* (2005) Human Fanconi anemia monoubiquitination pathway promotes homologous DNA repair. *Proc Natl Acad Sci USA* **102**: 1110–1115.
17. Pace P, Mosedale G, Hodskinson MR, *et al.* (2010) Ku70 corrupts DNA repair in the absence of the Fanconi anemia pathway. *Science* **329**: 219–223.
18. Adamo A, Collis SJ, Adelman CA, *et al.* (2010) Preventing nonhomologous end joining suppresses DNA repair defects of Fanconi anemia. *Mol Cell* **39**: 25–35.
19. Andreassen PR, D'Andrea AD, Taniguchi T. (2004) ATR couples FANCD2 monoubiquitination to the DNA-damage response. *Genes Dev* **18**: 1958–1963.
20. Garcia-Higuera I, Taniguchi T, Ganesan S, *et al.* (2001) Interaction of the Fanconi anemia proteins and BRCA1 in a common pathway. *Mol Cell* **7**: 249–262.
21. Smogorzewska A, Matsuoka S, Vinciguerra P, *et al.* (2007) Identification of the FANCI protein, a monoubiquitinated FANCD2 paralog required for DNA repair. *Cell* **129**: 289–301.
22. Meetei AR, Sechi S, Wallisch M, *et al.* (2003) A multiprotein nuclear complex connects Fanconi anemia and Bloom syndrome. *Mol Cell Biol* **23**: 3417–3426.
23. Whitby MC. (2010) The FANCM family of DNA helicases/translocases. *DNA Repair (Amst)* **9**: 224–236.
24. Ishiai M, Kitao H, Smogorzewska A, *et al.* (2008) FANCI phosphorylation functions as a molecular switch to turn on the Fanconi anemia pathway. *Nat Struct Mol Biol* **15**: 1138–1146.
25. Pichierri P, Rosselli F. (2004) The DNA crosslink-induced S-phase checkpoint depends on ATR-CHK1 and ATR-NBS1-FANCD2 pathways. *EMBO J* **23**: 1178–1187.
26. Cohn MA, Kowal P, Yang K, *et al.* (2007) A UAF1-containing multisubunit protein complex regulates the Fanconi anemia pathway. *Mol Cell* **28**: 786–797.
27. Zhang F, Fan Q, Ren K, *et al.* (2010) FANCJ/BRIP1 recruitment and regulation of FANCD2 in DNA damage responses. *Chromosoma* **119**: 637–649.
28. Tischkowitz M, Xia B. (2010) PALB2/FANCN: recombining cancer and Fanconi anemia. *Cancer Res* **70**: 7353–7359.
29. Jensen RB, Carreira A, Kowalczykowski SC. (2010) Purified human BRCA2 stimulates RAD51-mediated recombination. *Nature* **467**: 678–683.
30. Dray E, Etchin J, Wiese C, *et al.* (2010) Enhancement of RAD51 recombinase activity by the tumor suppressor PALB2. *Nat Struct Mol Biol* **17**: 1255–1259.

31. Buisson R, Dion-Cote AM, Coulombe Y, *et al.* (2010) Cooperation of breast cancer proteins PALB2 and piccolo BRCA2 in stimulating homologous recombination. *Nat Struct Mol Biol* **17**: 1247–1254.
32. Cantor SB, Bell DW, Ganesan S, *et al.* (2001) BACH1, a novel helicase-like protein, interacts directly with BRCA1 and contributes to its DNA repair function. *Cell* **105**: 149–160.
33. Svendsen JM, Harper JW. (2010) GEN1/Yen1 and the SLX4 complex: solutions to the problem of Holliday junction resolution. *Genes Dev* **24**: 521–536.
34. Bhagwat N, Olsen AL, Wang AT, *et al.* (2009) XPF-ERCC1 participates in the Fanconi anemia pathway of cross-link repair. *Mol Cell Biol* **29**: 6427–6437.
35. Meetei AR, de Winter JP, Medhurst AL, *et al.* (2003) A novel ubiquitin ligase is deficient in Fanconi anemia. *Nat Genet* **35**: 165–170.
36. Machida YJ, Machida Y, Chen Y, *et al.* (2006) UBE2T is the E2 in the Fanconi anemia pathway and undergoes negative autoregulation. *Mol Cell* **23**: 589–596.
37. Alpi AF, Pace PE, Babu MM, *et al.* (2008) Mechanistic insight into site-restricted monoubiquitination of FANCD2 by Ube2t, FANCL, and FANCI. *Mol Cell* **32**: 767–777.
38. Matsushita N, Kitao H, Ishiai M, *et al.* (2005) A FancD2-monoubiquitin fusion reveals hidden functions of Fanconi anemia core complex in DNA repair. *Mol Cell* **19**: 841–847.
39. Mirchandani KD, McCaffrey RM, D'Andrea AD. (2008) The Fanconi anemia core complex is required for efficient point mutagenesis and Rev1 foci assembly. *DNA Repair (Amst)* **7**: 902–911.
40. Hicks JK, Chute CL, Paulsen MT, *et al.* (2010) Differential roles for DNA polymerases eta, zeta, and REV1 in lesion bypass of intrastrand versus interstrand DNA cross-links. *Mol Cell Biol* **30**: 1217–1230.
41. Yan Z, Delannoy M, Ling C, *et al.* (2010) A histone-fold complex and FANCM form a conserved DNA-remodeling complex to maintain genome stability. *Mol Cell* **37**: 865–878.
42. Singh TR, Saro D, Ali AM, *et al.* (2010) MHF1-MHF2, a histone-fold-containing protein complex, participates in the Fanconi anemia pathway via FANCM. *Mol Cell* **37**: 879–886.
43. Singh TR, Bakker ST, Agarwal S, *et al.* (2009) Impaired FANCD2 monoubiquitination and hypersensitivity to camptothecin uniquely characterize Fanconi anemia complementation group M. *Blood* **114**: 174–180.
44. Ciccia A, Ling C, Coulthard R, *et al.* (2007) Identification of FAAP24, a Fanconi anemia core complex protein that interacts with FANCM. *Mol Cell* **25**: 331–343.

45. Gari K, Decaillet C, Delannoy M, *et al.* (2008) Remodeling of DNA replication structures by the branch point translocase FANCM. *Proc Natl Acad Sci USA* **105**: 16107–16112.
46. Gari K, Decaillet C, Stasiak AZ, *et al.* (2008) The Fanconi anemia protein FANCM can promote branch migration of Holliday junctions and replication forks. *Mol Cell* **29**: 141–148.
47. Luke-Glaser S, Luke B, Grossi S, *et al.* (2010) FANCM regulates DNA chain elongation and is stabilized by S-phase checkpoint signalling. *EMBO J* **29**: 795–805.
48. Xue Y, Li Y, Guo R, *et al.* (2008) FANCM of the Fanconi anemia core complex is required for both monoubiquitination and DNA repair. *Hum Mol Genet* **17**: 1641–1652.
49. Collis SJ, Ciccia A, Deans AJ, *et al.* (2008) FANCM and FAAP24 function in ATR-mediated checkpoint signaling independently of the Fanconi anemia core complex. *Mol Cell* **32**: 313–324.
50. Schwab RA, Blackford AN, Niedzwiedz W. (2010) ATR activation and replication fork restart are defective in FANCM-deficient cells. *EMBO J* **29**: 806–818.
51. Huang M, Kim JM, Shiotani B, *et al.* (2010) The FANCM/FAAP24 complex is required for the DNA interstrand crosslink-induced checkpoint response. *Mol Cell* **39**: 259–268.
52. Zhang YW, Otterness DM, Chiang GG, *et al.* (2005) Genotoxic stress targets human Chk1 for degradation by the ubiquitin-proteasome pathway. *Mol Cell* **19**: 607–618.
53. Collis SJ, Barber LJ, Clark AJ, *et al.* (2007) HCLK2 is essential for the mammalian S-phase checkpoint and impacts on Chk1 stability. *Nat Cell Biol* **9**: 391–401.
54. Deans AJ, West SC. (2009) FANCM connects the genome instability disorders Bloom's syndrome and Fanconi anemia. *Mol Cell* **36**: 943–953.
55. Taniguchi T, Garcia-Higuera I, Xu B, *et al.* (2002) Convergence of the Fanconi anemia and ataxia telangiectasia signaling pathways. *Cell* **109**: 459–472.
56. Guervilly JH, Mace-Aime G, Rosselli F. (2008) Loss of CHK1 function impedes DNA damage-induced FANCD2 monoubiquitination but normalizes the abnormal G2 arrest in Fanconi anemia. *Hum Mol Genet* **17**: 679–689.
57. Wang X, Kennedy RD, Ray K, *et al.* (2007) Chk1-mediated phosphorylation of FANCE is required for the Fanconi anemia/BRCA pathway. *Mol Cell Biol* **27**: 3098–3108.

58. Zhi G, Wilson JB, Chen X, *et al.* (2009) Fanconi anemia complementation group FANCD2 protein serine 331 phosphorylation is important for Fanconi anemia pathway function and BRCA2 interaction. *Cancer Res* **69**: 8775–8783.
59. Roques C, Coulombe Y, Delannoy M, *et al.* (2009) MRE11-RAD50-NBS1 is a critical regulator of FANCD2 stability and function during DNA double-strand break repair. *EMBO J* **28**: 2400–2413.
60. Wang X, Andreassen PR, D'Andrea AD. (2004) Functional interaction of monoubiquitinated FANCD2 and BRCA2/FANCD1 in chromatin. *Mol Cell Biol* **24**: 5850–5862.
61. Smogorzewska A, Desetty R, Saito TT, *et al.* (2010) A genetic screen identifies FAN1, a Fanconi anemia-associated nuclease necessary for DNA interstrand crosslink repair. *Mol Cell* **39**: 36–47.
62. Liu T, Ghosal G, Yuan J, *et al.* (2010) FAN1 acts with FANCI-FANCD2 to promote DNA interstrand cross-link repair. *Science* **329**: 693–696.
63. MacKay C, Declais AC, Lundin C, *et al.* (2010) Identification of KIAA1018/FAN1, a DNA repair nuclease recruited to DNA damage by monoubiquitinated FANCD2. *Cell* **142**: 65–76.
64. Kratz K, Schopf B, Kaden S, *et al.* (2010) Deficiency of FANCD2-associated nuclease KIAA1018/FAN1 sensitizes cells to interstrand crosslinking agents. *Cell* **142**: 77–88.
65. Bergstralh DT, Sekelsky J. (2008) Interstrand crosslink repair: can XPF-ERCC1 be let off the hook? *Trends Genet* **24**: 70–76.
66. Fekairi S, Scaglione S, Chahwan C, *et al.* (2009) Human SLX4 is a Holliday junction resolvase subunit that binds multiple DNA repair/recombination endonucleases. *Cell* **138**: 78–89.
67. Yoshikiyo K, Kratz K, Hirota K, *et al.* (2010) KIAA1018/FAN1 nuclease protects cells against genomic instability induced by interstrand cross-linking agents. *Proc Natl Acad Sci USA* **107**: 21553–21557.
68. Yamamoto KN, Kobayashi S, Tsuda M, *et al.* (2011) Involvement of SLX4 in interstrand cross-link repair is regulated by the Fanconi anemia pathway. *Proc Natl Acad Sci USA* **108**: 6492–6496
69. Kannouche PL, Wing J, Lehmann AR. (2004) Interaction of human DNA polymerase eta with monoubiquitinated PCNA: a possible mechanism for the polymerase switch in response to DNA damage. *Mol Cell* **14**: 491–500.
70. Oestergaard VH, Langevin F, Kuiken HJ, *et al.* (2007) Deubiquitination of FANCD2 is required for DNA crosslink repair. *Mol Cell* **28**: 798–809.
71. Niedzwiedz W, Mosedale G, Johnson M, *et al.* (2004) The Fanconi anaemia gene FANCC promotes homologous recombination and error-prone DNA repair. *Mol Cell* **15**: 607–620.

72. Howlett NG, Harney JA, Rego MA, *et al.* (2009) Functional interaction between the Fanconi anemia D2 protein and proliferating cell nuclear antigen (PCNA) via a conserved putative PCNA interaction motif. *J Biol Chem* **284**: 28935–28942.
73. Geng L, Huntoon CJ, Karnitz LM. (2010) RAD18-mediated ubiquitination of PCNA activates the Fanconi anemia DNA repair network. *J Cell Biol* **191**: 249–257.
74. Park HK, Wang H, Zhang J, *et al.* (2010) Convergence of Rad6/Rad18 and Fanconi anemia tumor suppressor pathways upon DNA damage. *PLoS ONE* **5**: e13313.
75. Song IY, Palle K, Gurkar A, *et al.* (2010) Rad18-mediated translesion synthesis of bulky DNA adducts is coupled to activation of the Fanconi anemia DNA repair pathway. *J Biol Chem* **285**: 31525–31536.
76. Takata M, Ishiai M, Kitao H. (2009) The Fanconi anemia pathway: insights from somatic cell genetics using DT40 cell line. *Mutat Res* **668**: 92–102.
77. McVey M, Lee SE. (2008) MMEJ repair of double-strand breaks (director's cut): deleted sequences and alternative endings. *Trends Genet* **24**: 529–538.
78. Sung P, Krejci L, Van Komen S, *et al.* (2003) Rad51 recombinase and recombination mediators. *J Biol Chem* **278**: 42729–42732.
79. Liu J, Doty T, Gibson B, *et al.* (2010) Human BRCA2 protein promotes RAD51 filament formation on RPA-covered single-stranded DNA. *Nat Struct Mol Biol* **17**: 1260–1262.
80. Sigurdsson S, Van Komen S, Bussen W, *et al.* (2001) Mediator function of the human Rad51B-Rad51C complex in Rad51/RPA-catalyzed DNA strand exchange. *Genes Dev* **15**: 3308–3318.
81. Badie S, Liao C, Thanasoula M, *et al.* (2009) RAD51C facilitates checkpoint signaling by promoting CHK2 phosphorylation. *J Cell Biol* **185**: 587–600.
82. French CA, Masson JY, Griffin CS, *et al.* (2002) Role of mammalian RAD51L2 (RAD51C) in recombination and genetic stability. *J Biol Chem* **277**: 19322–19330.
83. Liu Y, Masson JY, Shah R, *et al.* (2004) RAD51C is required for Holliday junction processing in mammalian cells. *Science* **303**: 243–246.
84. Huang J, Huen MS, Kim H, *et al.* (2009) RAD18 transmits DNA damage signalling to elicit homologous recombination repair. *Nat Cell Biol* **11**: 592–603.
85. Hiom K. (2010) FANCJ: solving problems in DNA replication. *DNA Repair (Amst)* **9**: 250–256.
86. Wu Y, Suhasini AN, Brosh RM, Jr. (2009) Welcome the family of FANCJ-like helicases to the block of genome stability maintenance proteins. *Cell Mol Life Sci* **66**: 1209–1222.

87. Xie J, Litman R, Wang S, *et al.* (2010) Targeting the FANCJ-BRCA1 interaction promotes a switch from recombination to poleta-dependent bypass. *Oncogene* **29**: 2499–2508.
88. Peng M, Litman R, Xie J, *et al.* (2007) The FANCJ/MutLalpha interaction is required for correction of the cross-link response in FA-J cells. *EMBO J* **26**: 3238–3249.
89. Gupta R, Sharma S, Sommers JA, *et al.* (2007) FANCJ (BACH1) helicase forms DNA damage inducible foci with replication protein A and interacts physically and functionally with the single-stranded DNA-binding protein. *Blood* **110**: 2390–2398.
90. Wu Y, Shin-ya K, Brosh RM, Jr. (2008) FANCJ helicase defective in Fanconia anemia and breast cancer unwinds G-quadruplex DNA to defend genomic stability. *Mol Cell Biol* **28**: 4116–4128.
91. Gong Z, Kim JE, Leung CC, *et al.* (2010) BACH1/FANCJ acts with TopBP1 and participates early in DNA replication checkpoint control. *Mol Cell* **37**: 438–446.
92. Suhasini AN, Rawtani NA, Wu Y, *et al.* (2011) Interaction between the helicases genetically linked to Fanconi anemia group J and Bloom's syndrome. *EMBO J* **30**: 692–705.
93. Houghtaling S, Newell A, Akkari Y, *et al.* (2005) Fancd2 functions in a double strand break repair pathway that is distinct from non-homologous end joining. *Hum Mol Genet* **14**: 3027–3033.
94. Lundberg R, Mavinakere M, Campbell C. (2001) Deficient DNA end joining activity in extracts from Fanconi anemia fibroblasts. *J Biol Chem* **276**: 9543–9549.
95. Zhang J, Zhao D, Park HK, *et al.* (2010) FAVL elevation in human tumors disrupts Fanconi anemia pathway signaling and promotes genomic instability and tumor growth. *J Clin Invest* **120**: 1524–1534.
96. Ashworth A. (2008) A synthetic lethal therapeutic approach: poly(ADP) ribose polymerase inhibitors for the treatment of cancers deficient in DNA double-strand break repair. *J Clin Oncol* **26**: 3785–3790.
97. Kennedy RD, Chen CC, Stuckert P, *et al.* (2007) Fanconi anemia pathway-deficient tumor cells are hypersensitive to inhibition of ataxia telangiectasia mutated. *J Clin Invest* **117**: 1440–1449.
98. Taniguchi T, Tischkowitz M, Ameziane N, *et al.* (2003) Disruption of the Fanconi anemia-BRCA pathway in cisplatin-sensitive ovarian tumors. *Nat Med* **9**: 568–574.
99. Dhillon KK, Swisher EM, Taniguchi T. (2011) Secondary mutations of BRCA1/2 and drug resistance. *Cancer Sci* **102**: 663–669.

100. Sakai W, Swisher EM, Karlan BY, *et al.* (2008) Secondary mutations as a mechanism of cisplatin resistance in BRCA2-mutated cancers. *Nature* **451**: 1116–1120.
101. Swisher EM, Sakai W, Karlan BY, *et al.* (2008) Secondary BRCA1 mutations in BRCA1-mutated ovarian carcinomas with platinum resistance. *Cancer Res* **68**: 2581–2586.
102. Sakai W, Swisher EM, Jacquemont C, *et al.* (2009) Functional restoration of BRCA2 protein by secondary BRCA2 mutations in BRCA2-mutated ovarian carcinoma. *Cancer Res* **69**: 6381–6386.
103. Norquist B, Wurz K, Pennil CC, *et al.* (2011) Secondary somatic mutations restoring BRCA1/2 predict chemotherapy resistance in hereditary ovarian carcinomas. *J Clin Oncol*, in press.
104. Edwards SL, Brough R, Lord CJ, *et al.* (2008) Resistance to therapy caused by intragenic deletion in BRCA2. *Nature* **451**: 1111–1115.
105. Chu WK, Hickson ID. (2009) RecQ helicases: multifunctional genome caretakers. *Nat Rev Cancer* **9**: 644–654.
106. Bachrati CZ, Borts RH, Hickson ID. (2006) Mobile D-loops are a preferred substrate for the Bloom's syndrome helicase. *Nucleic Acids Res* **34**: 2269–2279.
107. Wu L. (2007) Role of the BLM helicase in replication fork management. *DNA Repair (Amst)* **6**: 936–944.
108. Chan KL, North PS, Hickson ID. (2007) BLM is required for faithful chromosome segregation and its localization defines a class of ultrafine anaphase bridges. *EMBO J* **26**: 3397–3409.
109. Chan KL, Palmai-Pallag T, Ying S, *et al.* (2009) Replication stress induces sister-chromatid bridging at fragile site loci in mitosis. *Nat Cell Biol* **11**: 753–760.
110. Naim V, Rosselli F. (2009) The FANC pathway and BLM collaborate during mitosis to prevent micro-nucleation and chromosome abnormalities. *Nat Cell Biol* **11**: 761–768.
111. Vinciguerra P, Godinho SA, Parmar K, *et al.* (2010) Cytokinesis failure occurs in Fanconi anemia pathway-deficient murine and human bone marrow hematopoietic cells. *J Clin Invest* **120**: 3834–3842.

CHAPTER 7

Lysosomopathies: Pathophysiology and Treatment of Lysosomal Storage Diseases

*Beth L. Thurberg, MD, PhD**

INTRODUCTION

Over the past 10–20 years, there have been significant advances in understanding the pathophysiology, clinical course, and treatment of several lysosomal storage diseases (LSDs). Rather than attempt a comprehensive review of all known storage disorders, as already represented in many general pathology and internal medicine textbooks,[1,2] this chapter will focus on those disorders for which treatment is now available (Table 1). Lysosomopathies are a diverse group of diseases, but several unifying pathophysiological principles have emerged, which will be emphasized in this chapter. Although many challenges remain, lysosomopathies represent an excellent example of how advances in our understanding of the genetic and cell-biological basis of a disease can lead to rapid and significant improvements in diagnosis and therapy, and in turn, how clinical studies including clinical trials can add to our understanding of disease pathophysiology. This latest information should be most useful and clinically

*Department of Pathology, Genzyme Corporation, Five Mountain Road, Framingham, MA 01701–9322, USA. Tel.: 508-271-2739; Fax: 508-820-7664; E-mail: Beth.Thurberg@genzyme.com.

Table 1 Summary of the Lysosomal Storage Diseases for Which Therapy is Available or Currently Under Investigation

Disease	Enzyme deficiency	Accumulating substrate	Treatment type	Name	Status
Gaucher disease type 1	Glucocerebrosidase	Glucosylceramide (GC) 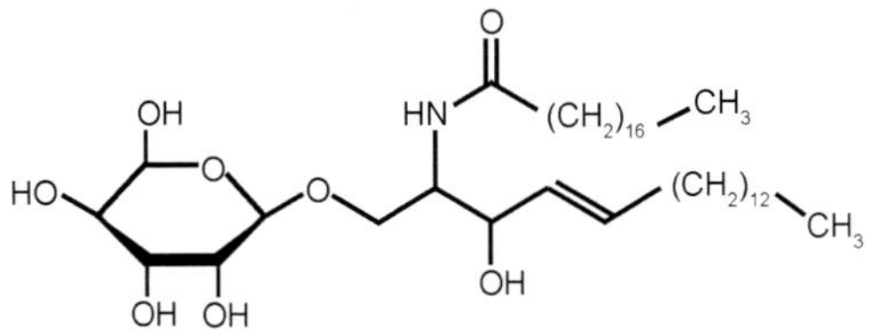	ERT	Alglucerase	Commercially available
				Imiglucerase	Commercially available
				Velaglucerase alfa	Commercially available
				Taliglucerase alfa	Under investigation
			SRT	Miglustat	Commercially available
				Eliglustat tartrate	Under investigation
			CT	isofagomine	Under investigation
Fabry disease	α-galactosidase A	Globotriaosylceramide (GL3) 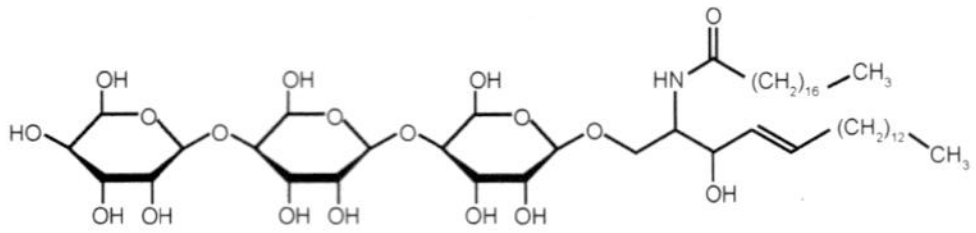	ERT	Agalsidase-β	Commercially available
				Agalsidase-α	Commercially available
			SRT	Eliglustat tartrate	Under investigation
			CT	Migalastat HCl	Under investigation

(*Continued*)

Table 1 (*Continued*)

Disease	Enzyme deficiency	Accumulating substrate	Treatment type	Name	Status
Mucopolysaccharidosis I	α-L-iduronidase	Glycosaminoglycans	ERT	Laronidase	Commercially available
Mucopolysaccharidosis II	Iduronate 2-sulfate sulfatase	heparan sulfate			
Mucopolysaccharidosis VI	Arylsulfatase B	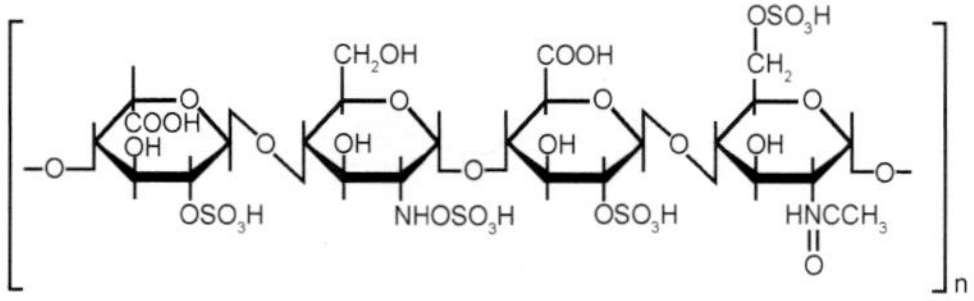	ERT	Idursulfase	Commercially available
			ERT	galsulfase	Commercially available
		dermatan sulfate			
		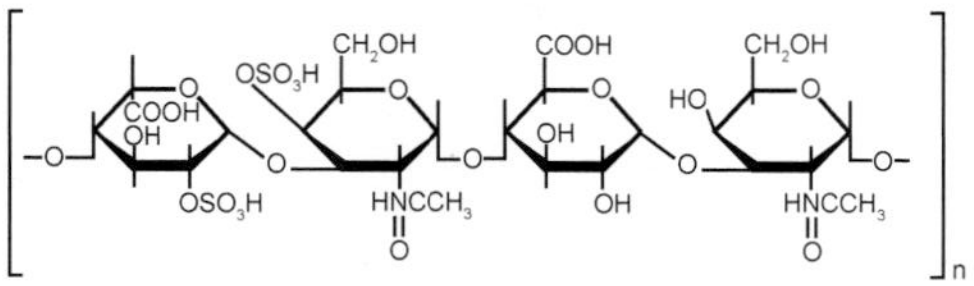			

(*Continued*)

Table 1 (Continued)

Disease	Enzyme deficiency	Accumulating substrate	Treatment type	Name	Status
Pompe disease	Acid α-glucosidase	Lysosomal glycogen 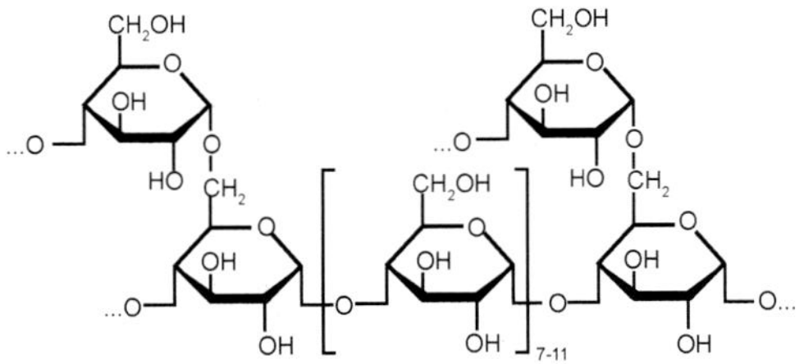	ERT	Alglucosidase-α	Commercially available
			CT	1-deoxynojirimycin HCl	Under investigation
Niemann–Pick disease type B	Acid sphingomyelinase	Sphingomyelin (SM) 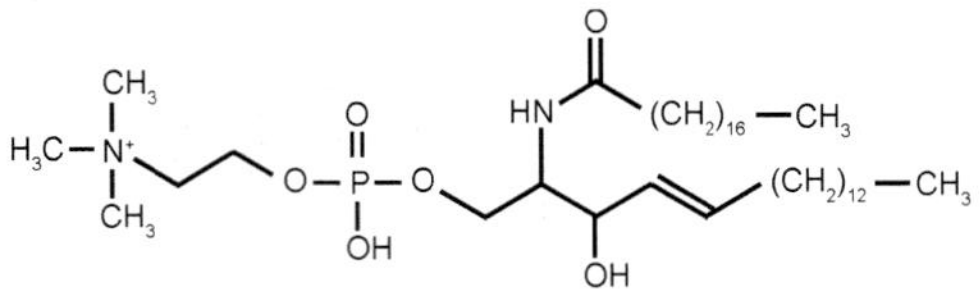	ERT	Recombinant human acid sphingomyelinase	Under investigation

ERT, enzyme replacement therapy; SRT, substrate reduction therapy; CT, chaperone therapy.

relevant to practicing pathologists and physicians for the diagnosis and treatment of such patients.

Each of the LSDs is caused by a genetic mutation leading to a deficiency of a critical lysosomal enzyme. The resulting substrate accumulation leads to tissue damage and clinical disease. Lysosomal enzymes are normally produced in the rough endoplasmic reticulum and undergo glycosylation with mannose-6-phosphate (M6P) residues in the Golgi. The M6P residues ensure proper delivery of the enzyme to the lysosome by targeting via the cation-independent mannose-6-phosphate receptor (CIMPR), present on all normal mammalian cells. The CIMPR cycles continuously from the cell surface, where it also binds and endocytoses M6P-containing ligands (such as newly synthesized lysosomal enzymes that have escaped sorting in the Golgi) to the lysosome, where these ligands are released. In this manner, the CIMPR can also transport extracellular M6P-containing ligands back to the lysosome.[3,4]

It is this secretion–recapture process which permits delivery of exogenous lysosomal enzymes to cells throughout the body following intravenous delivery and makes enzyme replacement therapy (ERT) possible.[5] ERT involves the administration of a purified, recombinant enzyme, produced in cell culture systems, and is available for a number of LSDs (Table 1). It has successfully treated the visceral manifestations of LSDs; however, due to the large size of these therapeutic proteins, treating neuropathic manifestations has proven more difficult, since this requires molecules capable of crossing the blood–brain barrier. In addition, long-term delivery of exogenous protein also carries the risk of antibody formation, which may be clinically insignificant, cause neutralization of enzyme activity, or result in formation of immune complexes.[6]

To address some of these challenges, current research is focusing on alternative approaches to treatment such as substrate reduction therapy (SRT) and pharmacological chaperone therapy (CT) (Table 1). SRT involves the use of a small-molecule inhibitor of a key enzyme in the substrate synthetic pathway, thereby halting further synthesis and accumulation, and has the potential advantage of oral delivery.[5] For many of the LSDs, the enzyme defect results not from a single characteristic mutation, but from a host of heterogeneous mutations present in the patient population, including missense mutations, null mutations, frame shift mutations, insertions, and splice site mutations. Each mutation encodes a mutant enzyme with a different size and residual activity. For the subset of patients with missense mutations, pharmacological chaperones may represent a new therapeutic approach. These small molecules bind to the mutant enzyme to provide physical stability, and promote proper folding, improved activity, and lysosomal trafficking. They also have the potential advantage of oral administration, and agents that cross the blood–brain barrier could be used to treat diseases with a neurological component. However, not all genotypes are responsive to these molecules, particularly mutations that occur in the active site, or are destabilizing to protein structure, such as those involved in disulfide bonds,[7] so their use would be limited to patients with responsive mutations. Thus, optimal treatment of LSDs should be highly individualized — a long-term goal that will require a deep understanding of how allelic variants impact protein function in each lysosomopathy.

GAUCHER DISEASE TYPE 1

Gaucher disease, an autosomal recessive LSD caused by a deficiency of lysosomal acid β-glucosidase (glucocerebrosidase), leads to the accumulation of glucosylceramide (GC) in cells of the monocyte/macrophage lineage. Gaucher disease type 1 (GD1) is the nonneuropathic form of the disease, and was the first LSD for which ERT was successfully developed. Accumulation of GC in macrophages produces enlarged, 20–100 nm "Gaucher cells" with characteristic crinkles and striations under routine light microscopy[8] (Fig. 1). Under electron microscopy, the accumulated GC forms tubular structures within numerous lysosomes which engorge the cell cytoplasm.[9] The involvement of Kupffer cells and splenic macrophages results in prominent hepatosplenomegaly. Gaucher cells also infiltrate the bone marrow, resulting in anemia and thrombocytopenia. The accumulation of substrate within osteoclasts contributes to osteopenia, osteoporosis, bone pain, and growth retardation. In more severe cases, pulmonary disease can also develop.[10] Gaucher cells infiltrate the alveolar spaces and lobular septa, leading to airspace and interstitial lung disease, which can be seen as reticulonodular infiltrates and intralobular thickening on chest X-ray and high-resolution CT scans.[11,12]

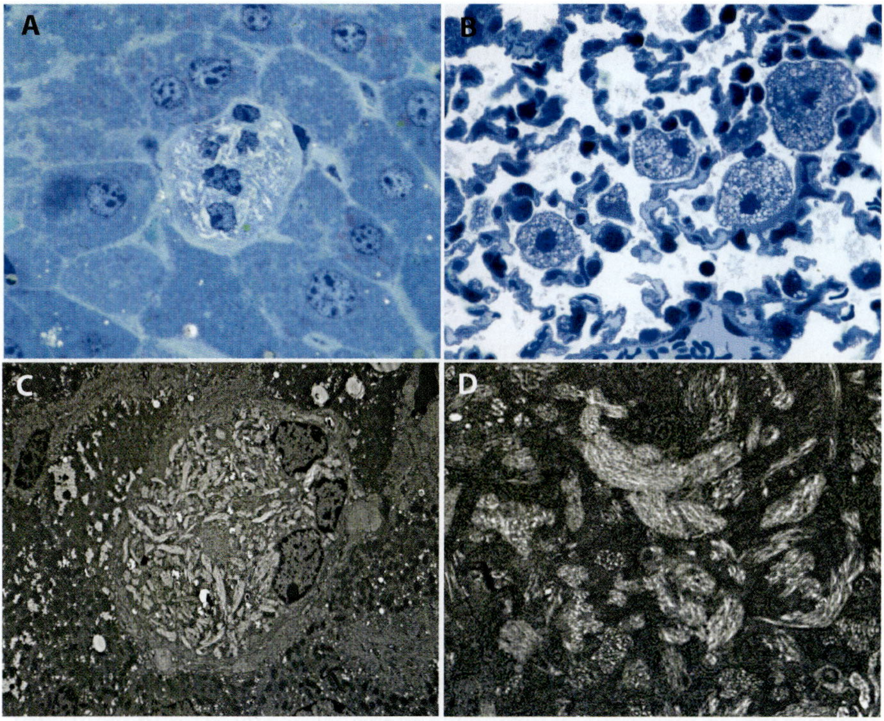

Fig. 1 Light and electron microscopy demonstrate the characteristic appearance of "Gaucher cells" in the liver and lung from a mouse model of Gaucher disease. **(A)** A semithin epoxy resin section of the liver shows the typical crinkled appearance of Kupffer cells engorged with the accumulating substrate, glucosylceramide (epoxy resin, Richardson's stain, 100× oil objective). **(B)** Semithin section of the lung containing clusters of enlarged, foamy pulmonary macrophages within alveolar spaces (epoxy resin, Richardson's stain, 100× oil objective). **(C, D)** Electron microscopy of the Kupffer cell inclusions (4000×). Panel D illustrates the enlarged, elongated lysosomes containing bundles of tubular structures characteristic of Gaucher disease (16,000×).

These patients are at risk for recurrent pulmonary infection and progressive respiratory insufficiency.[11]

The precise mechanism by which GC accumulation contributes to the pathophysiology of disease is not well understood. However, studies suggest that accumulation of GC alters normal macrophage function, which may contribute to tissue and clinical disease, and that clearance of GC from these cells may help to restore normal cell and tissue function, thereby reversing or halting further progression of disease. Elevations of macrophage-derived factors and cytokines in Gaucher patient serum have been found to correlate with disease severity.[13,14] Serum levels of the macrophage-derived cytokines IL8 and M-CSF, as well as the monocyte/macrophage activation marker sCD14, were elevated in one patient study and appeared to correlate with disease severity scores of liver and spleen volume.[13] After enzyme replacement treatment, the levels of IL8 remained unchanged; however, investigators did observe declines in serum M-CSF and sCD14 after treatment, suggesting that the biological activity of these factors may contribute to disease. A similar study observed correlations between elevated IL1-β, TNF-α and IL6 and the severity of the bone and hematological manifestations of the disease.[14] Elevations in reactive oxygen species and APE/Ref-1 levels have been observed in primary Gaucher fibroblast cell cultures, suggesting dysregulation in the adaptive response to oxidative stress. These elevations were prevented by treatment of the cell cultures with a recombinant enzyme.[15]

The first form of ERT developed for GD1, alglucerase, was isolated and purified from human placentas, and modified to expose mannose residues which target the macrophage mannose receptor.[8,16] In early clinical trials, the enzyme was administered intravenously to patients every two weeks, resulting in reduced hepatosplenomegaly, improved hemoglobin and platelet counts, and some skeletal improvement. The current form of ERT available today in the US is a recombinant human enzyme, imiglucerase, produced in mammalian cell culture. A more recent, histological study compared the appearance of pre- and posttreatment trephine bone marrow biopsies from patients before and after ERT. This demonstrated a reduction in Gaucher cell number and size with imiglucerase, accompanied by an increase in normal hematopoiesis and fat.[17] In addition, years of clinical experience with ERT in Gaucher children have shown that treatment can avert the growth retardation and skeletal abnormalities associated with the disease.[10] Today there are a number of ERTs available or under development for Gaucher disease (Table 1).

Substrate reduction therapies (SRTs) are orally available, small molecules which limit the accumulation of GC by inhibiting glucosylceramide synthase, the enzyme responsible for its synthesis. The first available SRT was the iminosugar miglustat, which has been available for mild-to-moderate GD1 patients intolerant of ERT. Recently, a more potent small-molecule substrate inhibitor, eliglustat tartrate, has been tested in patients and has not only been shown to improve hepatosplenomegaly and hematological parameters, but has also resulted in significant improvement of bone mineral density in the spine.[18] A small-molecule chaperone, isofagomine, is currently being tested and has shown activity in GD1 patient cell lines and beneficial effects on the liver and spleen in mouse models[19] by stabilizing missense mutations of β-glucosidase, thereby restoring partial enzyme activity. All of these small molecules are able to cross the blood–brain barrier, and have the potential to address the neuropathic aspects of the disease.

FABRY DISEASE

Fabry disease is an X-linked recessive disorder in which affected males are deficient in the lysosomal enzyme α-galactosidase A, leading to accumulation of the neutral glycosphingolipid, globotriaosylceramide (GL-3), in tissues throughout the body. Progressive GL-3 accumulation results in clinical disease, primarily in affected males, but some female heterozygotes may be symptomatic, depending on the pattern of X-chromosome inactivation.[20–22] Accumulation of GL-3 begins *in utero* and has been observed in renal epithelial cells of second trimester male fetuses.[23,24] In addition, substrate accumulation can be found in both fetal and maternal cells of placentas from heterozygous female Fabry patients, including intermediate trophoblasts (Fig. 2) maternal vasculature and umbilical cord vasculature (Thurberg and Politei, *Human Pathology*, in press; and Ref. 25).

Involvement of the skin in adolescent and adult patients results in angiokeratomas, hypohidrosis (abnormal lack of sweat) and acroparesthesias (the sensation of "pins and needles" in the extremities).[26] Examination of biopsies from these patients shows accumulation in endothelial cells (Fig. 3) and vascular smooth muscle cells of both superficial and deep dermis, as well as within the perineurium.[27] ERT aids in the removal of GL-3 from these cells.[27]

The involvement of the heart by Fabry disease is associated with a number of cardiac signs and symptoms, including left ventricular hypertrophy

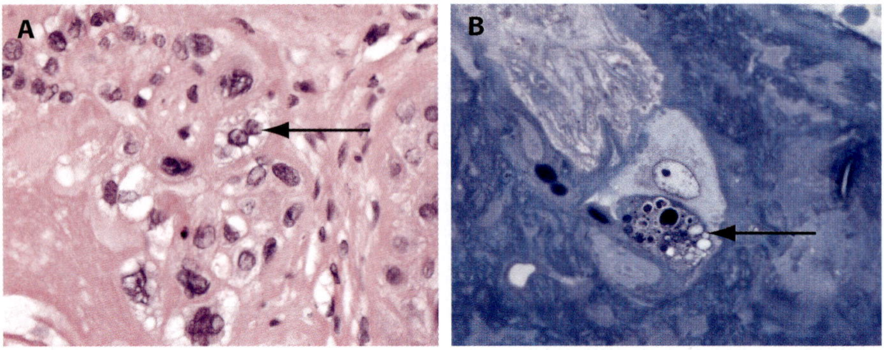

Fig. 2 GL-3 accumulates in intermediate trophoblasts of the Fabry placenta. (**A**) In paraffin sections, intermediate trophoblasts appear enlarged and foamy, due to loss of GL-3 during routine tissue processing (H&E, 60× objective). (**B**) In glutaraldehyde-fixed, epoxy-resin-embedded tissues, GL-3 is preserved and appears as dense, blue inclusions of various sizes (1 μm epoxy resin section, Richardson's stain, 100× oil objective).

(LVH), mitral insufficiency, conduction abnormalities, and ischemic heart disease.[28] The prominent GL-3 accumulation present in cardiomyocytes (Fig. 4, asterix) can occupy greater than 50% of the total cell volume, and likely contributes to LVH.[29] Survey studies indicate that cardiac involvement by Fabry disease can go undiagnosed when initially labeled as hypertrophic cardiomyopathy (HCM).[29,30] In addition, female Fabry patients, who until recently had been considered "asymptomatic carriers," experience a delay of 16–19 years between the onset of symptoms and diagnosis.[31–33]

There is a high incidence of thrombotic events in Fabry patients (both men and women),[26] which contribute to acute events such as myocardial infarction and stroke. Light- and electron-microscopic examination of cardiac biopsies from Fabry patients reveals GL-3 accumulation in the microvasculature,[34] which has been associated with elevations in markers of endothelial injury and activation, leukocyte activation, and coagulation.[35,36] Such proinflammatory and procoagulant alterations in normal endothelial homeostasis are well-characterized components of atherosclerotic disease in general,[37,38] and are likely responsible for similar clinical manifestations observed in Fabry patients. The clearance of GL-3 from the vasculature (Fig. 4), therefore, would be expected to prevent or reduce the occurrence of these events.

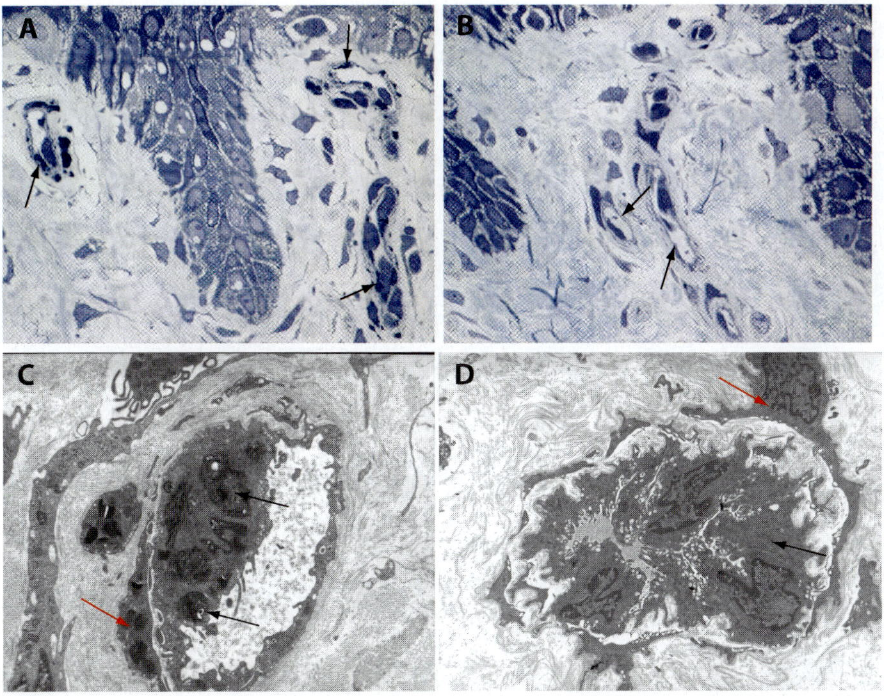

Fig. 3 GL-3 is cleared from vascular endothelial cells and pericytes of the skin after enzyme replacement therapy. **(A)** Pretreatment skin biopsy. GL-3 is present adjacent to endothelial cell nuclei and scattered throughout the cytoplasm. The larger granules protrude into the capillary lumen. **(B)** Posttreatment biopsy. Capillary endothelium is clear of GL-3 (1 μm epoxy resin section, Richardson's stain, magnification of A and B, 100× oil objective). **(C)** Pretreatment skin biopsy, electron microscopy. Black arrows indicate capillary endothelial cell GL-3; red arrows indicate pericyte GL-3 (magnification 6000×) **(D)** Posttreatment skin biopsy, electron microscopy. GL-3 has been removed from both endothelial cells (black) and pericytes (red) (magnification 3300×). Reproduced with permission from: Thurberg *et al.*, *J Invest Dermatol* (2004).

Vascular injury has also been implicated in the kidney damage which leads to progressive renal failure in Fabry disease.[26,36] The vascular endothelium plays an important normal physiological role in glomerular perfusion, filtration, and maintenance of a continuous anticoagulant vascular lining. GL-3 accumulation in glomerular (Fig. 4) and interstitial capillary endothelium interferes with this function. In addition, GL-3 accumulation occurs in vascular smooth muscle cells, mesangial cells,

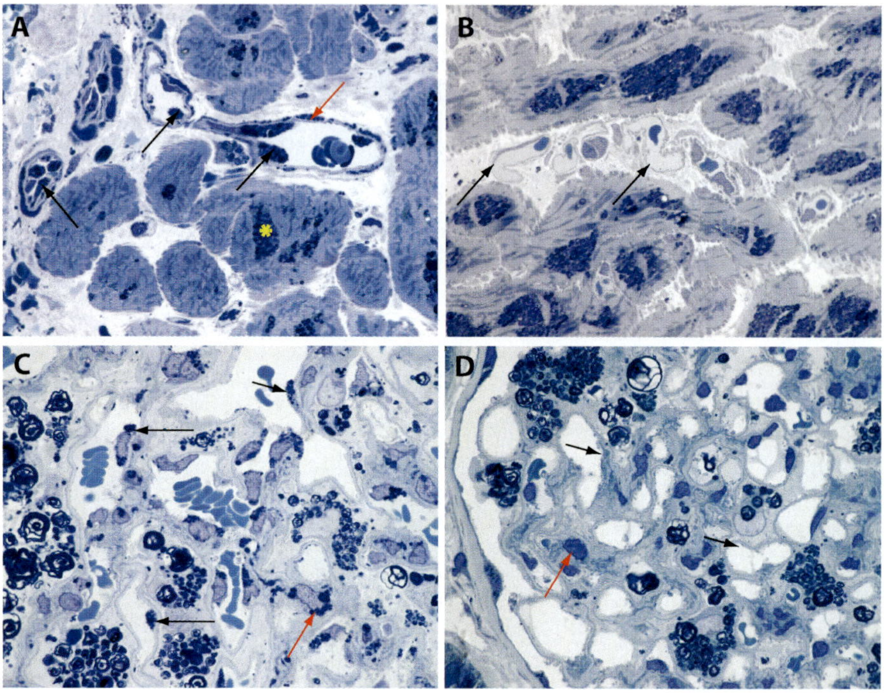

Fig. 4 GL-3 is cleared from cells of the heart and kidney after enzyme replacement therapy. **(A)** Cardiac biopsy, pretreatment: prominent, clustered inclusions of GL-3 protrude into the vascular lumen of interstitial capillaries (black arrow). Accumulation of GL-3 is also apparent in pericytes (red arrow). **(B)** Cardiac biopsy, posttreatment: GL-3 has been cleared from capillaries after ERT. Cardiomyocyte GL-3 (yellow asterix in panel A) was present at baseline and persisted posttreatment. **(C)** GL-3 is cleared from glomerular capillaries (black arrows) and mesangial cells (red arrows) after ERT, as shown in panel **D**. Many large myelin figures remain in podocytes. Cardiomyocytes and podocytes are terminally differentiated cells; this lack of cell turnover may contribute to the large GL-3 burden in these cells and their slow response to ERT (1 μm epoxy resin sections, Richardson's stain, 100× oil objective). Renal images reproduced with permission from: Thurberg *et al.*, *Kidney Int* (2002).

podocytes, and interstitial cells (fibroblasts and macrophages).[39] Recent ultrastructural studies examined the relationships between GL-3 load, podocyte injury, and the development of Fabry nephropathy.[40] The longstanding accumulation of GL-3 ultimately results in the irreversible secondary pathology, specifically global glomerulosclerosis and interstitial fibrosis (Fig. 5), responsible for progressive renal failure.[41] This is a common

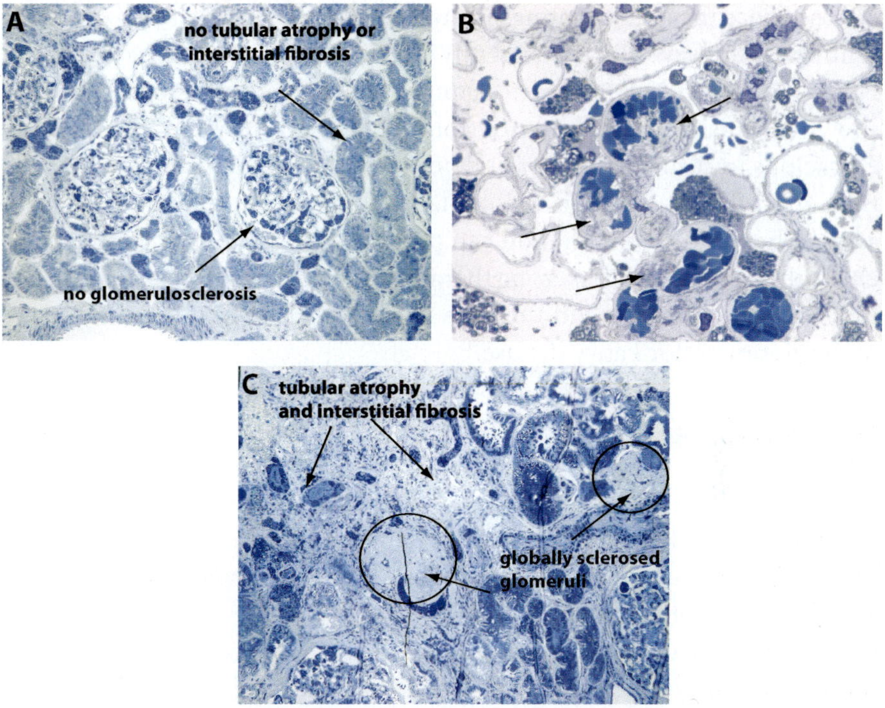

Fig. 5 Secondary pathology contributes to renal failure in Fabry patients. Early in Fabry disease, GL-3 accumulation is the predominant pathology observed, as seen in a biopsy from a young adult Fabry patient, in panel **A** (Richardson's stain, 20× objective). Progressive renal vascular thrombosis (see fibrin thrombi occluding glomerular capillaries in panel **B**, 100× oil objective) is a likely contributor to the irreversible secondary pathology which develops over time and contributes to renal decline. Panel **C** shows the biopsy of a patient in the mid-40s. Note the multiple, globally sclerosed glomeruli, the prominent interstitial scarring, and the widespread tubular atrophy (Richardson's stain, 20× objective).

cause of morbidity and mortality in male patients in their third-to-fifth decades of life.

There are two commercially available enzyme replacement treatments for Fabry disease: agalsidase-β and agalsidase-α. Both are administered intravenously, once every two weeks, and represent lifelong therapy required to maintain the clearance of substrate. SRT with the glucosylceramide synthase inhibitor, eliglustat tartrate, which has been successfully tested in Gaucher patients, is currently being examined for use in Fabry disease, since the accumulating substrates, GC and GL-3, share a common synthetic pathway. SRT is currently being tested in a mouse model of Fabry disease and compared to standard ERT.[42] In this model, SRT has shown promising GL-3 clearance from the kidney, whereas ERT was more efficacious in clearing the liver and heart. Combination therapy with both ERT and SRT, therefore, is being planned for future testing in order to assess whether the combination can provide superior results compared to either therapy alone. Chaperone therapy, which can restore partial enzyme function to misfolding and unstable mutations, is also in the early stages of development for Fabry disease.[43]

MUCOPOLYSACCHARIDOSES

Mucopolysaccharidoses (MPSs) are caused by deficiencies in lysosomal enzymes required for degradation of glycosaminoglycans (GAGs). Because disease can be caused by mutations in any one of several enzymes required for GAG metabolism, MPSs are subclassified into types I through VI, each correlating with a specific enzyme deficiency. GAG accumulation occurs in numerous cell types, particularly those of connective tissues. This tissue pathology translates into a wide variety of clinical signs and symptoms, which can include skeletal abnormalities, joint contractures, facial dysmorphism, restrictive pulmonary disease, hepatosplenomegaly, cardiomegaly, valvular disease, macroglossia, corneal clouding, deafness, meningeal thickening and, in severe forms, mental retardation. Patients typically excrete high levels of GAGs into the urine.[5]

Cellular accumulation begins *in utero*, and has been observed in 20-week-old fetuses within heart valves, vascular endothelial cells, dura mater, anterior horn cells of the spinal cord, and neurons of the dorsal root ganglia.[44,45] In a histopathological study of MPS-I and MPS-II pediatric skin biopsies (patient age range 1–11 years), GAG inclusions were observed in keratinocytes, dermal fibroblasts, macrophages, smooth muscle cells, and Schwann cells.[46] The brains of MPS-I children also show engorgement of neurons, likely responsible for the severe mental retardation in these patients.[47] A pediatric autopsy study seeking to understand the conductive and sensorineural deafness in these patients revealed deformation of the auditory ossicles and the presence of vacuolated cells in nearby vascular spaces. The study also observed replacement of the spiral ganglion and disruption of the vestibulocochlear nerves by swaths of foamy cells which were not otherwise identified but presumed to be affected cells of mesenchymal origin engorged with GAGs.[48] The muscle biopsy of an MPS-II patient with progressive gait disturbance revealed GAG accumulation in skeletal myocytes, satellite cells, vascular endothelial cells, and fibroblasts of the endomysium.[49]

Under electron microscopy, the inclusions are heterogeneous and can appear as a mixture of open, vacuolated spaces (type A vacuoles), along with others containing fluffy, filamentous, granular, or dense inclusions, as well as inclusions with loose whorled membranes and stacks of membranes (zebra bodies, also known as type B vacuoles).[9,46,47] Similar heterogeneous lysosomal inclusions can be observed in a mouse model of MPS-I (Hurler's disease) in multiple tissues (Fig. 6). It has been suggested that the abnormal storage of GAGs may produce a secondary defect in lysosomal metabolism affecting glycolipid degradation, resulting in heterogeneous inclusions.[44,50] The authors proposed that the accumulated GAGs can form complexes with other lysosomal enzymes, such as α-galactosidase, which diminish their activity, and result in the additional accumulation of glycolipids, which can form the laminated and zebra body inclusions (type B vacuoles). Recently, a canine model of MPS-I examined the arterial pathology which develops in this disease, and shares morphological similarities with atherosclerosis, such as the formation of lesions in areas of low shear stress. The lesions were composed of foamy, GAG-laden macrophages, smooth muscle cells, and fibroblasts, which were reduced with recombinant enzyme treatment.[51]

ERT is currently available for three of the MPSs, I, II, and VI, in which heparan sulfate and dermatan sulfate are the major accumulating GAG species (Table 1). Laronidase treatment for MPS-I (Hurler's, Hurler–Scheie's, and Scheie's disease) has been shown to lower urinary GAG levels and reduce liver and spleen volumes, and resulted in improvements in pulmonary function and the six-minute walk test (6MWT). The 6MWT test is a useful clinical measure of mobility and physical function in such patients, since the accumulation of GAGs in mesenchymal cells of joint tissues affects range of motion.[52] Similarly, treatment of MPS-II (Hunter's disease) with idursulfase results in reduced urinary GAGs, and

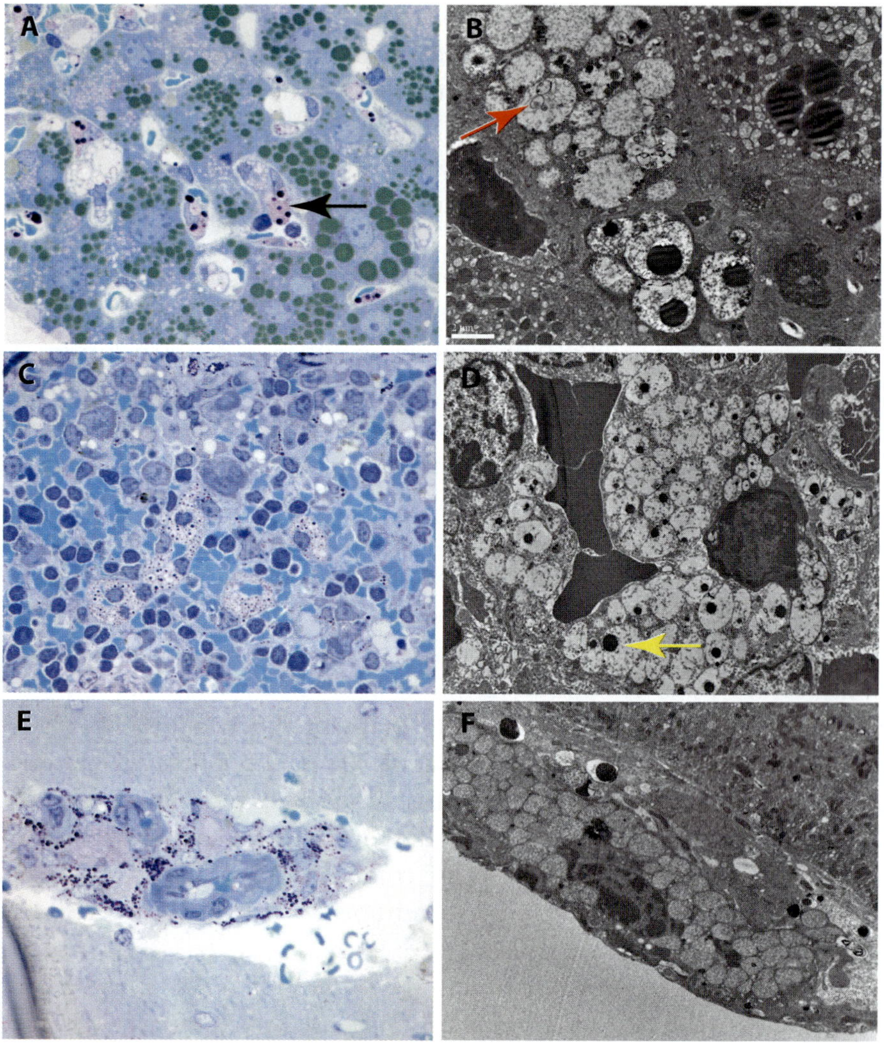

Fig. 6 Heterogeneous lysosomal inclusions are present in multiple tissues of the MPS-I (Hurler's disease) mouse. The variation in lysosomal inclusions can be seen in both high-resolution light microscopy sections (**A, C, E**; epoxy resin, PAS/Richardson's stain, 100× oil objective) and by electron microscopy (**B, D, F**). In light microscopy sections, the accumulated GAGs stain pinkish-purple with PAS, due to their high sugar moiety content (black arrow, panel A). A few lysosomes contain loose myelin figures (red arrow, panel B), while others contain a granular, flocculent material, either alone or in combination with electron-dense, lipid-like droplets (yellow arrow, panel D), which gives them the appearance of black-eyed peas. Panels A and B — liver; panels C and D — spleen; panels E and F — pia mater of cerebellum.

improvements in hepatosplenomegaly and the 6MWT.[5] MPS-VI (Maroteaux–Lamy's disease) is currently treated with galsulfase, which has shown reduction in urinary GAGs, improvement in joint pain and range of motion, and improvements in both six- and twelve-minute walk tests in clinical trials.[5,53]

POMPE DISEASE (GLYCOGEN STORAGE DISEASE TYPE 2)

Pompe disease (glycogen storage disease type II, acid maltase deficiency) is an autosomal recessive, lysosomal storage disorder in which affected individuals are deficient in the lysosomal enzyme acid α-glucosidase (acid maltase). Abnormal lysosomal glycogen accumulates in multiple cell types, particularly in the myocytes of skeletal, cardiac, and smooth muscle, and has been detected in fetuses at as early as 16–18 weeks' gestation.[54,55] In the most severe, classical infantile form of Pompe disease, macroglossia, cardiomyopathy, hypotonia, and respiratory insufficiency dominate, with death occurring around the first year of life due to cardiorespiratory failure.[56–58] In the juvenile- and adult-onset disease, an insidious, slowly evolving skeletal muscle weakness predominates. Overall, there is an inverse correlation between disease severity and the level of residual enzyme activity, with the most severely affected infants having no detectable enzyme activity.[56] Until the recent development of an enzyme replacement, little — other than symptomatic — treatment was available for such patients.

Examination of autopsy tissue samples from a 14-month-old infantile patient revealed multiorgan involvement, including the limbs, diaphragm, heart, bladder, and CNS (Fig. 7). The patient had clinically evident cardiac hypertrophy, consistent with a greater-than-two fold increase over normal organ weight at autopsy, and prominent lysosomal glycogen in cardiomyocytes (Figs. 7D and 7J). Infiltration of septal myocytes has been reported in association with the shortened P–R intervals observed in some Pompe patients,[59] and a corrective increase in the P–R interval has been noted in conjunction with ERT.[60] Involvement of vascular smooth muscle (Fig. 7F) noted in the juvenile and adult forms of Pompe disease was reportedly responsible for vascular aneurysms resulting in severe headaches, cerebellar infarction, and fatal rupture.[61] Glycogen accumulation in motor neurons of the spinal cord (Fig. 7I) has also been reported previously in

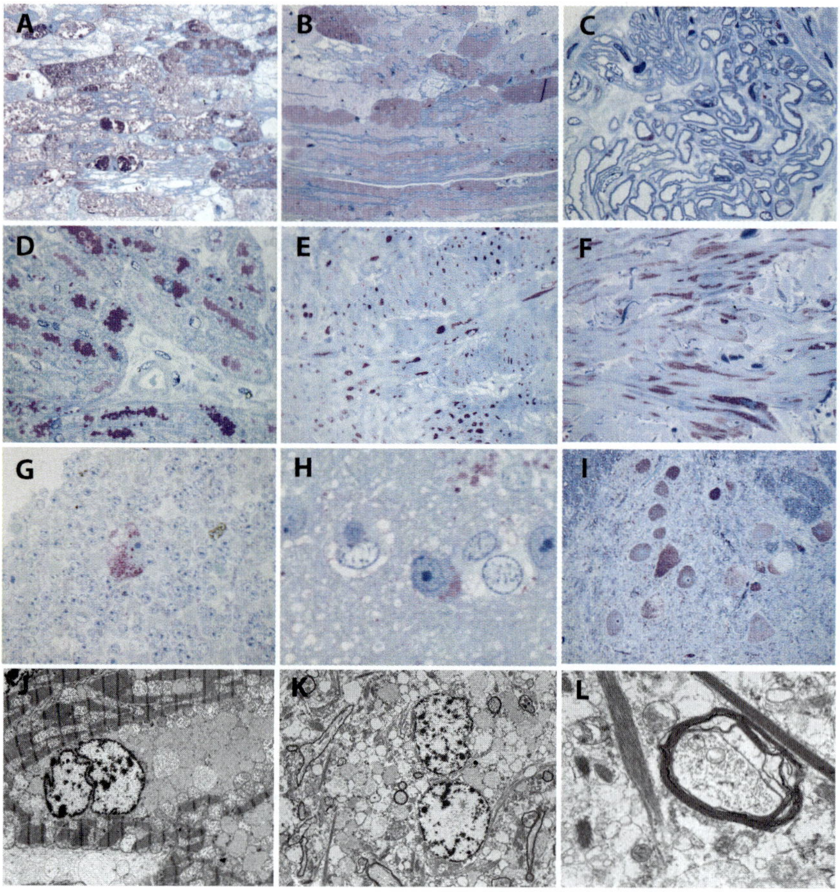

Fig. 7 Pompe autopsy tissues demonstrate abnormal glycogen accumulation in multiple organs and cell types. **(A)** Skeletal muscle, deltoid (40× objective); **(B)** skeletal muscle, diaphragm (40× objective); **(C)** nerve from the diaphragm: glycogen can be seen in Schwann cell cytoplasm (100× oil objective); **(D)** heart, interventricular septum (60× objective); **(E)** bladder, smooth muscle cells of the muscularis propria (40× objective); **(F)** large artery, smooth muscle cells of the media (100× oil objective); **(G)** cerebellum — glycogen accumulation can be seen in a Purkinje cell (60× objective); **(H)** frontal lobe — glycogen accumulation in a neuron (100× oil objective), **(I)** spinal cord — lysosomal glycogen can be seen in motor neurons of the ventral horn (40× objective); panels **A–I** — 1 μm epoxy resin sections stained with Richardson's/PAS stain, **(J)** cardiomyocyte of the interventricular septum containing multiple glycogen-filled lysosomes adjacent to the nucleus. (EM: 28,000×), **(K)** clusters of membrane-bound glycogen vacuoles are present in motor neurons of the ventral horn (EM: 8800×), **(L)** myelin sheaths of nerve fibers exhibit splitting (EM: 44,000×). Reproduced with permission from: Thurberg *et al.*, *Lab Invest* (2006).

infantile patients[62,63] and raises the question as to whether neuronal involvement is a secondary component contributing to muscle weakness. This area requires further investigation.

However, it is the involvement of the skeletal muscle that is most clinically evident in all forms of the disease. Skeletal muscle biopsy is a relatively straightforward way to assess disease severity and response to treatment (Fig. 8). However, in Pompe disease, the response to treatment amongst patients can vary significantly. It is important, therefore, to understand the cellular pathophysiology that evolves within skeletal myocytes and underlies this variation in clinical response.

Variation in the disposition of intracellular glycogen (membrane-bound versus cytoplasmic predominance) has been observed and can be explained as part of several progressive stages (Fig. 9) of ultrastructural change that occur in skeletal myocytes.[64] Essential to the understanding of these stages of disease is the "lysosomal rupture hypothesis," which asserts that during normal contraction the unique contractile nature of myocytes subjects the glycogen-containing lysosomes to stress forces during the increased rigidity of surrounding myofibrils. Once lysosomes reach a critical size, these forces cause lysosomal rupture and release of

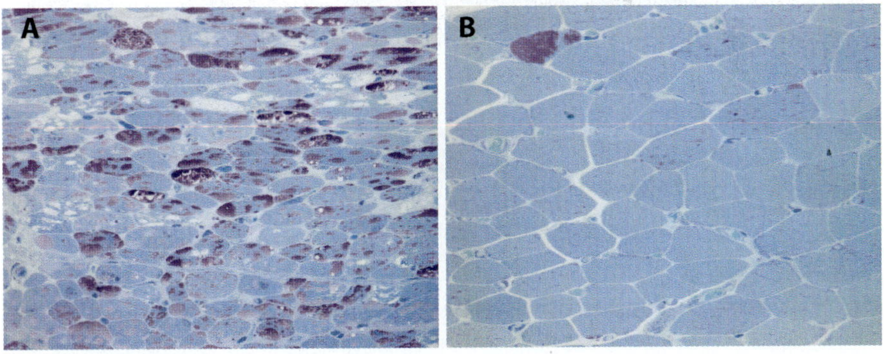

Fig. 8 Light-microscopic examination of Pompe quadriceps biopsies demonstrates the histologic response to enzyme replacement therapy. The glycogen accumulation in infantile patient A at baseline (panel **A**) has been cleared in the majority of myocytes after one year; a rare myocyte appears completely replaced by glycogen and remains unaffected by ERT (asterix, panel **B**) (1 μm epoxy resin sections stained with Richardson's/PAS stain, magnification 40× objective). Reproduced with permission from: Thurberg *et al.*, *Lab Invest* (2006).

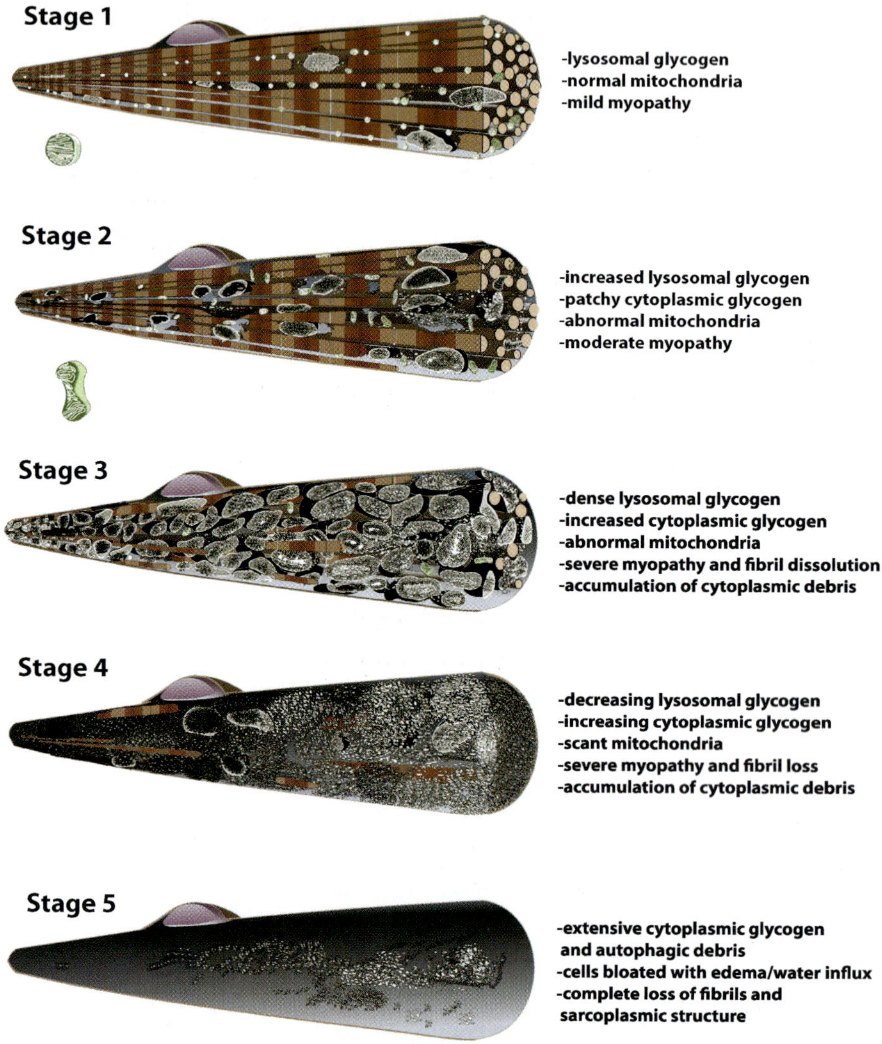

Fig. 9 Five stages of ultrastructural disease progression in Pompe myocytes. Reproduced with permission from: Thurberg *et al.*, *Lab Invest* (2006).

glycogen and lytic enzymes into the cytoplasm. The lytic enzymes cause damage to myofibrils, leading to loss of myofibrils and loss of contractile function. In more advanced stages (stage 4 and stage 5), myocytes filled with lakes of cytoplasmic glycogen and debris, and devoid of contractile elements, are considered to represent end stage disease.[65–68] These changes are particularly relevant to the use of ERT, since the cytoplasmic glycogen released from lysosomes is likely inaccessible to the membrane-receptor-dependent targeting mechanism required for enzyme uptake.[69] Even if delivery to the cytoplasm were possible, the neutral pH of the cytoplasm would render exogenously delivered recombinant enzyme less active. In clinical studies which examined the efficacy of ERT, it was observed that muscle biopsies containing a predominance of early stage disease cells (Fig. 10) were more responsive to treatment than those with a predominance of late stage diseased cells (stages 3–5) where the effects of lysosomal rupture had become evident.[70]

As lysosomal rupture gradually erodes the interior cellular architecture and contributes to the accumulation of cytoplasmic debris, the evolution of autophagic vacuoles becomes visible, as myocytes sequester the debris in the center of the cell. Autophagic buildup — a normal cellular process historically referred to in the field of pathology for decades as lipofuscin, "wear-and-tear pigment," or "age pigment" — is a common, age-related, incidental finding at autopsy in cardiomyocytes, hepatocytes, and

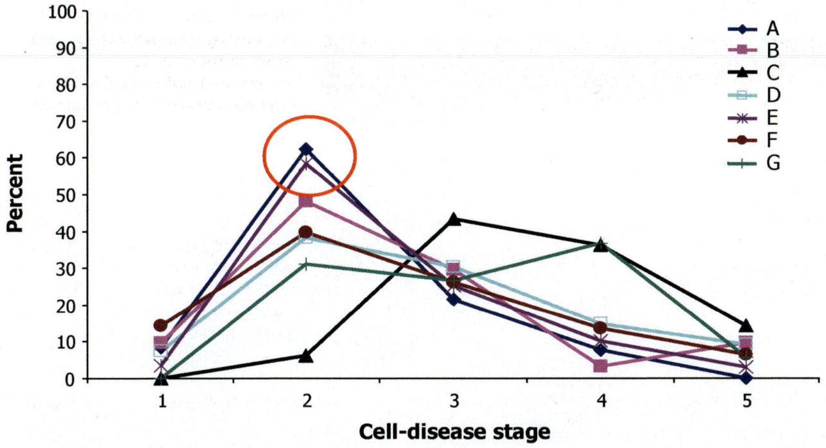

Fig. 10 Comparison of the distribution of the cell disease stage in baseline biopsies from seven infantile Pompe patients. All cells in a representative field were counted and graded according to the stages shown in Fig. 9. Patients with the highest percentage of early stage cells had the best clinical and biopsy response to treatment. Patients A and E had high proportions of stage 1 and 2 cells prior to treatment (circle) and both did well after treatment. Pre- and posttreatment biopsy for patient A was shown in Fig. 8. Reproduced with permission from: Thurberg *et al.*, *Lab Invest* (2006).

neurons of the brain.[1,71] In normal circumstances, it is not injurious to the cell or its functions, which suggests the body's tolerance for the long-term storage of a certain level of debris with no adverse effects.[1] Figure 11B shows a myocyte containing autophagic debris confined to the central core of the cell while intact, contiguous contractile elements straddle the autophagic bundle on either side of the cell, an arrangement that appears most biomechanically favorable for the continued function of the cell. Skeletal myocytes with such neatly sequestered debris can be observed in otherwise healthy, posttreatment biopsies from patients who are highly responsive to ERT and walk independently. This suggests that there is a certain level of cellular tolerance of the storage of autophagic debris, which in such cases represents a kind of cellular scar from past injury secondary to disease, and prior to initiation of treatment.

If left untreated, however, diseases such as Pompe disease may accelerate the normal ageing function of autophagy. The observation of autophagic vacuoles in skeletal myocytes of both infants and adults with Pompe disease has prompted a larger study of the possible role that alterations in autophagic function might play in the evolution of lysosomal storage diseases, and their response to treatment. In a mouse model of Pompe disease, muscle fibers show upregulation of genes involved in the formation of autophagosomes and downregulation of proteins for endocytosis

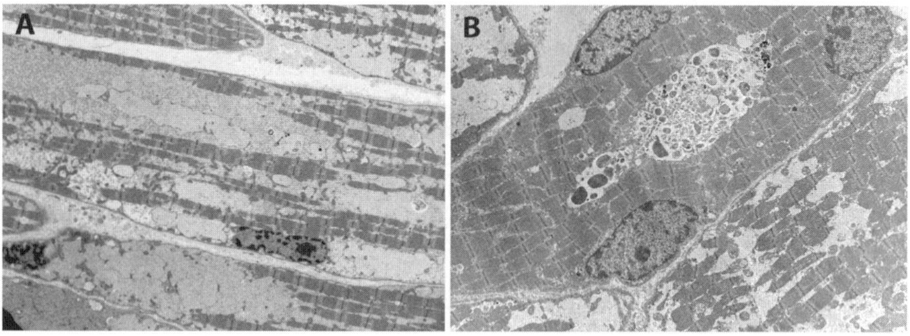

Fig. 11 Electron-microscopic appearance of skeletal myocytes in Pompe disease. In panel **A**, myofibrils are frayed and discontinuous, interrupted by a mixture of lysosomal and cytoplasmic glycogen, mixed with cell debris. Panel **B** shows a cell with centrally sequestered autophagic debris, straddled by continuous myofibrils on either side — a configuration which appears to be biomechanically favorable for continued cell function.

and lysosomal trafficking (CIMPR, clathrin, AP-2 complex).[72] In addition, some fibers have been found to have an increased number of lysosomes with a pH above the normal range, suggesting a possible acidification defect.[73] What might account for these changes in autophagy? Nutritional deprivation in general leads to autophagy,[71] and the failure of cellular glycogen to be metabolized to glucose in Pompe disease may deprive the cell of essential or needed energy. The resulting increase in autophagosomes and lipofuscin debris may affect the vesicular trafficking needed for enzyme delivery. In addition, autophagic buildup forms masses that may ultimately interrupt the contractile proteins of the myofibrils and adversely affect muscle contraction. Thus, autophagy represents one feature of the disease, which if significantly advanced could influence a patient's response to treatment.

Other features of advancing disease include secondary muscle pathology (particularly evident in adult patient biopsies), inflammation, fibrosis, and fatty replacement of skeletal myocytes (Fig. 12) — irreversible changes not amenable to current therapy. Figure 12 shows adult Pompe muscle tissue in which the majority of myocytes have been lost. Therefore, early diagnosis and intervention is essential. Many adult patients experience an

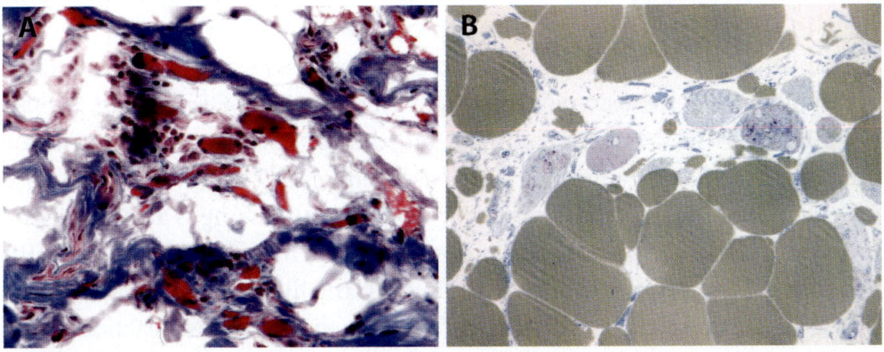

Fig. 12 Secondary pathology is common in adult-onset Pompe disease. (**A**) Biceps tissue from an adult patient which shows severely atrophic myocytes with marked fibrosis and scattered inflammatory cells on trichrome-stained sections (40× objective). (**B**) Semithin epoxy resin sections show glycogen accumulation (purple) in the few remaining myocytes, surrounded by fatty replacement (1 μm epoxy resin section stained with Richardson's and PAS, 100× oil objective).

over-10-years delay between the onset of symptoms and diagnosis,[74] and the rapidly progressive nature of the infantile form makes timely diagnosis and treatment a challenge. Fortunately, recent advances in newborn screening have been made in Taiwan, leading to early diagnosis and significantly improved patient response rates compared to earlier trials.[75,76]

NIEMANN–PICK DISEASE TYPE B (ACID SPHINGOMYELINASE DEFICIENCY)

Niemann–Pick disease type B (NPB) is an autosomal recessive storage disorder caused by a deficiency of lysosomal acid sphingomyelinase. The disease is characterized by prominent sphingomyelin accumulation in macrophages throughout the body, leading to hepatosplenomegaly, interstitial lung disease, and low platelet levels secondary to bone marrow involvement.[77] Lesser amounts of sphingomyelin can also be observed in hepatocytes, endothelial cells, fibroblasts, and other cell types.

In the liver, hepatomegaly is due to the dramatic accumulation of sphingomyelin in hepatocytes and Kupffer cells, which gives the tissue a foamy cell appearance on traditional paraffin-processed, H&E-stained sections (Fig. 13A). Sphingomyelin is better preserved in the gluaraldehyde-based fixatives used for electron microscopy. Semithin sections from epoxy-resin-embedded tissues demonstrate dense substrate accumulation in these cells (Fig. 14A), and under electron microscopy the sphingomyelin appears as tight myelin figures (Fig. 14B). As the disease progresses, hepatic fibrosis, and eventually frank cirrhosis, develop (Fig. 15), resulting in GI bleeding, ascites, peripheral edema,[78] hepatic failure,[79] and death.[80] Cirrhosis is a risk factor for hepatocellular carcinoma,[81] and has been reported in a few NPB case reports.[82] Accumulation of foamy macrophages in the lung is also accompanied by interstitial fibrosis, as demonstrated in lung biopsies from NPB patients.[83] This common observation in both the liver and the lung suggests that long-standing sphingomyelin storage may elicit a fibrotic reaction in the surrounding tissues.

Elevations in sphingomyelin up to 20 times the normal levels have been measured biochemically in NPB liver tissue. These measurements were associated with elevations in hepatic cholesterol and total phospholipids as well.[84] Elevated sphingomyelin is a risk factor for vascular disease in the general population and, interestingly, NPB patients also have abnormal lipoprotein profiles with low HDL levels and elevated LDL levels, leading to early signs and symptoms of heart disease.[77] Unstable angina[83] and vascular involvement have been reported in these patients. The autopsy of a 19-year-old female patient who died of acute heart failure revealed an enlarged and dilated left ventricle, with histological

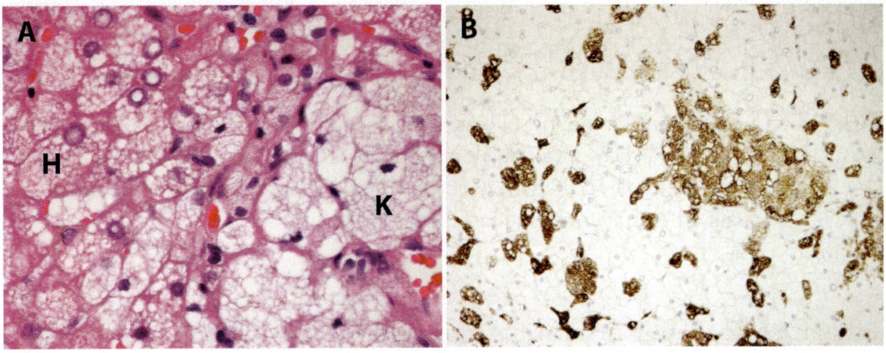

Fig. 13 Dramatic accumulation of sphingomyelin is present in the Niemann–Pick B liver. (**A**) Both hepatocytes (H) and Kupffer cells (K) appear foamy and vacuolated throughout the field. (H&E, 60× objective). (**B**) Immunohistochemistry for the macrophage marker CD68 highlights the Kupffer cell enlargement (20× objective).

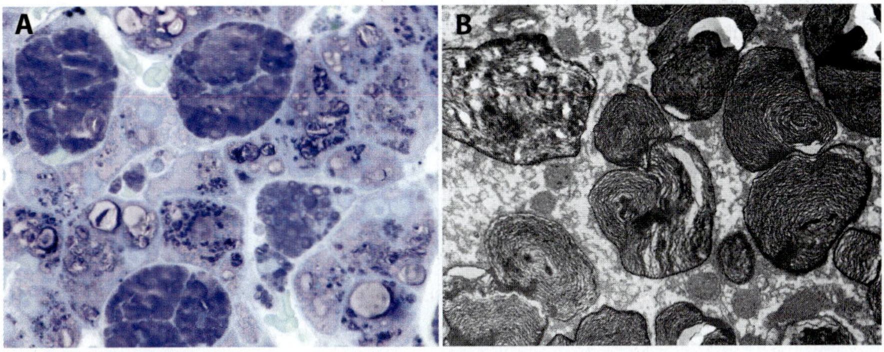

Fig. 14 Sphingomyelin is preserved in biopsies processed into epoxy resin. (**A**) Dark-blue–purple staining of sphingomyelin within Kupffer cells and hepatocytes (1 μm epoxy resin section, modified toluidene blue from the author's laboratory, 100× oil objective). (**B**) Ultrathin sections can be cut from the same epoxy resin block to examine under electron microscopy. Sphingomyelin appears as laminated whorls.

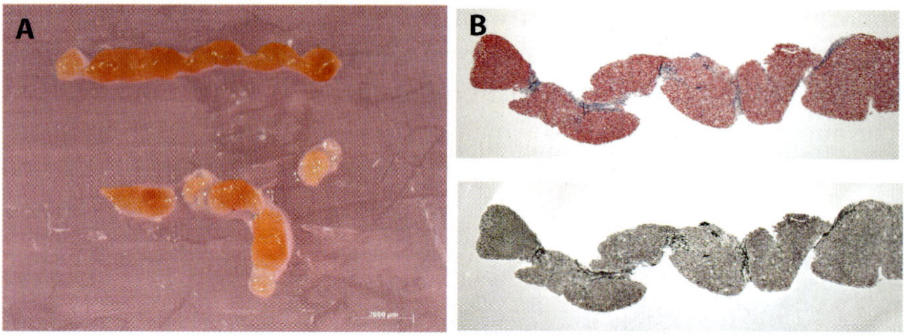

Fig. 15 Cirrhosis can develop in Niemann–Pick B disease. (**A**) Gross photo of a liver needle biopsy prior to processing. The biopsy fragments and the irregular surface of cirrhotic nodules can be appreciated grossly. (**B**) Trichrome and reticulin stains highlight the bridging fibrosis and nodule formation across the entire field (2× objective).

evidence of ischemic myocardial injury and narrowing of the distal coronary arteries by swollen smooth muscle cells.[85]

Some patients present with xanthomatous skin lesions containing foam cell infiltration,[82] and myelin figures.[86] Dermal sphingomyelin can accumulate in multiple cell types, including fibroblasts, macrophages, vascular endothelial cells, vascular smooth muscle cells, perineurium, and Schwann cells (Fig. 16).

Treatment for NPB has included splenectomy,[87] bone marrow transplantation,[88] liver transplantation,[89] and subcutaneous implantation of epithelial amniotic cells isolated from human placentas,[90] with variable results. Bone marrow transplantation resulted in modest reductions in hepatic and splenic sphingomyelin levels with regression of pulmonary infiltrates; however, long-term bone marrow transplantation failed to halt the progression of liver disease to cirrhosis.[88] Isolated epithelial amniotic cells are believed to contain high levels of lysosomal enzymes and to be nonimmunogenic. The subcutaneous implantation of these cells served as a transient reservoir of enzyme replacement which resulted in decreased hepatomegaly; however, the implantation procedure must be repeated every 1–4 months. Because of the limitations of the current treatment modalities, an ERT for NPB patients, consisting of recombinant acid sphingomyelinase, is presently under development in early clinical trials.

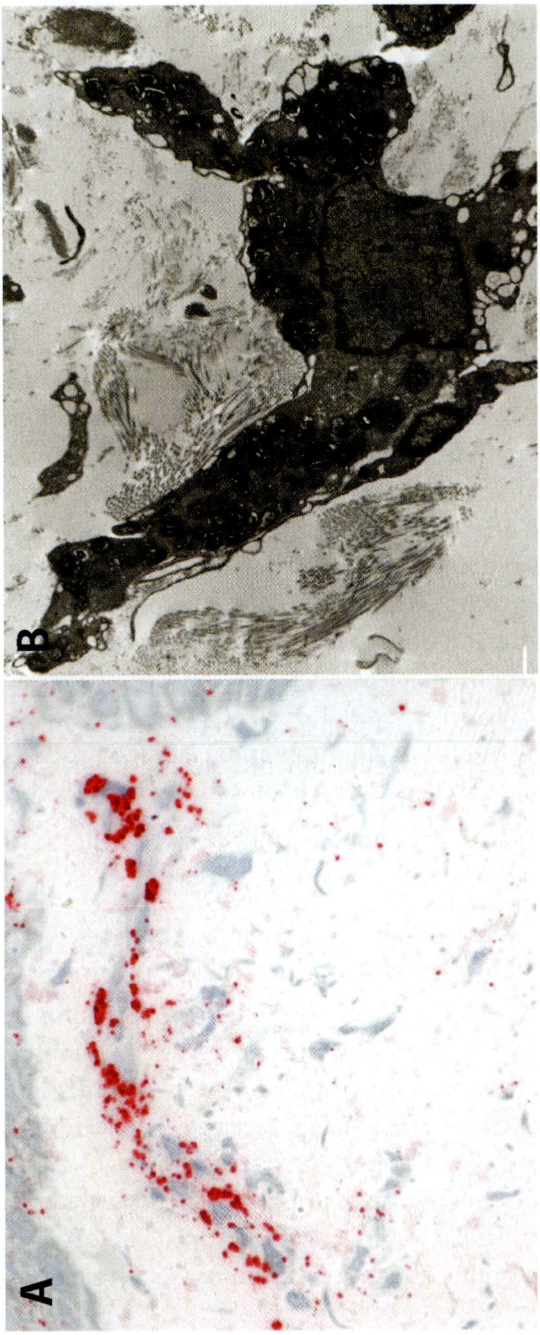

Fig. 16 Sphingomyelin also accumulates in cells of the skin. (**A**) A lysenin affinity stain shows sphingomyelin in the superficial dermis (red staining, 100× oil objective). (**B**) Electron-microscopic image of a dermal macrophage engorged with dense myelin figures of sphingomyelin (8000×).

APPENDIX: RECOMMENDATIONS ON TISSUE COLLECTION, HANDLING, AND PROCESSING

In hospitals' histology laboratories, 10% neutral buffered formalin (NBF) is the standard fixative used for tissue preservation, followed by paraffin processing and embedding,[91] and this is typically how most biopsies are processed in cases of undiagnosed and unsuspected storage disorders. For tissues from LSD patients, these routine procedures yield tissue sections filled with clear, foamy, or vacuolated cells, artifacts caused by poor cell preservation and loss of storage products during processing. For example, glycogen is a water-soluble molecule and is often lost by diffusion from the tissue when aqueous-based solutions such as NBF are used.[92–94] Conversely, lipids are lost during subsequent steps in tissue processing where organic solvents are used. As the availability of treatment increases awareness of storage disorders and inclusion in the differential diagnosis, clinicians and pathologists will need to work together to obtain optimally processed tissue for assessment of disease. Whenever possible, separate pieces of tissue should be collected for frozen sections, and embedding in both epoxy resin and paraffin.

For optimal morphological assessment, collection of biopsies in a glutaraldehyde-based fixative used for electron microscopy preserves the cellular integrity of tissues containing most storage products. Epoxy-resin-embedded tissues can be cut into 1 micron, semithin sections which can be stained and examined under light microscopy, as well as ultrathin sections for electron microscopy preparation and examination.[39,70,95] Novel staining protocols which "color-separate" the accumulating substrate from the surrounding tissue have been developed for use on semithin sections of Pompe and Niemann–Pick tissues (Ref. 95; the author's laboratory's novel tissue preparation and stain). Such color separation allows the use of computer-assisted morphometry for quantification of the substrate load. The preservation of CNS tissues presents unique challenges, but improved protocols have recently been developed that employ specific resins.[96] The use of epoxy resin restricts the size of the tissue to 1 mm cubes, whereas the use of glycol methacrylate permits the production of larger blocks which can accommodate an entire cross-section of the mouse brain. The preservation of CNS tissue in 10% NBF followed by a secondary fixation in osmium tetroxide/potassium dichromate prior to embedding in glycol methacrylate produces crisp, well-preserved cell morphology. Fresh tissue snap-frozen in optimal cutting temperature (OCT) media should be obtained for immunohistochemical and enzymatic staining, particularly for muscle biopsies where muscle fiber typing analysis may be desired. Any additional tissue should be preserved in NBF and processed into paraffin for archival purposes and the application of routine stains such as H&E, trichrome, reticulin, and others, used to assess LSDs with secondary pathology effects on target organs.

ACKNOWLEDGMENTS

I would like to thank members of the histology laboratory in the Department of Pathology at Genzyme for their diverse contributions to our understanding of lysosomopathies.

REFERENCES

1. Cotran RS, Kumar V, Collins T, eds. (1999) *Robbins Pathologic Basis of Disease.* WB Saunders, Philadelphia.
2. Fauci AS, Braunwald E, Isselbacher KJ, *et al.*, eds. (1998) *Harrison's Principles of Internal Medicine.* McGraw-Hill, New York.
3. Hille-Rehfeld A. (1995) Mannose-6-phosphate receptors in sorting and transport of lysosomal enzymes. *Biochim Biophys Acta* **1141**: 177–194.
4. Braulke T. (1999) Type-2 IGF receptor: a multi-ligand binding protein. *Horm Metab Res* **31**: 242–246.
5. Rohrbach M, Clarke JT. (2007) Treatment of lysosomal storage disorders: progress with enzyme replacement therapy. *Drugs* **67**(18): 2697–2716.
6. Hunley TE, Corzo E, Dudek M, *et al.* (2004) Nephrotic syndrome complicating α-glucosidase replacement therapy for Pompe disease. *Pediatrics* **114**: e532–e535.
7. Andreotti G, Guarracino MR, Cammisa M, *et al.* (2010) Prediction of the responsiveness to pharmacological chaperones: lysosomal human alpha-galactosidase, a case of study. *Orphanet J Rare Dis* **5**(36): 1–11.

8. Hughes DA, Pastores GM. (2010) The pathophysiology of GD — current understanding and rationale for existing and emerging therapeutic approaches. *Wien Med Wochenschr* **160**: 594–599.
9. Dickersin GR. (2000) *Diagnostic Electron Microscopy: A Text/Atlas*. Springer-Verlag, New York.
10. Mikosch P, Hughes D. (2010) An overview on bone manifestations in Gaucher disease. *Wien Med Wochenschr* **160**: 609–624.
11. Santamaria F, Parenti G, Guidi G, *et al.* (1998) Pulmonary manifestations of Gaucher disease. *Am J Respir Crit Care Med* **157**(3): 985–989.
12. Amir G, Ron N. (1999) Pulmonary pathology in Gaucher disease. *Hum Pathol* **30**(2): 666–670.
13. Hollak CEM, Evers L, Aerts J, Van Oers M. (1997) Elevated levels of M-CSF, sCD14 and IL8 in type 1 Gaucher disease. *Blood Cell Mol Dis* **23**(2): 201–212.
14. Barak V, Acker M, Kalickman I, *et al.* (1999) Cytokines in Gaucher's disease. *Eur Cytokine Netw* **10**(2): 205–210.
15. Deganuto M, Pittis MG, Pines A, *et al.* (2007) Altered intracellular redox status in Gaucher disease fibroblasts and impairment of adaptive response against oxidative stress. *J Cell Physiol* **12**(1): 223–235.
16. Zhang XS, Brondyk W, Lydon JT, *et al.* (2011) Biotherapeutic target or sink: analysis of the macrophage mannose receptor tissue distribution in murine models of lysosomal storage diseases. *J Inherit Metab Dis*, in press.
17. Rudzki A, Okon K, Machaczka M, *et al.* (2003) Enzyme replacement therapy reduces Gaucher cells burden but may accelerate osteopenia in patients with type I disease — histological study. *Eur J Haematol* **70**: 273–281.
18. Lukina E, Watman N, Arreguin EA, *et al.* (2010) Improvement in hematologic, visceral, and skeletal manifestations of Gaucher disease type 1 with oral eliglustat tartrate (Genz-112638) treatment: 2-year results of a phase 2 study. *Blood* **116**(20): 4095–4098.
19. Khanna R, Benjamin ER, Pellegrino L, *et al.* (2010) The pharmacological chaperone isofagomine increases the activity of the Gaucher disease L444P mutant form of beta-glucosidase. *FEBS J* **277**(7): 1618–1638.
20. Gubler MC, Lenoir G, Grunfeld JP, *et al.* (1978) Early renal changes in hemizygous and heterozygous patients with Fabry's disease. *Kidney Int* **13**: 223–235.
21. Farge D, Nadler S, Wolfe LS, *et al.* (1985) Diagnostic value of kidney biopsy in heterozygous Fabry's disease. *Arch Pathol Lab Med* **109**: 85–88.
22. Marguery MC, Giordano F, Parant M, *et al.* (1993) Fabry's disease: heterozygous form of different expression in two monozygous twin sisters. *Dermatology* **187**: 9–15.

23. Malouf M, Kirkman H, Buchanan P. (1976) Ultrastructure changes in antenatal Fabry's disease. *Am J Path* **82**: 13a.
24. Tsutsumi O, Sato M, Sata K, *et al.* (1985) Early prenatal diagnosis of inborn error of metabolism: a case report of a fetus affected with Fabry's disease. *Asia Oceania J Obstet Gynaecol* **11**(1): 39–45.
25. Bouwman MG, Hollak CE, van den Bergh Weerman MA, *et al.* (2010) Analysis of placental tissue in Fabry disease with and without enzyme replacement therapy. *Placenta* **31**(4): 344–346.
26. Desnick RJ, Ionnou YA, Eng CM. (2001) α-galactosidase A deficiency: Fabry disease. In: Scriver CR, Beaudet AL, Sly WS, Valle D (eds.), The Metabolic and Molecular Bases of Inherited Disease, 8th edn. McGraw-Hill, New York, Vol. 3, pp. 3733–3774.
27. Thurberg BL, Byers HR, Granter SR, *et al.* (2004) Monitoring the three year efficacy of enzyme replacement therapy in Fabry disease by repeated skin biopsies. *J Investig Dermatol* **122**(4): 900–908.
28. Bodary PF, Shayman JA, Eitzman DT. (2007) Alpha galactosidase A in vascular disease. *Trends Cardiovasc Med* **17**: 129–133.
29. Sachdev B, Takenaka T, Teraguchi H, *et al.* (2002) Prevalence of Anderson–Fabry disease in male patients with late onset hypertrophic cardiomyopathy. *Circulation* **105**: 1407–1411.
30. Montserrat L, Gimeno-Blanes JR, Marin F, *et al.* (2007) Prevalence of Fabry disease in a cohort of 508 unrelated patients with hypertrophic cardiomyopathy. *J Am Coll Cardiol* **50**: 2399–2403.
31. Wang RY, Lelis A, Mirocha J, Wilcox W. (2007) Heterozygous Fabry women are not just carriers, but have a significant burden of disease and impaired quality of life. *Genet Med* **9**: 34–45.
32. Eng CM, Fletcher J, Wilcox WR, *et al.* (2007) Fabry disease: baseline medical characteristics of a cohort of 1765 males and females in the Fabry Registry. *J Inherit Metab Dis* **30**: 184–192.
33. Wilcox WR, Oliveira JP, Hopkin RJ, *et al.* (2008) Females with Fabry disease frequently have major organ involvement: lessons from the Fabry registry. *Mol Genet Met* **93**: 112–128.
34. Thurberg BL, Fallon JT, Mitchell R, *et al.* (2009) Cardiac microvascular pathology in Fabry disease: evaluation of endomyocardial biopsies before and after enzyme replacement therapy. *Circulation* **118**: 2561–2567.
35. Sakuraba H, Igarashi T, Shibata T, Suzuki Y. (1987) Effect of vitamin E and ticlopidine on platelet aggregation in Fabry's disease. *Clin Genet* **31**: 349–354.
36. DeGraba T, Azhar S, Dignat-George F, *et al.* (2000) Profile of endothelial and leukocyte activation in Fabry patients. *Ann Neurol* **47**: 229–233.

37. Kannel WB. (2005) Overview of hemostatic factors involved in atherosclerotic cardiovascular disease. *Lipids* **40**: 1215–1250.
38. Galkina E, Ley K. (2007) Vascular adhesion molecules in atherosclerosis. *Arterioscler Thromb Vasc Biol* **27**: 2292–2301.
39. Thurberg BL, Rennke H, Colvin RB, *et al.* (2002) Globotriaosylceramide accumulation in the Fabry kidney is cleared from multiple cell types after enzyme replacement therapy. *Kidney Int* **62**: 1933–1946.
40. Najafian B, Svarstad E, Bostad L, *et al.* (2011) Progressive podocyte injury and globotriaosylceramide (GL-3) accumulation in young patients with Fabry disease. *Kid Int* **79**(6): 663–670.
41. Germain DP, Waldek S, Banikazemi M, *et al.* (2007) Sustained, long-term renal stabilization after 54 months of agalsidase-β therapy in patients with Fabry disease. *J Am Soc Nephrol* **18**: 1547–1557.
42. Marshall J, Ashe KM, Bangari D, *et al.* (2010) Substrate therapy reduction augments the efficacy of enzyme therapy in a mouse model of Fabry disease. *PLoS One* **5**(11): 1–10.
43. Motabar O, Sidransky E, Goldin E, *et al.* (2010) Fabry disease — current treatment and new drug development. *Curr Chem Genom* **4**: 50–56.
44. Crow J, Gibbs DA, Cozens W, *et al.* (1983) Biochemical and histopathological studies on patients with mucopolysaccharidoses, two of whom had been treated by fibroblast transplantation. *J Clin Pathol* **36**: 415–430.
45. Martin JJ, Ceuterick C. (1983) Prenatal pathology in mucopolysaccharidoses: a comparison with postnatal cases. *Clin Neuropathol* **2**(3): 122–127.
46. Bioulac P, Mercier M, Beylot C, *et al.* (1975) The diagnosis of mucopolysaccharidoses by electron microscopy of skin biopsies. *J Cutan Pathol* **2**(4): 179–190.
47. Dekaban AS, Constantopoulos G, Herman MM, *et al.* (1976) Mucopolysaccaridosis type V (Scheie syndrome): a postmortem study by multidisciplinary techniques with emphasis on the brain. *Arch Pathol Lab Med* **100**(5): 237–245.
48. Friedman I, Spellacy E, Crow J, *et al.* (1985) Histopathological studies of the temporal bones in Hurler's disease (MPS 1H). *J Laryngol Otol* **99**(1): 29–41.
49. Wakai S, Minami R, Kameda K, *et al.* (1988) Skeletal muscle involvement in mucopolysaccharidosis type IIA: severe type of Hunter syndrome. *Pediatr Neurol* **4**(3): 178–180.
50. Kint JA, Dacremont G, Carton D, *et al.* (1973) Mucopolysaccharidosis: secondarily induced abnormal distribution of lysosomal isoenzymes. *Science* **181**(97):352–354.

51. Lyons JA, Dickson Pi, Wall JS, *et al.* (2011) Arterial pathology in canine mucopolysaccharidosis-I and response to therapy. *Lab Invest* **91**: 665–674.
52. Kakkis ED, Muenzer J, Tiller GE, *et al.* (2001) Enzyme replacement therapy in muccopolysaccharidosis I. *N Engl J Med* **18**:344(3): 182–188.
53. Harmatz P, Whitley CB, Waber L, *et al.* (2004) Enzyme replacement therapy in mucopolysaccharidosis VI (Maroteaux–Lamy syndrome). *J Pediatr* **144**: 574–580.
54. Hug G. (1979) Pre- and postnatal pathology, enzyme treatment, and unresolved issues in five lysosomal disorders. *Pharmacol Rev* **90**: 565–591.
55. Phupong V, Shuangshoti S, Sutthiruangwong P, *et al.* (2005) Prenatal diagnosis of Pompe disease by electron microscopy. *Arch Gynecol Obstet* **271**: 259–261.
56. Hirschhorn R, Reuser AJJ. (2001) Glycogen storage disease type II: acid α-glucosidase (acid maltase) deficiency. In: Scriver CR, Beaudet AL, Sly WS, Valle D (eds.), The Metabolic and Molecular Basis of Inherited Disease, 8th edn. McGraw-Hill, New York, pp. 3389–3420.
57. Van den Hout HM, Hop W, van Diggelen OP, *et al.* (2003) The natural course of infantile Pompe's disease: 20 original cases compared with 133 cases from the literature. *Pediatrics* **112**: 332–340.
58. Kishnani PS, Hwu P, Mandel H, *et al.* (2006) A retrospective, multinational, multicenter study of the natural history of infantile Pompe disease. *J Pediatr* **148**: 671–676.
59. Bharati S, Serratto M, DuBrow I, *et al.* (1982) The conduction system in Pompe's disease. *Ped Cardiol* **2**: 25–32.
60. Ansong A, Li JS, Nozik-Grayck E, *et al.* (2006) Electrocardiographic response to enzyme replacement therapy for Pompe disease. *Genet Med* **8**(5): 297–301.
61. Makos MM, McComb RD, Hart MN, *et al.* (1987) Alpha-glucosidase deficiency and basilar artery aneurysm: report of a sibship. *Ann Neurol* **22**(5): 629–633.
62. Martin JJ, de Barsy T, Van Hoof F, *et al.* (1973) Pompe's disease: an inborn lysosomal disorder with storage of glycogen — a study of brain and striated muscle. *Acta Neuropath* **23**: 229–244.
63. Martini C, Ciana G, Benettoni A, *et al.* (2001) Intractable fever and cortical neuronal glycogen storage in glycogenosis type 2. *Neurology* **57**: 906–908.
64. Griffin JL. (1984) Infantile acid maltase deficiency. *Virchows Arch* **45**: 23–36.
65. Cardiff RD. (1966) A histochemical and electron microscopic study of skeletal muscle in a case of Pompe's disease (glycogenosis II). *Pediatrics* **37**(2): 249–259.
66. Hug G, Garancis JC, Schubert WK, *et al.* (1966) Glycogen storage disease, type II, III, VIII, and IX. *Amer J Dis Child* **111**: 457–474.
67. Garancis JC. (1968) Type II glycogenosis: biochemical and electron microscopic study. *Am J Med* **44**: 289–300.

68. Hudgson P, Fulthorpe JJ. (1975) The pathology of type II skeletal muscle glycogenosis: a light and electron-microscopic study. *J Path* **116**: 139–147.
69. Hesselink RP, Wagenmakers AJM, Drost MR, *et al.* (2003) Lysosomal dysfunction in muscle with special reference to glycogen storage disease type II. *Biochem Biophys Acta* **1637**: 164–170.
70. Thurberg BL, Lynch Maloney C, Vaccaro C, *et al.* (2006) Characterization of pre- and post-treatment pathology after enzyme replacement for Pompe disease. *Lab Invest* **86**(12): 1208–1220.
71. Rajawat YS, Hilioti Z, Bossis I. (2009) Aging: central role of autophagy and the lysosomal degradative system. *Ageing Res Rev* **8**: 199–213.
72. Raben N, Gilbert AL, Fukuda T, *et al.* (2005) Replacing a deficient enzyme in Pompe disease: recombinant human acid α-glucosidase (rhGAA) and endogenous transgenic human GAA are equipotent, but neither completely clears stored glycogen from type II skeletal muscle. *Mol Ther* **11**(1): 48–56.
73. Fukada T, Ewan L, Bauer M, *et al.* (2006) Dysfunction of endocytic and autophagic pathways in a lysosomal storage disease. *Ann Neurol* **59**(4): 700–708.
74. Kobayashi H, Shimada Y, Ikegami M, *et al.* (2010) Prognostic factors for the late onset Pompe disease with enzyme replacement therapy: from our experience of 4 cases including an autopsy case. *Mol Genet Metab* **100**(1): 14–19.
75. Chien Y-H, Lee N-C, Thurberg BL, *et al.* (2009) Pompe disease in infants: improving the prognosis by newborn screening and early treatment. *Pediatrics* **124**(6): e1116–e1125.
76. Kishnani PS, Nicolino M, Voit T, *et al.* (2006) Chinese hamster ovary cell–derived recombinant human acid alpha-glucosidase in infantile-onset Pompe disease. *J Pediatr* **149**(1): 89–97.
77. Wasserstein MP, Desnick RJ, Schuchmann EH, *et al.* (2004) The natural history of type B Niemann–Pick disease: results from a 10-year longitudinal study. *Pediatrics* **114**: e672–e677.
78. Tassoni JP, Fawaz KA, Johnson DE. (1991) Cirrhosis and portal hypertension in a patient with adult Niemann–Pick disease. *Gastroenterology* **100**: 567–569.
79. Putterman C, Zelingher J, Shouval D. (1992) Liver failure and the sea-blue histiocyte/adult Niemann–Pick disease. *J Clin Gastroenterol* **15**(2): 146–149.
80. Labrune P, Bedossa P, Huguet P, *et al.* (1991) Fatal liver failure in two children with Niemann–Pick disease type B. *J Pediatr Gastroenterol Nutr* **13**: 104–109.
81. Odze RD, Goldblum JR, Crawford JM, eds. (2004) *Surgical Pathology of the GI Tract, Liver, Billiary Tract and Pancreas.* Saunders, Philadelphia, Pennsylvania, pp. 878–882.

82. Crocker AC, Farber S. (1958) Niemann–Pick disease: a review of eighteen patients. *Medicine* **37**(1): 1–95.
83. Nicholson AG, Florio R, Hansell DM, *et al.* (2006) Pulmonary involvement by Niemann–Pick disease: a report of six cases. *Histopathology* **48**: 596–603.
84. Vanier MT. (1983) Biochemical studies in Niemann–Pick disease: major sphingolipids of liver and spleen. *Biochim Biophys Acta* **750:** 178–184.
85. Ishii TT, Toyono M, Tamure M, *et al.* (2006) Acid sphingomyelinase deficiency: cardiac dysfunction and characteristic findings of the coronary arteries. *J Inherit Metab Dis* **29**(1): 232–234.
86. Mardini MK, Gergen P, Akhtar M, *et al.* (1982) Niemann–Pick disease: report of a case with skin involvement. *Am J Dis Child* **136**: 650–651.
87. Mylla Neto G, Costa R, Fernandes PM, *et al.* (1983) Niemann–Pick disease in adult: report of a case surgically treated. *Rev Hosp Clin Fac Med Sao Paulo* **38**(2): 83–85.
88. Victor S, Coulter JBS, Besley TN, *et al.* (2003) Niemann–Pick disease: sixteen-year follow-up of allogeneic bone marrow transplantation in a type B variant. *J Inherit Metab Dis* **26**: 775–785.
89. Smanik EJ, Tavill AS, Jacobs GH, *et al.* (1993) Orthotopic liver transplantation in two adults with Niemann–Pick and Gaucher's disease: implications for the treatment of inherited metabolic disease. *Hepatology* **17**: 42–49.
90. Scaggiante B, Pineschi A, Sustersich M, *et al.* (1987) Successful therapy of Niemann–Pick disease by implantation of human amniotic membrane. *Transplantation* **44**(1): 59–61.
91. Carson F. (1997) *Histotechnology: A Self-Instructional Text*, 2nd edn. ASCP Press, Chicago.
92. Hug G. (1976) Glycogen storage diseases. *Birth Defects*, Original Article Series **12**(6): 145–175.
93. McAdams AJ, Hug G, Bove KE. (1974) Glycogen storage disease, type I to X: criteria for morphologic diagnosis. *Hum Pathol* **5**: 463–487.
94. Premasiri MK, Lee Y-S. (2003) The myopathology of floppy and hypotonic infants in Singapore. *Pathology* **35**: 409–413.
95. Lynch CM, Johnson J, Vaccaro C, Thurberg BL. (2005) High resolution light microscopy (HRLM) and digital analysis of Pompe disease pathology. *J Histochem Cytochem* **53**(1): 63–73.
96. Taksir TV, Griffiths D, Johnson J, *et al.* (2007) Optimized preservation of CNS morphology for the identification of glycogen in the Pompe mouse model. *J Histochem Cytochem* **55**(10): 991–998.

CHAPTER 8

Phosphatopathies

*Makoto Kuro-o**

INTRODUCTION

Inorganic phosphate (Pi) is an essential mineral that forms cell and body structure (DNA, membrane phospholipids, and bone). Pi also mediates numerous cellular functions, including ATP synthesis and kinase-mediated signal transduction. It is so fundamental to life that its deficiency can be fatal, which may explain the evolution of a behavioral adaptation called "phosphate appetite": animals in a Pi-deficient state seek out Pi-rich foodstuffs, resulting in osteophagic behavior (licking/eating bones of dead animals). This has been observed even in herbivores.[1,2]

Unlike animals in the wild, people on a western diet take in too much phosphorus. According to a recent USDA report,[3] phosphorus consumption per capita has been continuously increasing beyond the recommended intake level for the past three decades. Importantly, actual increases in phosphorus consumption in the diet are likely much greater, because the USDA statistics have not included large amounts of Pi added to processed foods and soft drinks as additives and preservatives, which are now estimated to contribute more than 30% of the total phosphorus intake.[4]

*Department of Pathology, University of Texas Southwestern Medical Center at Dallas, 5323 Harry Hines Blvd., Dallas, TX 75390-9072, USA. E-mail: makoto.kuro-o@utsouthwestern.edu, Tel.: 214-648-4018, Fax: 214-648-4070.

The increasing trend of Pi intake in our current diet is a serious health concern, particularly for patients with chronic kidney disease (CKD), who have reduced ability to excrete excess Pi into their urine. CKD is defined as a state of reduced renal function over months or years. It can be caused by many chronic diseases that affect the kidney (most notably hypertension and diabetes) or can occur as a consequence of natural aging. More than 26 million Americans, or 13% of the total population, have CKD, and 600,000 Americans suffer from advanced CKD, also known as end-stage renal disease (ESRD), which can only be treated by dialysis or renal transplantation.[5] Because of its broad impact and burden in health care, developing therapeutic strategies to prevent CKD progression has been designated as a key priority of "Healthy People 2020" by the United Stated Department of Health and Human Services (http://www.healthypeople.gov).

Considering the increasing trend of Pi consumption and the grave impact of CKD on public health, there is an urgent need for our aging society to clarify potential harms of Pi overload and elucidate the underlying mechanisms. The purposes of this chapter are to overview recent progress in the understanding of endocrine regulation of Pi homeostasis and discuss various pathologies associated with Pi retention, which are now referred to as "phosphatopathies." Phosphatopathy is a novel concept which we propose here for the first time, prompted by recent findings that have forged unexpected links between vitamin D, parathyroid hormone, fibroblast growth factor-23, and a molecule known as Klotho, leading to critical new insights into Pi metabolism.

ENDOCRINE REGULATION OF PHOSPHATE HOMEOSTASIS

Pi homeostasis is maintained by a balance between absorption of dietary Pi from the intestine and renal excretion of blood Pi. These processes are regulated by endocrine factors. The active form of vitamin D (1,25-dihydroxyvitamin D_3, or calcitriol) and parathyroid hormone (PTH), which have been extensively studied as hormones that regulate calcium (Ca) metabolism, are also involved in Pi metabolism.[6–8] PTH is secreted when a G-protein-coupled Ca-sensing receptor expressed on the surface of parathyroid chief cells detects decreases in blood Ca levels. PTH acts on the kidney to increase calcitriol synthesis in proximal tubules by inducing expression of 1α-hydroxylase, an enzyme essential for biosynthesis of calcitriol. Calcitriol secreted from the kidney in turn acts on the intestine to increase absorption of Ca and Pi, thereby restoring blood Ca levels to normal. Unlike calcitriol, PTH does not cause a net increase in blood Pi levels, because PTH has an activity that induces Pi excretion into urine (phosphaturia). Calcitriol also acts on the parathyroid gland and suppresses PTH production to close a negative feedback loop. This kidney–parathyroid endocrine axis mediated by PTH and calcitriol was until quite recently the only known mechanism of endocrine regulation of Pi homeostasis.

This classic view has been transformed by the discovery of novel endocrine axes mediated by fibroblast growth factor-23 (FGF23) and Klotho. FGF23 is one of the 22 members of the fibroblast growth factor (FGF) family.[9] The function of FGF23 was unknown until the *FGF23* gene was identified as the genetic basis for autosomal dominant hypophosphatemic rickets (ADHR).[10] ADHR is characterized by Pi wasting into urine, hypophosphatemia, inappropriately low serum calcitriol levels, and defects in bone mineralization (rickets). ADHR patients harbor missense mutations in the *FGF23* gene at the ^{176}RXXR^{179} motif, where FGF23 protein is cleaved and inactivated by a protease(s) yet to be identified. The ADHR mutations identified to date (R176Q, R179W, and R179Q) confer resistance to the proteolytic cleavage on FGF23 protein.[11] As a result, blood levels of intact FGF23 are elevated in ADHR patients. Animal studies have demonstrated that injection of intact FGF23 induces phosphaturia and lowers blood Pi levels. FGF23 also suppresses calcitriol synthesis in the kidney and lowers blood calcitriol levels as well.[12] Thus, FGF23 has emerged as a novel phosphaturic hormone as well as a counterregulatory hormone for calcitriol.[13] It does not bind to any known FGF receptor with high affinity. Thus, the identity of the FGF23 receptor had been an open question.

Klotho, on the other hand, was originally identified as an aging suppressor gene in mice that extended the lifespan when overexpressed[14] and induced complex phenotypes resembling human premature aging syndromes when disrupted.[15] The *Klotho* gene encodes a single-pass transmembrane protein and is expressed primarily in the kidney.[15] The precise biological function of Klotho was not clear until Klotho-deficient mice (*Klotho*$^{-/-}$ mice) and FGF23-deficient mice (*Fgf23*$^{-/-}$ mice) were found to be identical in terms of phenotypes. *Fgf23*$^{-/-}$ mice exhibited high serum Pi and calcitriol levels associated with ectopic calcification in arteries (vascular calcification).[16] In addition to these predictable phenotypes, *Fgf23*$^{-/-}$ mice unexpectedly developed complex phenotypes resembling aging, including growth arrest, emphysematous lung, osteopenia, and atrophy of gonads, skin, muscle, and intestine, resulting in a shortened lifespan.[17] These aging-like phenotypes were highly reminiscent of *Klotho*$^{-/-}$ mice. Conversely, *Klotho*$^{-/-}$ mice had been known to exhibit high serum Pi and calcitriol levels[18] and ectopic calcification in addition to the aging-like phenotypes.

These observations led to the identification of Klotho protein function: Klotho protein forms a constitutive binary complex with the FGF receptor (FGFR), thereby creating a *de novo* high-affinity binding site for FGF23.[19–21] Thus, it is not FGFR but the Klotho–FGFR complex that functions as a high-affinity receptor for FGF23. The fact that Klotho functions as an obligate coreceptor for FGF23 explains why FGF23 does not bind to FGFR alone with high affinity. It also explains why *Klotho*$^{-/-}$ mice and *Fgf23*$^{-/-}$ mice developed identical phenotypes. Furthermore, the relatively high kidney-specific expression of Klotho explains why FGF23 acts primarily on the kidney among the many other tissues that express multiple FGFR isoforms.

The parathyroid gland is one of the few organs that express Klotho endogenously, raising the possibility that the parathyroid may be another target organ of FGF23 besides the kidney. In fact, FGF23 can suppress PTH production and secretion.[22] Because PTH can increase FGF23,[23] FGF23 and PTH regulate each other in a negative feedback loop between bone and parathyroid. In addition, calcitriol can increase FGF23 production in the bone,[12] indicating that FGF23 and calcitriol also regulate each other in a negative feedback loop between bone and kidney. Thus, the discovery of novel endocrine axes mediated by FGF23 and Klotho has added new dimensions to and significantly advanced our understanding of endocrine regulation of Pi homeostasis (Fig. 1).

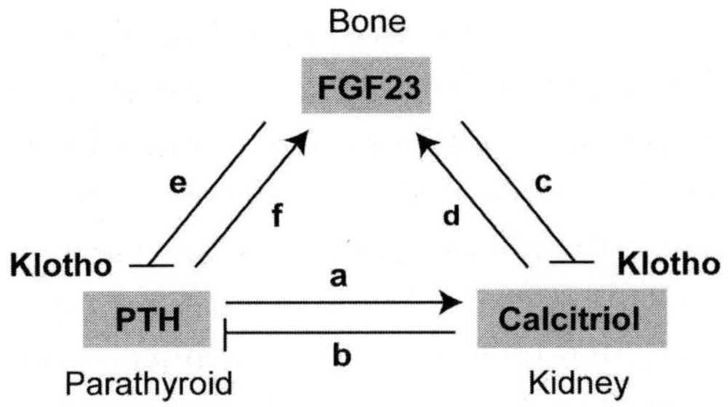

Fig. 1 Pi-regulating hormones (modified from Refs. 85 and 86). Parathyroid hormone (PTH) increases the synthesis of calcitriol (1,25-dihydroxyvitamin D_3) in the kidney (**a**). Calcitriol in turn decreases PTH (**b**) and closes a negative feedback loop. FGF23 is secreted from bone and acts on the kidney to reduce calcitriol synthesis in a Klotho-dependent manner (**c**). Because calcitriol increases FGF23 expression in bone (**d**), there exists a negative feedback loop between FGF23 and calcitriol. FGF23 also acts on the parathryroid to reduce PTH in a Klotho-dependent manner (**e**). Because PTH increases FGF23 expression (**f**), there exists another negative feedback loop between PTH and FGF23.

PHOSPHATOPATHIES — A NOVEL CONCEPT

Defects in either FGF23 (hormone) or Klotho (receptor) in mice increase blood Pi and calcitriol levels and induce complex phenotypes resembling aging.[24] Blood Ca levels are increased secondary to elevated calcitriol levels. To investigate the contribution of Pi, Ca, and calcitriol to the aging-like phenotypes, several laboratories have asked if resolving hyperphosphatemia or hypervitaminosis D by means of dietary or genetic interventions might rescue the aging-like phenotypes in *Klotho*$^{-/-}$ mice and *Fgf23*$^{-/-}$ mice.

First, a vitamin-D-deficient diet was reported to rescue many of the aging-like phenotypes in *Klotho*$^{-/-}$ mice[18] and *Fgf23*$^{-/-}$ mice.[25] Second, disruption of the *Cyp27b1* gene also rescued many of the aging-like phenotypes observed in *Klotho*$^{-/-}$ mice[26] and *Fgf23*$^{-/-}$ mice[17]: the *Cyp27b1* gene encodes 1α-hydroxylase, which converts an inactive form of vitamin D (25-hydroxyvitamin D_3) to calcitriol (1,25-dihydroxyvitamin D_3) in the kidney. Thus, disruption of the *Cyp27b1* gene reduces blood calcitriol to undetectable levels. *Cyp27b1*$^{-/-}$–*Klotho*$^{-/-}$ double-knockout mice and *Cyp27b1*$^{-/-}$–*Fgf23*$^{-/-}$ double-knockout mice were exempt from soft tissue calcification, growth arrest, emphysematous lung, osteopenia, and atrophy of gonads, skin, muscle, and intestine, resulting in significantly prolonged survival. Lastly, disruption of the *Vdr* (vitamin D receptor) gene also rescued accelerated aging phenotypes in *Fgf23*$^{-/-}$ mice.[27] These observations strongly suggest that the aging-like phenotypes in *Klotho*$^{-/-}$ mice and *Fgf23*$^{-/-}$ mice can be attributed to calcitriol intoxication. However, ablation of calcitriol activity reduced blood Pi and Ca levels at the same time, raising the possibility that the true culprit(s) might be Pi and/or Ca.

A low-Pi diet significantly reduced blood Pi levels, whereas it did not restore blood Ca and calcitriol levels in *Fgf23*$^{-/-}$ mice.[25] Rather, it exacerbated existing hypercalcemia and hypervitaminosis D. Increases in calcitriol in mice on a low-Pi diet are regarded as an attempt to maximize Pi absorption from the intestine and an adaptation to limited Pi availability. However, calcitriol promotes not only Pi absorption but also Ca absorption from the intestine. Because Ca was not restricted in the low-Pi diet, increases in calcitriol resulted in hypercalcemia. Despite persistent hypercalcemia and hypervitaminosis D, a low-Pi diet rescued many aging-like phenotypes in *Klotho*$^{-/-}$ mice[28] and *Fgf23*$^{-/-}$ mice.[25] Moreover, these mutant mice were rescued by disruption of the *Npt2a* gene.[29] Na-dependent Pi cotransporter type 2a (Npt2a) is expressed on the brush border membrane of renal proximal tubules and is primarily responsible for transepithelial Pi reabsorption in the kidney. Thus, mice lacking Npt2a lose Pi into urine and develop hypophosphatemia. Like mice placed on low-Pi diet, *Npt2a*$^{-/-}$ mice have high blood calcitriol and Ca levels. Despite that, *Npt2a*$^{-/-}$–*Klotho*$^{-/-}$ double-knockout mice and *Npt2a*$^{-/-}$–*Fgf23*$^{-/-}$ double-knockout mice were exempt from many of the *Klotho*$^{-/-}$ and *Fgf23*$^{-/-}$ single-knockout phenotypes. In addition, a high-Pi diet induced aging-like phenotypes in *Npt2a*$^{-/-}$–*Klotho*$^{-/-}$ double-knockout mice.[29] Thus, it is not calcitriol or Ca but Pi that is primarily responsible for the aging-like phenotypes.

Based on these findings, we propose that the complex aging-like phenotypes displayed by *Klotho*$^{-/-}$ mice and *Fgf23*$^{-/-}$ mice be called "phosphatopathies." These animal studies have not excluded the possibility that hypercalcemia and/or hypervitaminosis D may be a prerequisite for Pi to induce phosphatopathies, because mice that exhibit hyperphosphatemia with low/normal blood Ca or calcitriol levels have not been tested for phosphatopathies. However, this possibility is unlikely, because patients with chronic renal failure develop phosphatopathies associated with hyperphosphatemia, low blood calcitriol, and low/normal blood Ca levels, which will be discussed in the following sections.

PHOSPHATOPATHIES IN HUMANS

Reduction-of-function mutations in the FGF23–Klotho endocrine system cause a rare hereditary disorder known as hyperphosphatemic familial tumoral calcinosis (FTC), which is characterized by high blood Pi levels and ectopic calcification in soft tissues such as skin and arteries.[30] Hyperphosphatemic FTC is caused by mutations in the *FGF23*, *GALNT3*, or *KLOTHO* gene. In contrast to the ADHR mutations, FTC mutations in the *FGF23* gene result in increased susceptibility to its proteolytic inactivation, thereby lowering blood levels of intact FGF23.[31,32] The *GALNT3* gene encodes a glycosyltransferase necessary for *O*-glycosylation of the FGF23 protein adjacent to its proteolytic cleavage site, which reduces the vulnerability of FGF23 to protease attack. Thus, loss-of-function mutations in the *GALNT3* gene also result in low blood levels of intact FGF23.[30] A missense mutation in the *KLOTHO* gene was reported in a 13-year-old girl who developed FTC-like symptoms, including hyperphosphatemia, osteopenia, and calcification in soft tissue and arteries.[33] Thus, defects in the FGF23–Klotho endocrine system clearly also cause phosphatopathies in humans.

Typical phosphatopathies are universally observed in patients with CKD. ESRD patients exhibit significant increases in blood FGF23 levels, hyperphosphatemia, and vascular calcification, which are reminiscent of Klotho-deficient mice.[34–36] In fact, Klotho expression is significantly decreased in CKD patients.[35,37] Since hyperphosphatemia was identified as a potent mortality risk factor for CKD patients,[38–40] controlling serum Pi levels below 4.5 mg/dL by a low-Pi diet and Pi binders (pills that chelate Pi in the gut and prevent Pi absorption) has been proposed as an important therapeutic goal in the management of CKD patients.[41] Several clinical trials have shown that Pi binders indeed lower blood Pi levels and delay progression of vascular calcification, resulting in improved survival of hemodialysis patients.[42,43]

Importantly, the vast majority of CKD patients die not of renal failure *per se*, but due to the early onset of common age-related diseases such as cardiovascular disease, cancer, or infection.[44,45] Consequently, the spectrum of causes of death in CKD patients is not very different from that of the general population. CKD patients also suffer many aging-like symptoms, including skin atrophy, hypogonadism, osteopenia, and cognitive impairment. Thus, CKD may be viewed as a state of accelerated aging and age-related diseases associated with Klotho deficiency and Pi retention.

PHOSPHATE METABOLISM IN CKD

As the number of functional nephrons decreases in CKD, the glomerular filtration rate (GFR) falls progressively. The GFR has been used for classifying the severity of CKD into five stages: stage 1 (GFR $\geq$ 90 ml/min/1.73m^2), stage 2 (GFR 89–60), stage 3 (GFR 59–30), stage 4 (GFR 29–15), and stage 5 (GFR $<$ 15). Epidemiological studies have demonstrated that hyperphosphatemia becomes evident only in advanced stages (stage 4 and stage 5) and that changes in Pi-regulating hormones precede hyperphosphatemia during CKD progression: Low calcitriol, high FGF23, and high PTH are observed in CKD stage 3 or earlier.[46]

To prevent Pi retention, ingested Pi must be excreted from the kidney into urine. When the number of intact nephrons is progressively decreased during CKD progression, each functional nephron must excrete increasing amounts of Pi unless Pi intake is restricted. This demand can be met by increasing the secretion of phosphaturic hormones: FGF23 or PTH, or both. In fact, the serum levels of FGF23 and PTH increase long before the serum Pi levels increase during CKD progression.[46] Although both FGF23 and PTH can function as phosphaturic hormones, FGF23 is apparently responsible for preventing Pi retention in the setting of CKD for the following reasons: First, the serum calcitriol levels also decrease the long before the serum Pi levels increase during CKD progression,[46] suggesting that the FGF23 effect (decreasing calcitriol) predominates over the PTH effect (increasing calcitriol). Second, reduction of PTH by parathyroidectomy or administration of calcimimetics does not increase the serum Pi levels in CKD patients.[47,48] Lastly, the serum Pi levels are increased after administration of an FGF23-neutralizing antibody to a rat model of early-stage CKD with normophosphatemia.[49] Treatment with the FGF23 antibody also increases the serum calcitriol levels, followed by increases in Ca and decreases in PTH. Thus, high PTH in CKD is regarded as secondary to FGF23-induced decreases in calcitriol.

These observations indicate that early-stage CKD patients maintain normal blood Pi levels primarily by increasing FGF23 to compensate for reduced ability of the kidney to excrete Pi *at the expense of calcitriol.* Decreases in calcitriol can lower Ca absorption from the intestine and blood Ca levels. In fact, many CKD patients exhibit low blood Ca levels.[5] Because calcitriol is a potent upregulator of Klotho,[18] decreases in calcitriol can reduce Klotho expression in the kidney and parathyroid. In fact, urine Klotho levels start decreasing in early-stage CKD long before hyperphosphatemia ensues.[37] Decreases in Klotho can increase resistance to FGF23 in its target organs (kidney and parathyroid) and further increase blood FGF23 levels. Resistance to FGF23 in the parathyroid may also attenuate its ability to suppress PTH and contribute to PTH increase.[50] Moreover, increases in PTH can in turn further increase FGF23.[23] Thus, these changes in the Pi-regulating hormones and Klotho form a vicious cycle that leads to high FGF23, high PTH, low calcitriol, and low Klotho, which denote characteristics of ESRD (Fig. 2). Hyperphosphatemia ensues

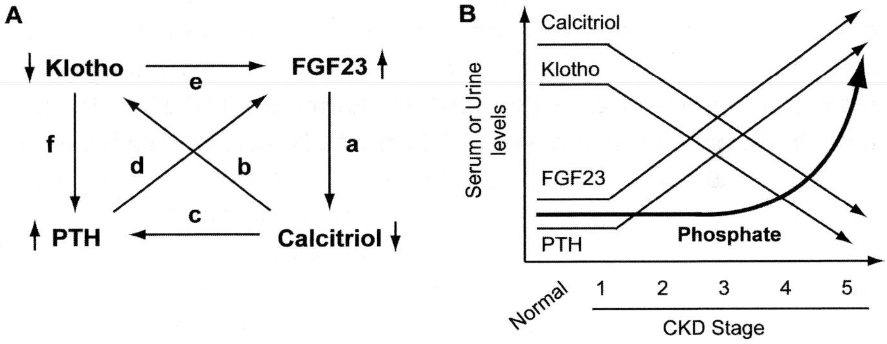

Fig. 2 Changes in the Pi-regulating hormones and Klotho during CKD progression (modified from Refs. 85 and 86). (**A**) Vicious cycles leading to high FGF23, high PTH, low vitamin D, and low Klotho in CKD. To meet an increasing demand for Pi excretion per nephron, CKD patients increase FGF23. High FGF23 suppresses calcitriol synthesis (**a**). Low calcitriol reduces Klotho expression (**b**) and increases PTH (**c**). High PTH further increases FGF23 (**d**). Low Klotho increases resistance to FGF23 and also increases FGF23 (**e**). Low Klotho in the parathyroid attenuates the ability of FGF23 to suppress PTH (**f**). (**B**) Increases in serum FGF23 and serum PTH levels and decreases in serum calcitriol and urine Klotho levels precede hyperphosphatemia during CKD progression from stage 1 to stage 5.

when the number of functional nephrons decreases to the extent that ingested Pi is not sufficiently excreted into urine. Thus, phosphatopathies in ESRD patients are associated with hyperphosphatemia, low blood calcitriol, and low blood Ca levels. These observations support the notion that Pi retention induces phosphatopathies even in the absence of Ca and calcitriol retention.

MECHANISM OF PHOSPHATOPATHIES

It has been known that cultured cells exposed to high Pi concentrations exhibit various responses that partly recapitulate phosphatopathies *in vivo*. Dulbecco's modified eagle medium (DMEM), the most commonly used cell culture medium, contains 1.8 mM Ca and 0.9 mM Pi, which are equivalent to blood levels in normal humans. Vascular smooth muscle cells cultured in DMEM never acquire CaPi deposits spontaneously. However, when Pi concentrations are increased to 2.0 mM (equivalent to those in advanced CKD patients), the cellular calcium content is increased and CaPi deposits are observed in a dose-dependent manner.[51] These changes induced by the high-Pi medium are accompanied by increased expression of Runx2 (also known as Cbfa1), a transcription factor that induces expression of bone markers including osteopontin, osteocalcin, and collagen-1, and functions as a master regulator of osteogenesis. In addition, these transdifferentiated vascular smooth muscle cells harbor cytoplasmic vesicles highly competent to uptake Ca and Pi like osteoblasts. These vesicles, called matrix vesicles, are extruded into the extracellular matrix. Once externalized, the matrix vesicles further uptake Ca and Pi and serve to nucleate hydroxyapatite crystals $[Ca_{10}(PO_4)_6(OH)_2]$, leading to medial calcification of arteries.[52] Thus, high ambient Pi may directly trigger the phenotypic transition in vascular smooth muscle cells to become "bone-like" and promote vascular calcification.

Although increases in extracellular Pi can activate the osteogenic cellular program, it is not necessary for Pi to enter the cell to trigger this program. Cellular Pi uptake is largely mediated by PiT-1, a type III Na-dependent Pi cotransporter that is ubiquitously expressed in various types of cells.[53] Pi-induced increases in Runx2 and osteopontin expression were abolished by phosphonoformic acid (PFA), a phosphate analog that competitively inhibits Npt2a and was once thought to inhibit PiT-1 as well.[51] Thus, it had been believed that Pi entry into the cell was required for the cellular responses to high ambient Pi. However, recent studies excluded this possibility: PFA did not inhibit cellular Pi uptake by PiT-1[54] but prevented formation of insoluble CaPi precipitation in the medium: PFA, like pyrophosphates or bisphosphonates, turned out to be a compound that can interfere with CaPi crystal growth.[55] These observations imply that the cellular responses to the high-Pi medium may be attributed to CaPi precipitates generated in the medium when the Ca and Pi concentrations exceed the solubility limit. Vascular smooth muscle cells may have intrinsic machinery that recognizes insoluble CaPi precipitates in the extracellular space and triggers Runx2 and osteopontin expression. On the other hand, knockdown of PiT-1 expression by siRNA abolished induction of Runx2 and osteopontin following exposure to the high-Pi medium,[56] which apparently contradicts the notion that Pi entry into the cell is not required for cellular responses to the high-Pi medium. However, considering the fact that PiT-1 was originally identified as a retroviral receptor,[57–59] it is intriguing to speculate that PiT-1 may contribute to Pi-induced cellular responses not by importing Pi into the cell but by participating in the recognition of CaPi precipitates and/or signal transduction triggered by CaPi precipitates.

Recent studies further support the notion that it is not soluble Pi *per se* but insoluble CaPi precipitates that are responsible for the Pi-induced cellular responses. In the presence of 10% fetal bovine serum (FBS) and 1.8 mM Ca, CaPi nanocrystals complexed with serum fetuin were generated in a high-Pi medium (2.0–4.0 mM Pi) spontaneously after incubation at 37°C for three days.[60] This medium containing CaPi nanocrystals induced osteopontin expression when applied to vascular smooth muscle cells. However, when the CaPi nanocrystals were precipitated by centrifugation, the supernatants did not induce osteopontin, whereas the pellet resuspensions did.[61] In addition to activation of the osteogenic program, a high-Pi medium can specifically activate the Raf/MEK/ERK pathway but not the other MAP kinase pathays (JNK and p38) and the AKT pathway.[62,63] Activation of ERK by a high-Pi medium is also attributed not to soluble Pi but to insoluble CaPi precipitates generated in the medium.[62] The other cellular responses induced by a high-Pi medium include induction of

TGF-β1 expression in vascular smooth muscle cells,[64] increases in cellular levels of reactive oxygen species (ROS), and apoptosis in vascular endothelial cells.[65,66] It remains to be determined whether these cellular responses to a high-Pi medium are also attributed not to soluble Pi but to CaPi precipitates.

The question remains whether such insoluble CaPi precipitates are just artifacts of cell culture experiments or are indeed generated *in vivo* despite the existence of many endogenous inhibitors for growth of CaPi crystals, such as fetuin, matrix Gla protein, osteopontin, and pyrophosphate.[61,67–71]

NANOBACTERIA AND PHOSPHATOPATHIES

"Nanobacteria" were originally described in the 1980s as self-propagating particles residing in serum, urine, and calcified tissues of animals and humans.[72,73] Although the concept was controversial even at its inception, nanobacteria were thought to be the smallest life form, visible only through electron microscopy.[72] However, recent studies have proven that nanobacteria are not living entities but CaPi nanocrystals (<500 nm) complexed with fetuin.[60,74–76] Thus, CaPi nanocrystals have turned out to be prevalent *in vivo* as nanobacteria. Because the term is now clearly misleading, we propose that these CaPi nanocrystals be called thus.

Recent studies have revealed the biophysical process of fetuin–mineral complex formation (Fig. 3).[77] When Ca and Pi concentrations exceed the solubility limit, tiny $Ca_9(PO_4)_6$ clusters called Posner clusters (7–9 Å in diameter) are generated. In the absence of fetuin, Posner clusters serve as mineral nuclei and building blocks of amorphous CaPi.[78] In the presence of fetuin, Posner clusters bind to fetuin and cannot nucleate amorphous CaPi.[79] One fetuin molecule can bind to up to ~120 Posner clusters, generating CaPi-laden monomeric or multimeric fetuin molecules called calciprotein monomers (CPMs) or primary calciprotein particles (primary CPPs), respectively. Primary CPPs are subject to a topological rearrangement to turn into stable structures, in which a densely packed fetuin monolayer covers a mineral core, thereby preventing further crystal growth.[80] These particles are referred to as secondary CPPs. All of these fetuin–mineral particles exist as a mixture in various ratios, depending

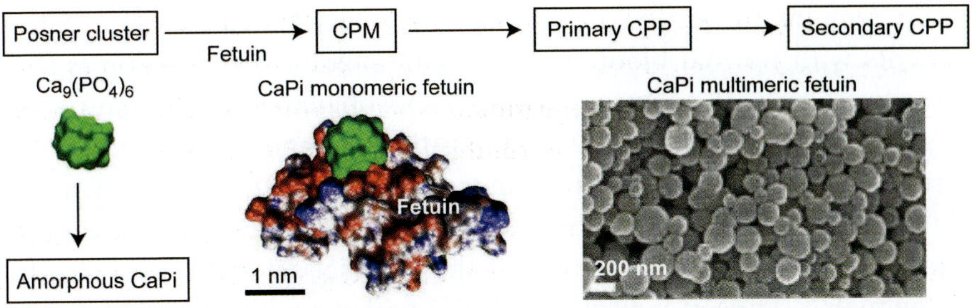

Fig. 3 Various forms of fetuin–mineral complexes. A Posner cluster is generated when Ca and Pi reach the saturating concentrations in solution. Fetuin does not inhibit formation of Posner clusters but prevents their aggregation and suppresses formation of amorphous CaPi. 3D views of a Posner cluster (green) and a CaPi monomeric fetuin (CPM) are shown. Negative and positive charges on the fetuin molecule are shown in red and blue, respectively (modified from Ref. 77). Scanning electron microscopy of CaPi multimeric fetuin particles (primary and secondary CPPs) is also shown (modified from Ref. 75).

on temperature, pH, incubation time (in the course of hours, days, weeks, and even months), and concentrations of Ca, Pi, and fetuin. These particles generated in test tubes are indistinguishable from nanobacteria isolated from blood, urine, or tissues in terms of morphology and composition determined by scanning and transmission electron microscopy, mass spectrometry, and energy-dispersive X-ray spectroscopy,[60,74–76] although actual nanobacteria can incorporate some additional ions (hydrogen, hydroxide, carbonate, magnesium, etc.) and proteins (albumin, apolipoprotein, etc.).[52,60,74–76]

Nanobacteria have been demonstrated in human blood.[81] When serum samples from CKD patients were centrifuged at high speed (16,000 × g for 2 h), the concentrations of fetuin in the supernatants were significantly lower than those measured before centrifugation, indicating that some serum fetuin was precipitated by centrifugation. In fact, the pellets consisted of fetuin–mineral complexes (i.e. nanobacteria). Moreover, the amount of nanobacteria was correlated with the severity of coronary artery calcification. Nanobacteria were barely detectable in serum samples from normal individuals but measurable in those from patients with stage 1–2 CKD and increased with disease progression. These findings indicate that nanobacteria can propagate in the blood even in the absence of overt hyperphosphatemia.

It remains to be determined why nanobacteria can grow in CKD patients with normal blood Pi levels. One possible explanation is that repetitive transient postprandial hyperphosphatemia may be sufficient to allow nucleation and gradual propagation of nanobacteria, even if the serum Pi levels return to normal under fasting conditions. In fact, a single Pi-rich meal transiently increases the serum Pi levels for a few hours and acutely impairs vasodilation of the brachial artery even in healthy individuals.[65] Thus, it is likely that CKD patients with reduced capacity to excrete Pi into urine may suffer prolonged and/or enhanced postprandial hyperphosphatemia. It should be noted that many CKD complications exemplified by vascular calcification occur in CKD patients with normal blood Pi levels, which is associated with an increased incidence of cardiovascular events.[44,82,83] Also, many otherwise healthy older individuals without hyperphosphatemia exhibit vascular calcification.[84] Because renal function declines in an age-dependent manner, the prevalence of CKD increases with age: more than a third of individuals over 70 years of age have CKD.[5] Thus, older individuals with reduced renal function likely have higher levels of nanobacteria in the blood relative to younger individuals with normal renal function. These "phosphatopathies without hyperphosphatemia" may be explained by nanobacteria.

CONCLUDING REMARKS

Studies on *Klotho*$^{-/-}$ mice and *Fgf23*$^{-/-}$ mice have identified novel endocrine axes that regulate Pi homeostasis and forged an unexpected link between Pi and aging. Pi retention causes complex phenotypes resembling aging, which can be collectively referred to as phosphatopathies. Phosphatopathies are universally observed in the elderly and are associated with a decline of renal function. Notably, hyperphosphatemia is sufficient but not required for the development of phosphatopathy. This may be explained by the fact that it is not soluble Pi but insoluble CaPi nanocrystals that are responsible for cellular responses to high ambient Pi. CaPi nanocrystals, which were previously known as nanobacteria, are prevalent *in vivo* and increase as renal function declines. Further studies on phosphatopathies are expected to provide new insights into the mechanism of aging.

REFERENCES

1. Blair-West JR, Denton DA, McKinley MJ, *et al.* (1992) Behavioral and tissue responses to severe phosphorus depletion in cattle. *Am J Physiol* **263**: R656–R663.
2. Sweeny JM, Seibert HE, Woda C, *et al.* (1998) Evidence for induction of a phosphate appetite in juvenile rats. *Am J Physiol* **275**: R1358–R1365.
3. Gerrior S, Bente L, Hiza H. (2004) Nutrient content of the U.S. food supply, 1909–2000. U.S. Department of Agriculture, Center for Nutritional Policy and Promotion Home Economics Research Report No. 56.
4. Calvo MS, Park YK. (1996) Changing phosphorus content of the U.S. diet: potential for adverse effects on bone. *J Nutr* **126**: 1168S–1180S.
5. Stevens LA, Li S, Wang C, *et al.* (2010) Prevalence of CKD and comorbid illness in elderly patients in the United States: results from the Kidney Early Evaluation Program (KEEP). *Am J Kidney Dis* **55**: S23–S33.
6. Dusso AS, Brown AJ, Slatopolsky E. (2005) Vitamin *D. Am J Physiol Renal Physiol* **289**: F8–F28.
7. Berndt T, Kumar R. (2007) Phosphatonins and the regulation of phosphate homeostasis. *Annu Rev Physiol* **69**: 341–359.
8. Tfelt-Hansen J, Brown EM. (2005) The calcium-sensing receptor in normal physiology and pathophysiology: a review. *Crit Rev Clin Lab Sci* **42**: 35–70.
9. Itoh N, Ornitz DM. (2008) Functional evolutionary history of the mouse Fgf gene family. *Dev Dyn* **237**: 18–27.
10. White KE, Evans WE, O'Rlordan JLH, *et al.* (2000) Autosomal dominant hypophosphataemic rickets is associated with mutations in FGF23. *Nat Genet* **26**: 345–348.
11. White KE, Carn G, Lorenz-Depiereux B, *et al.* (2001) Autosomal-dominant hypophosphatemic rickets (ADHR) mutations stabilize FGF-23. *Kidney Int* **60**: 2079–2086.
12. Shimada T, Hasegawa H, Yamazaki Y, *et al.* (2004) FGF-23 is a potent regulator of vitamin D metabolism and phosphate homeostasis. *J Bone Miner Res* **19**: 429–435.
13. Liu S, Tang W, Zhou J, *et al.* (2006) Fibroblast growth factor 23 is a counter-regulatory phosphaturic hormone for vitamin D. *J Am Soc Nephrol* **17**: 1305–1315.
14. Kurosu H, Yamamoto M, Clark JD, *et al.* (2005) Suppression of aging in mice by the hormone Klotho. *Science* **309**: 1829–1833.
15. Kuro-o M, Matsumura Y, Aizawa H, *et al.* (1997) Mutation of the mouse klotho gene leads to a syndrome resembling ageing. *Nature* **390**: 45–51.

16. Shimada T, Kakitani M, Yamazaki Y, *et al.* (2004) Targeted ablation of Fgf23 demonstrates an essential physiological role of FGF23 in phosphate and vitamin D metabolism. *J Clin Invest* **113**: 561–568.
17. Razzaque MS, Sitara D, Taguchi T, *et al.* (2006) Premature aging-like phenotype in fibroblast growth factor 23 null mice is a vitamin D-mediated process. *FASEB J* **20**: 720–722.
18. Tsujikawa H, Kurotaki Y, Fujimori T, *et al.* (2003) Klotho, a gene related to a syndrome resembling human premature aging, functions in a negative regulatory circuit of vitamin D endocrine system. *Mol Endocrinol* **17**: 2393–2403.
19. Kurosu H, Ogawa Y, Miyoshi M, *et al.* (2006) Regulation of fibroblast growth factor-23 signaling by klotho. *J Biol Chem* **281**: 6120–6123.
20. Urakawa I, Yamazaki Y, Shimada T, *et al.* (2006) Klotho converts canonical FGF receptor into a specific receptor for FGF23. *Nature* **444**: 770–774.
21. Goetz R, Beenken A, Ibrahimi OA, *et al.* (2007) Molecular Insights into the Klotho-dependent, endocrine mode of action of FGF19 subfamily members. *Mol Cell Biol* **27**: 3417–3428.
22. Ben-Dov IZ, Galitzer H, Lavi-Moshayoff V, *et al.* (2007) The parathyroid is a target organ for FGF23 in rats. *J Clin Invest* **117**: 4003–4008.
23. Lavi-Moshayoff V, Wasserman G, Meir T, *et al.* (2010) PTH increases FGF23 gene expression and mediates the high-FGF23 levels of experimental kidney failure: a bone parathyroid feedback loop. *Am J Physiol Renal Physiol* **299**: F882–F889.
24. Kuro-o M. (2010) Klotho. *Pflugers Arch* **459**: 333–343.
25. Stubbs JR, Liu S, Tang W, *et al.* (2007) Role of hyperphosphatemia and 1,25-dihydroxyvitamin D in vascular calcification and mortality in fibroblastic growth factor 23 null mice. *J Am Soc Nephrol* **18**: 2116–2124.
26. Ohnishi M, Nakatani T, Lanske B, Razzaque MS. (2009) Reversal of mineral ion homeostasis and soft-tissue calcification of klotho knockout mice by deletion of vitamin D 1alpha-hydroxylase. *Kidney Int* **75**: 1166–1172.
27. Hesse M, Frohlich LF, Zeitz U, *et al.* (2007) Ablation of vitamin D signaling rescues bone, mineral, and glucose homeostasis in Fgf-23 deficient mice. *Matrix Biol* **26**: 75–84.
28. Morishita K, Shirai A, Kubota M, *et al.* (2001) The progression of aging in klotho mutant mice can be modified by dietary phosphorus and zinc. *J Nutr* **131**: 3182–3188.
29. Ohnishi M, Razzaque MS. (2010) Dietary and genetic evidence for phosphate toxicity accelerating mammalian aging. *FASEB J* **24**: 3562–3571.
30. Sprecher E. (2010) Familial tumoral calcinosis: from characterization of a rare phenotype to the pathogenesis of ectopic calcification. *J Invest Dermatol* **130**: 652–660.

31. Benet-Pages A, Orlik P, Strom TM, Lorenz-Depiereux B. (2005) An FGF23 missense mutation causes familial tumoral calcinosis with hyperphosphatemia. *Hum Mol Genet* **14**: 385–390.
32. Larsson T, Davis SI, Garringer HJ, *et al.* (2005) Fibroblast growth factor-23 mutants causing familial tumoral calcinosis are differentially processed. *Endocrinology* **146**: 3883–3891.
33. Ichikawa S, Imel EA, Kreiter ML, *et al.* (2007) A homozygous missense mutation in human KLOTHO causes severe tumoral calcinosis. *J Clin Invest* **117**: 2692–2701.
34. Gutierrez OM, Mannstadt M, Isakova T, *et al.* (2008) Fibroblast growth factor 23 and mortality among patients undergoing hemodialysis. *N Engl J Med* **359**: 584–592.
35. Koh N, Fujimori T, Nishiguchi S, *et al.* (2001) Severely reduced production of klotho in human chronic renal failure kidney. *Biochem Biophys Res Commun* **280**: 1015–1020.
36. El-Abbadi M, Giachelli CM. (2005) Arteriosclerosis, calcium phosphate deposition and cardiovascular disease in uremia: current concepts at the bench. *Curr Opin Nephrol Hypertens* **14**: 519–524.
37. Hu MC, Shi M, Zhang J, *et al.* (2011) Klotho deficiency causes vascular calcification in chronic kidney disease. *J Am Soc Nephrol* **22**: 124–136.
38. Kestenbaum B, Sampson JN, Rudser KD, *et al.* (2005) Serum phosphate levels and mortality risk among people with chronic kidney disease. *J Am Soc Nephrol* **16**: 520–528.
39. Tonelli M, Sacks F, Pfeffer M, *et al.* (2005) Relation between serum phosphate level and cardiovascular event rate in people with coronary disease. *Circulation* **112**: 2627–2633.
40. Ganesh SK, Stack AG, Levin NW, *et al.* (2001) Association of elevated serum PO(4), Ca × PO(4) product, and parathyroid hormone with cardiac mortality risk in chronic hemodialysis patients. *J Am Soc Nephrol* **12**: 2131–2138.
41. Kasiske BL. (2003) K/DOQI clinical practice guidelines for bone metabolism and disease in chronic kidney disease. *Am J Kidney Dis* **42**: S1–S201.
42. Martin KJ, Gonzalez EA. (2011) Prevention and Control of phosphate retention/hyperphosphatemia in CKD–MBD: what is normal, when to start, and how to treat? *Clin J Am Soc Nephrol* **6**: 440–446.
43. Isakova T, Gutierrez OM, Chang Y, *et al.* (2009) Phosphorus binders and survival on hemodialysis. *J Am Soc Nephrol* **20**: 388–396.
44. Sarnak MJ, Levey AS, Schoolwerth AC, *et al.* (2003) Kidney disease as a risk factor for development of cardiovascular disease: a statement from the American Heart Association Councils on Kidney in Cardiovascular Disease,

High Blood Pressure Research, Clinical Cardiology, and Epidemiology and Prevention. *Circulation* **108**: 2154–2169.
45. Go AS, Chertow GM, Fan D, *et al.* (2004) Chronic kidney disease and the risks of death, cardiovascular events, and hospitalization. *N Engl J Med* **351**: 1296–1305.
46. Gutierrez O, Isakova T, Rhee E, *et al.* (2005) Fibroblast growth factor-23 mitigates hyperphosphatemia but accentuates calcitriol deficiency in chronic kidney disease. *J Am Soc Nephrol* **16**: 2205–2215.
47. Sato T, Tominaga Y, Ueki T, *et al.* (2004) Total parathyroidectomy reduces elevated circulating fibroblast growth factor 23 in advanced secondary hyperparathyroidism. *Am J Kidney Dis* **44**: 481–487.
48. Torres PU, Prie D, Beck L, Friedlander G. (2006) New therapies for uremic secondary hyperparathyroidism. *J Ren Nutr* **16**: 87–99.
49. Hasegawa H, Nagano N, Urakawa I, *et al.* (2010) Direct evidence for a causative role of FGF23 in the abnormal renal phosphate handling and vitamin D metabolism in rats with early-stage chronic kidney disease. *Kidney Int* **78**: 975–980.
50. Komaba H, Goto S, Fujii H, *et al.* (2010) Depressed expression of Klotho and FGF receptor 1 in hyperplastic parathyroid glands from uremic patients. *Kidney Int* **77**: 232–238.
51. Jono S, McKee MD, Murry CE, *et al.* (2000) Phosphate regulation of vascular smooth muscle cell calcification. *Circ Res* **87**: E10–E17.
52. Villa-Bellosta R, Millan A, Sorribas V. (2010) Role of calcium-phosphate deposition in vascular smooth muscle cell calcification. *Am J Physiol Cell Physiol* **300**: C210–C220.
53. Werner A, Dehmelt L, Nalbant P. (1998) Na^+-dependent phosphate cotransporters: the NaPi protein families. *J Exp Biol* **201**: 3135–3142.
54. Ravera S, Virkki LV, Murer H, Forster IC. (2007) Deciphering PiT transport kinetics and substrate specificity using electrophysiology and flux measurements. *Am J Physiol Cell Physiol* **293**: C606–C620.
55. Villa-Bellosta R, Sorribas V. (2009) Phosphonoformic acid prevents vascular smooth muscle cell calcification by inhibiting calcium-phosphate deposition. *Arterioscler Thromb Vasc Biol* **29**: 761–766.
56. Li X, Yang HY, Giachelli CM. (2006) Role of the sodium-dependent phosphate cotransporter, Pit-1, in vascular smooth muscle cell calcification. *Circ Res* **98**: 905–912.
57. Johann SV, Gibbons JJ, O'Hara B. (1992) GLVR1, a receptor for gibbon ape leukemia virus, is homologous to a phosphate permease of *Neurospora crassa* and is expressed at high levels in the brain and thymus. *J Virol* **66**: 1635–1640.

58. Kavanaugh MP, Miller DG, Zhang W, *et al.* (1994) Cell-surface receptors for gibbon ape leukemia virus and amphotropic murine retrovirus are inducible sodium-dependent phosphate symporters. *Proc Natl Acad Sci USA* **91**: 7071–7075.
59. Miller DG, Edwards RH, Miller AD. (1994) Cloning of the cellular receptor for amphotropic murine retroviruses reveals homology to that for gibbon ape leukemia virus. *Proc Natl Acad Sci USA* **91**: 78–82.
60. Young JD, Martel J, Young D, *et al.* (2009) Characterization of granulations of calcium and apatite in serum as pleomorphic mineralo-protein complexes and as precursors of putative nanobacteria. *PLoS One* **4**: e5421.
61. Sage AP, Lu J, Tintut Y, Demer LL. (2010) Hyperphosphatemia-induced nanocrystals upregulate the expression of bone morphogenetic protein-2 and osteopontin genes in mouse smooth muscle cells *in vitro. Kidney Int* **79**: 414–422.
62. Khoshniat S, Bourgine A, Julien M, *et al.* (2010) Phosphate-dependent stimulation of MGP and OPN expression in osteoblasts via the ERK1/2 pathway is modulated by calcium. *Bone* **48**: 894–902.
63. Yamazaki M, Ozonoa K, Okada T, *et al.* (2010) Both FGF23 and extracellular phosphate activate Raf/MEK/ERK pathway via FGF receptors in HEK293 cells. *J Cell Biochem* **111**: 1210–1221.
64. Wang N, Wang X, Xing C, *et al.* (2010) Role of TGF-beta(1) in bone matrix production in vascular smooth muscle cells induced by a high-phosphate environment. *Nephron Exp Nephrol* **115**: e60–e68.
65. Shuto E, Taketani Y, Tanaka R, *et al.* (2009) Dietary phosphorus acutely impairs endothelial function. *J Am Soc Nephrol* **20**: 1504–1512.
66. Di Marco GS, Hausberg M, Hillebrand U, *et al.* (2008) Increased inorganic phosphate induces human endothelial cell apoptosis *in vitro. Am J Physiol Renal Physiol* **294**: F1381–F1387.
67. Ewence AE, Bootman M, Roderick HL, *et al.* (2008) Calcium phosphate crystals induce cell death in human vascular smooth muscle cells: a potential mechanism in atherosclerotic plaque destabilization. *Circ Res* **103**: e28–e34.
68. Aihara K, Byer KJ, Khan SR. (2003) Calcium phosphate-induced renal epithelial injury and stone formation: involvement of reactive oxygen species. *Kidney Int* **64**: 1283–1291.
69. Lieske JC, Norris R, Toback FG. (1997) Adhesion of hydroxyapatite crystals to anionic sites on the surface of renal epithelial cells. *Am J Physiol* **273**: F224–F233.

70. Molloy ES, McCarthy GM. (2006) Basic calcium phosphate crystals: pathways to joint degeneration. *Curr Opin Rheumatol* **18**: 187–192.
71. Sommer AP. (2010) Cytotoxicity of calcium phosphate crystals and human-derived nanoparticles: an overlooked link. *Circ Res* **106**: e10.
72. Kajander EO, Ciftcioglu N. (1998) Nanobacteria: an alternative mechanism for pathogenic intra- and extracellular calcification and stone formation. *Proc Natl Acad Sci USA* **95**: 8274–8279.
73. Kajander EO. (2006) Nanobacteria — propagating calcifying nanoparticles. *Lett Appl Microbiol* **42**: 549–552.
74. Young JD, Martel J, Young L, *et al.* (2009) Putative nanobacteria represent physiological remnants and culture by-products of normal calcium homeostasis. *PLoS One* **4**: e4417.
75. Wu CY, Martel J, Young D, Young JD. (2009) Fetuin-A/albumin–mineral complexes resembling serum calcium granules and putative nanobacteria: demonstration of a dual inhibition-seeding concept. *PLoS One* **4**: e8058.
76. Raoult D, Drancourt M, Azza S, *et al.* (2008) Nanobacteria are mineralo-fetuin complexes. *PLoS Pathog* **4**: e41.
77. Heiss A, Pipich V, Jahnen-Dechent W, Schwahn D. (2010) Fetuin-A is a mineral carrier protein: small angle neutron scattering provides new insight on fetuin-A controlled calcification inhibition. *Biophys J* **99**: 3986–3995.
78. Betts F, Blumenthal NC, Posner AS, *et al.* (1975) Atomic structure of intracellular amorphous calcium phosphate deposits. *Proc Natl Acad Sci USA* **72**: 2088–2090.
79. Rochette CN, Rosenfeldt S, Heiss A, *et al.* (2009) A shielding topology stabilizes the early stage protein–mineral complexes of fetuin-A and calcium phosphate: a time-resolved small-angle X-ray study. *Chembiochem* **10**: 735–740.
80. Heiss A, Jahnen-Dechent W, Endo H, Schwahn D. (2007) Structural dynamics of a colloidal protein–mineral complex bestowing on calcium phosphate a high solubility in biological fluids. *Biointerphases* **2**: 16–20.
81. Hamano T, Matsui I, Mikami S, *et al.* (2010) Fetuin–mineral complex reflects extraosseous calcification stress in CKD. *J Am Soc Nephrol* **21**: 1998–2007.
82. Hruska KA, Choi ET, Memon I, *et al.* (2010) Cardiovascular risk in chronic kidney disease (CKD): the CKD–mineral bone disorder (CKD–MBD). *Pediatr Nephrol* **25**: 769–778.
83. Meyer KB, Levey AS. (1998) Controlling the epidemic of cardiovascular disease in chronic renal disease: report from the National Kidney Foundation Task Force on cardiovascular disease. *J Am Soc Nephrol* **9**: S31–S42.

84. Rodriguez Garcia M, Naves Diaz M, Cannata Andia JB. (2005) Bone metabolism, vascular calcifications and mortality: associations beyond mere coincidence. *J Nephrol* **18**: 458–463.
85. John GB, Cheng CY, Kuro-o M. (2011) Role of Klotho in aging, phosphate metabolism, and chronic kidney disease. *Am J Kidney Dis* **58**: 127–134.
86. Kuro-o M. (2011) Phosphate and Klotho. *Kidney Int* **79**: S20–S23.

Index